Quantum Reality and Theory of Śūnya

Siddheshwar Rameshwar Bhatt
Editor

Quantum Reality and Theory of Śūnya

Springer

Editor
Siddheshwar Rameshwar Bhatt
Indian Council of Philosophical Research
New Delhi, Delhi, India

ISBN 978-981-13-1956-3 ISBN 978-981-13-1957-0 (eBook)
https://doi.org/10.1007/978-981-13-1957-0

Library of Congress Control Number: 2019933413

This Springer imprint is published by the registered company Springer Nature Singapore Pte Ltd.
The registered company address is: 152 Beach Road, #21-01/04 Gateway East, Singapore 189721, Singapore

Dedicated to
Seers, sages and thinkers
Who have shown the way to Reality

Foreword

As I assumed the office of the President of the ICCR, I received a challenging letter from Dr. Peter Gobets to record concrete historic evidence on the precise genesis of the zero as a digit. The date of the origin of zero deserves study. Its presence in grammar and metrics as a semantic category goes far back in history. He sent a note of Dr. Elizabeth P. den Boer of the Leiden University on the uncanny convergence of the world views of Indian philosophies and quantum physics. The practical utility of the zero has been adopted in the West, but its deeper meaning remains to be grasped. As all signs refer both to a concrete presence and an intangible realm, the duality of the vestigial and ineffable is the reality of the *pūrṇa* and *śūnya*. Sanskrit stands out in having the dual number besides the singular and plural. Creation was viewed in India as a binary phenomenon, in consonance with the androgynous birth of humans and animals. The duality of consciousness and intellection was denoted by analogous verbal roots. The fundament of consciousness was the root *cit* and its nominal form *cetana*. It became the verb *cint* 'to think' with the addition of the infix *n* (from *cit* to *cint*) and the noun *cintana*. Creativity was binary and so was its grammatical expression in the dual number. The concept of empirical *pūrṇa* suggested the subtle *śūnya* 'void'. The *nāsadīya* hymn of the Ṛgveda on creation speaks of sad 'existence' and a sad 'non-existence'. The binary principle of India's approach to reality, phenomena or vacuity induced the emergence of *śūnya* both as (i) zero and (ii) as a void vacuum where subatomic phenomena arise and decay.

The monocentric approach of Western philosophy, strengthened by monotheism, has led to the ontological quandary of quantum physics as to why subatomic phenomena arise and decay in an empty vacuum. The zero-based philosophy of nonism (no philosophy) has been missing in the West. The vast Buddhist literature of root texts and commentaries of the last two millennia on 'negation' and 'sūnyatā' can provide insights into the subatomic vacuum. Is this subatomic vacuum 'empty'. No, it is a 'creative void'.

The word *śūnya* goes back to the Ṛgveda where it means 'emptiness, lack, want, absence' from *śūna* 'swollen', or the state of hollowness. The verbal form *śvayati* means 'swells, increases and grows' as both increase and emptiness are ingrained in the word *śūnya* itself. *Śūnya* means 'sky' in the Śabdacandrikā. Jaṭādhara's lexicon translates *śūnyavādin* 'one who propounds śūnya' as a Buddhist. *Śūnya* developed as a philosophical term in Buddhism, e.g. *paribhāvitāśūnyatādīrgharātram* 'we have meditated long on the void' in the Lotus Sutra 117.7 (gāthā). There are different lists of *śūnyatā*: 18 in the Mahāvyutpatti 933–951 and ŚatasāhasrikāPrajñāpāramitā and 20 in the Dharma-saṅgraha. The different śūnyayās have to be interpreted by physicists and mathematicians.

Buddhist thought was based on the negative concepts of *anātmavāda* denying the existence of soul and secondly the theory of *anityatā* 'impermanence' denying the eternal dimension. Laṅkāvatāra-sūtra 22.10 speaks of *śāśvatoccheda* 'denying the eternal'. Both were to lead to śūnyatā. Nāgārjuna the founder of the Mādhyamikaschool was associated with the Prajñāpāramitā and gave rise to the doctrine of śūnyatā which should be translated as 'Creative Void'. It is not emptiness related to the empirical volume. He defined śūnyatā as *pratītya-samutpāda* which can be rendered as 'interdependent origination' to state that things are not self-arisen but come into existence in mutual dependence on conditions. The conventional truth (*saṁvṛti-satya*) is a necessary means for the ultimate truth (*paramārtha-satya*), and the ultimate makes the conventional possible. The language of Buddhism and that of quantum physics have to overcome dualism of value and validity. Scientists have started to think more subtly than they had hitherto been accustomed. To cite Heisenberg, 'Although I am now convinced that scientific truth is unassailable in its own field, I have never found it possible to dismiss the content of religious thinking as simply part of an outmoded phase in the consciousness of mankind'.

In the increasingly busy crossroads of artificial intelligence, neuroscience, consciousness, quantum physics and computing sciences, scientists have begun thinking deep below the level of consciousness, beyond the intellectual formalism of logic but with fluidity. The contemporary philosophy of science is becoming the spoilsport of modern logic, like Kurt Gödel's incompleteness theorem which declares that any attempt to build a complete and consistent logical theorem will inevitably be ruined by undecidable proportions. Modern understanding of how the glories of the mind spring from pure matter is a challenge. The mind has been viewed as creative and self-aware, how can we think of it as nothing but electrical impulses and biological tissue.

Intuition gave rise to Vedic Mathematics and to the unique mathematical genius of Ramanujam at Cambridge who was hailed as the 'man who knew infinity'. Semiotics, the science of signs and mathematical thought, invites us to study new realms of abstraction and meditational insights of Buddhist *śūnyatā* to co-contribute to progress, away from the deep-seated Greek aversion to the concept of emptiness. It will be *pāramitā* or trans-going to comprehend the universe.

These proceedings of the seminar on *śūnya* theory and quantum reality held in December 2016 are a milestone in the study of scientists. Philosopher and meditator Prof. S. R. Bhatt has provided insights into the India perceptions as well as put in hard labour to edit the presentations. The *śūnya* and *pūrṇa* are the elemental binary eternals, and the dedicated efforts of Prof. Bhatt enshrined in this volume remind me of the words of Poet Rabindranath Tagore: 'The flower's sacrifice ripens in the sweetness of the fruit'.

Former President, Indian Council for Cultural Relations, New Delhi, India
Lokesh Chandra

Preface

Quantum Reality and Intimations of *Śūnya*

The history of human kind reveals constant recurrence of some fundamental problems which have agitated reflective human mind right from the dawn of civilization. One of such perennial problems is the nature of Reality. The demand for some kind of universally acceptable conception of Reality is one which the rational human mind cannot help making. But no two minds have ever agreed with one another in their pronouncements. Another cognate problem which perplexes is 'How and why the mutative world has come to exist?' along with its corollaries of 'whence' and 'whither' of this universe.

There have been twofold approaches to Reality and hence to knowledge: the fundamental and the derivative, the ultimate and the proximate, the transcendental and the immanent, the ectype and the archetype and the *pāramārthika* and the *vyāvahārika*. There has been search for the transcendent and attempts to understand the immanent. For this, both the intuitive and the ratiocinative modes have been adopted. The inward exploration in the form of *pragyā* and outward observation in the form of *pratibhā* need to be taken as complimentary. It should be a synthesis of experience and expression, of course maintaining clear distinction between the two as Kaṭhopaniṣad rightly emphasizes. This is the message of the Īśopaniṣad also which talks of symbiosis (*ubhayosaha*) of *vidyā* and *avidyā*.

The subject matter of our deliberations consists of two subthemes, viz. search for the nature of quantum reality and the corresponding theory of *Śūnya* as a possible approach.

We begin with the Indian scene and expound the theory of *Śūnya* which is to be equated with Brahman, as the two have similar etymology and semantics. The roots *siv* and *bṛh* stand for 'to manifest' and 'to grow'. *ParamaŚiva* or Brahman is regarded as the ultimate ontological substratum of the entire cosmos. It is at once both transcendent and immanent. It is unitary and unifying (*Tadekam*) and has a natural tendency to manifest and diversify (*Eko'hambahusyām*). It is self-existing and self-sustaining (*svadhā*). It is luminous and illuminating (*Tasyabhasasarvamidamvibhāti*). It is peaceful and tranquil like deep sea and vibrant and turbulent like sea surface. It is to be noted that *Śūnya* is not void or vacuity but, in its original form, it means devoid of multiplicity, emptied of plurality. It is not absence of anything but abode of everything. It is self-luminous and therefore pure consciousness. It is fullness (*pūrṇa*) and therefore bliss (*Yadvābhūmotatsukham*). It is *Cidānandrūpa* but also *spandarūpa*. It is alogical, suprarational and intuitively realizable through *sādhanā*. It is comparable to 'quantum vacuum' of quantum physics.

Śūnya/Brahman is vibration-less vibration, unmoved mover. In the cosmic process, the *ParamaŚiva/Śūnya Brahma* assumes the role of *Śiva-Śaktisāmarasya* (symbiosis of pure consciousness and consciousness in manifestation). Then its dynamism gets manifested. This is known as *unmeṣa* (fulguration) or *spandana* (vibration). This is the process of stir of consciousness, the *vimaśa* of *prakāśa*. Śiva is *asti* (being), and Śakti is *bhavati* (becoming). When Śakti is in action from here, individuation and multiplication get manifested. It is not creation but manifestation. But it should be remembered that in this kinetic process the *ParamaŚiva/Brahman* is not affected. *That is why it is described as vibration-less vibration.* In this process, there is dependent origination and interdependent existence of all phenomena. This is comparable to principles of interconnection, entanglement, implicate and explicate orders, exclusion principle, etc. of quantum mechanics. This is how phenomenon of teleportation, non-locality, etc. can be understood.

There is a very poignant word for the start of cosmic process which is rich in meaning. This is an adjective of *Śūnya/ParamaŚiva/Brahman*. It is characterized as *svātantrya*. It means (a) *svamtanute* (manifests itself), (b) *svasmāttanute* (manifests from within itself), (c) *svecchayātanute* (out of one's sweet will), (d) *svaśaktyātanute* (from its own power without being controlled by anyone outside as there is nothing outside), (e) *svacchnadatayātanute* (without any causal constraint) and (f) *svasmintanute* (manifests within itself as it is second to none, *advaya* or *advaita*).

It is to be emphasized that all manifestations here are of the same nature of perfection as from perfection only perfection can arise (*pūrṇādpūrnamudachyate*). Due to improper understanding and due to individuation and consequent 'I-ness' or egoity, we feel separated and limited. So there is a need for self-recognition (*Pratyabhijñā*) or ātmalābha (self-realization). For this, the material world which is also a part and parcel of the same ultimate Reality provides an arena. This is the implication of the prayer *Mṛtyorma'mṛtamgamaya.*This is how we can have symbiosis of Vedanta and Tantra.

The multifaceted concept of *Śūnya* has been an ingenious thought of fertile and innovative Indian mind. It has been a unique contribution of India to world culture and civilization. This seminal and pivotal concept has its ramifications in various fields like metaphysics, cosmology and cosmogony, physics, mathematics, religion, yoga, etc. There are several implications of the theory of Śūnya. Some of them are philosophical (Vedanta and Buddhism), mathematical (concept of zero, decimal system, etc.), spiritual (concept of *Pūrṇa*), religious (Mahimā cult), in *sādhanāpakṣa* (in Yoga and Tantra), in astrophysics (pertaining to *Ākaśa* or *loka*), etc. Several new dimensions can also be discerned. In different contexts, it has different meaning. It would be a worthwhile and rewarding exercise if its different facets are attended to and analysed. They are all correlated, and only a holistic and integral approach can bring out their significance and value.

The concept of *Śūnya* has a profound metaphysical connotation which stands for the totality of Reality. In this sense, it is equated with *Pūrṇa* (completeness, infinitude and boundlessness). The Reality is a Whole which comprehends all iota of the universe as its intrinsic parts. But each part (*kośa* or *khaṇḍa*) is also a Whole (*piṇḍa*) within this widest Whole (*brahmāṇḍa*). It is represented as a Supreme Circle, and in the cosmos, there are circles within circles presenting a picture of concentric Wholes. The ultimate reality is full circle, an overarching circle. The invocation of *ĪśaUpaniṣad* remarkably puts forth this intuitive vision. It states that the Reality is a Whole or totality comprising all that was, that is and that will be, an idea available in the *PuruṣaSūkta* of the Ṛgveda. From this Whole, only Whole can spring forth even though we may not be aware of it. This Whole is infinite. Infinite cannot be *finitized*, and therefore the Advaita Vedanta regards all differentiations as finite appearances which are in ultimate analysis infinite only. Bhaskaracharya, a later mathematician, also avers the same. He opines that no change takes place in the infinite and immutable Brahman when worlds are evolved or withdrawn even though in these processes numerous orders of beings are put forth and absorbed. The equation of *Śūnya* and *pūrṇa* was a wonderful feat of Indian logical acumen which could be apprehended only by a *mantradṛṣṭāṛiṣi*. This concept of *Śūnya* is not emptiness or voidness but 'devoidness' in the sense that in its proto-nature the Reality is devoid of all differentiations. In itself, it is emptied of all manifest diversity. It is a non-dual realm, a quantum vacuum. In logical terms, it is referred to as 'null set', but this null set is not devoid of membership but only devoid of manifest membership. It is not to be regarded as 'nothingness' as nothing can come out of nothing. The *NāsadīyaSūkta* of the Ṛgveda begins with this intuitive realization. All quantum phenomena arise within it and get dissolved in it. The modern quantum physics endorses this understanding, and it is struggling to explain how this cosmic event takes place. The Big Bang theory is only an indication of this. The search for 'God-particle' or 'Boson' seems to be a futile exercise of the Western scientists. The *impartite* approach to Reality is a unique and ingenious gift of the Indian mind, and Western scientist should pay heed to it.

Deriving inspiration from this intuitive realization the Indian mathematicians like Bodhāyana, Brahmagupta, and many others have invented the idea of zero (cipher) and the decimal system. The mathematical zero has its obverse as infinity. All numbers or numerical signs act as tangible reference to finite, but zero represents the non-dual realm, the infinity. The Western mind has adopted zero only functionally for practical utility; it could not grasp its deeper metaphysical meaning. The concept of zero did have tremendous impact on Western science and mathematics in terms of 'decimal system' and 'quantum vacuum', but they could not reach to its metaphysical heights and depths because of the limitations of their empirical methodology. Only through *ṛtambharāpragyā* (to use Indian terminology) this realization is possible in a state of *samādhi*.

Apart from metaphysics, physics and mathematics, this concept of *Śūnya* has been profitably utilized by schools of Śaivism wherein Lord *ParamaŚiva* is referred to as *Śūnya* or *Bindu*. The *Śūnya* is described as *Aśūnya* which means that it is beyond one and many but supreme source of one and many. The worship of Lord Jagannātha in Puri and the text ŚūnyaSamhitā followed by the Mahimā school advocated by Pañcasakhas of Orissa are the elaborations of this idea synthesizing it with the Buddhist concept of *Śūnyata*. Their concept of *Śūnya Brahma* or *ŚūnyaPuruṣa* is a remarkable idea, the implications of which need to be brought out shorn of its religious connotation. *ŚūnyaPuruṣa* is *Anādimaṇḍala*, also described as *ŚūnyaMaṇḍala* which is the source of all creation. It is proto cause as well as the effects. The Buddhist concepts of 'dependent origination' and 'interdependent existence' coupled with this idea of *Śūnya* can help in revealing the nature of empirical and trans-empirical Reality as Nāgārjuna has pointed out. Nāgārjuna averred that there are two levels of approaching Reality. There is transcendental Reality underlying the world of phenomena.

A depth analysis of this seminal concept of *Śūnya* can open up new horizons and intellectual vistas and help in enriching improvised Western science and mathematics. Indian contributions in terms of place value decimal system, domestication of fire, invention of wheel and original gifts to science and mathematics are certainly recognized, but they have not been adequately utilized. Now that Indian contributions are gradually appreciated and the saturated Western mind is looking towards India for newer insights and fresh approaches, in-depth delineation of this concept will go a long way in spiritualizing science and making Indian spirituality more scientific.

Coming to Western scenario with this background, we may begin with the Greek and medieval periods to trace the development of scientific thought. There has been separation between physics and metaphysics after Aristotle resulting in divorce between science and spirituality. These days a need is being felt to reconcile the findings of the two for their mutual benefit. It seems that a holistic approach may be more rewarding and fruitful. In post-Newtonian physics, there is a transition from solid state to quantum. Limitations of classical physics have been realized. It is a good development. It is hoped that this may lead to the twofold approaches referred to earlier. Of course, some hard scientists may not find it easy or convenient to reach to the transcendent. However, in modern times, there is gradual realization of primacy of consciousness which transcends matter. Studies have been undertaken both in space-time bound local communication and also non-local communication independent of space and time. Non- local communication may be characterized by super non-locality and super-super-non-locality (David Bohm). The space-time bound local communication is measurable in terms of speed of light (and we are now celebrating 340 anniversary of determination of speed of light in the west), but non-local communication is not so measurable. So in this respect, the present science has to go beyond the recourse to measurability.

There is a welcome shift from Newtonian to quantum physics as it enlarges our vision, but it has to join hands with spirituality. Max Plank in 1900 and Einstein in 1905 spearheaded this movement, and this revolutionized our understanding of the basic nature of Reality. Subsequently, Niels Bohr, Heisenberg, Schrodinger, Pauli, David Bohm and many others have developed this. Many of them have acknowledged Indian and Chinese influences on their thinking. Apart from the famous 'Exclusion' principle, Pauli discovered the existence of 'Neutrino', an uncharged particle. Niel Bohr, Heisenberg and Schrodinger advanced ideas of complementary opposites and acausality. Synchronicity as an acausal connecting principle was recognized. The dual descriptions of subatomic micro-entities in terms of waves and particles were accepted. Schrodinger's theory of oneness of all existences is an outstanding contribution. In spite of controversy between 'Copenhagen interpretation' and EPR thought experiment regarding the nature of quanta, the Wholeness and implicate order and interconnectedness of all phenomena remain undisputed. There may be difference of opinion about 'uncertainty principle', 'collapse of wave function', etc., but the 'non-locality' and 'entanglement' at the quantum level, verified and established by Bell's theorem, find general acceptance. The same is the case with the theory of overcoming 'viewing subject'-'viewed object' distinction at this level. The point is that space-time-causality operation is applicable only at the physical level, and at the deep down transcendental level, non-locality prevails. There are no 'basic building blocks' of the universe but only complicated web of relations deeply interconnected between various parts of the Whole.

To conclude, contemporary quantum physics is driven to posit dual levels of reality, the empirical and the trans-empirical, and yet many theoretical scientists do not take this seriously. They still cherish the idea of ultimate state of matter. But no account of any reductionism—behaviourist or naturalistic or any sort of physicalism—can be free from logical flaws and therefore untenable. There is incurable limitation of theoretical reason, and the trans-empirical is not accessible to current positivistic methodologies.

New Delhi, Delhi, India Siddheshwar Rameshwar Bhatt

Acknowledgements

This book comprises papers presented in an international seminar jointly organized by the Indian Council for Cultural Relations and Indian Council of Philosophical Research. Scholars from different parts of the world participated in the deliberations on behalf of these two organizations. We extend our heartfelt thanks for their erudite contributions.

Ms. Shinjini, Ms. Priya Vyas and other staff have graciously undertaken its publication. I express my gratitude to them.

The theme is of topical interest and significance. It is hoped that it will be useful to enlightened readers.

Contents

About the Editor and Contributors

Editor

Siddheshwar Rameshwar Bhatt is an eminent philosopher and Sanskritist. Presently, he is Chairman of the Indian Council of Philosophical Research. He is also Chairman of the Indian Philosophical Congress and Asian Congress of Philosophy. He retired as Professor and Head of the Department of Philosophy at the University of Delhi, Delhi. He is internationally known as an authority on ancient Indian culture, Buddhism, Jainism and Vedanta. His research areas include Indian philosophy, logic, epistemology, ethics, value theory, philosophy of education, philosophy of religion, comparative religion, social and political thought, etc. He has lectured in many universities and research institutes of the USA, Canada, Finland, Trinidad, North Korea, South Korea, Japan, China, Vietnam, Sri Lanka, Singapore and Thailand. He is a member of many national and international associations. He is a Member of Board of Advisors of Council for Research in Values and Philosophy, Washington DC, USA, which has brought out 300 volumes on world cultures and civilizations.

Contributors

Rajendra Prasad Bajpai worked as lecturer and reader of physics in Himachal Pradesh University, Shimla, for 15 years and then as professor for 25 years in North-Eastern Hill University (NEHU), Shillong. He became a visiting professor in the Leiden University, the Netherlands, for a year after superannuation. He was a member of the International Institute of Biophysics, Neuss, Germany, from 1995 to 2011, where he provided definitive evidence of the quantum nature of life and developed procedures for extracting holistic properties from observed biophoton signals.

Herbert J. Bernstein Since 1971, he is professor of physics at Hampshire College. He has been a consultant to the World Bank and AAAS and US President's Science Adviser on science policy. He served as an MIT Visiting Scientist from 1984 through 2004. In 1986, he took over as PI from Cliff Shull (Nobel Prize in Physics 1994) to head an international team of research physicists. His research interests include science and society, the effects of modern knowledge, quantum information and teleportation and theoretical physics. He was a Mina Shaughnessy Scholar, a Kellogg National Leadership Fellow and recipient of the 1984 Sigma Xi Procter Prize (with Victor F. Weisskopf). He is a Five College '40th Anniversary' Professor and winner of their Jackie Pritzen Prize for public scholarship.

Michel Bitbol is *Directeur de Recherche* at the Centre National de la Recherche Scientifique, in Paris, France. He is presently based at the *Archives Husserl*, a centre of research in phenomenology. He worked as a research scientist from 1978 to 1990, specializing in biophysics. From 1990 onwards, he turned to the philosophy of physics, working on a neo-Kantian interpretation of quantum mechanics. In 1997, he was the recipient of an award from the *Académie Des Sciences Morales et Politiques* for his work in the philosophy of quantum mechanics.

Jayant Burde graduated with a degree in physics and received his M.Sc. degree in mathematics from Bombay University. He also holds a degree in law from Bangalore University and is a Certificated Associate of the Indian Institute of Bankers. His published papers contain mathematical models in finance, costing and organizational structure. He has authored many books, namely, *Rituals, Mantras and Science*, *The Mystique of Om*, *The World of Rhythm Called Rituals*, *Buddhist Logic and Quantum Dilemma*, *Śūnya and Nothingness in Science, Philosophy and Religion*, *Philosophy of Numbers* and *Jaina Relativism and Relativity Physics*. Most of his books are interdisciplinary having a strong philosophical core.

Geshe Dorji Damdul is Director of Tibet House, Cultural Center of H. H. the Dalai Lama, New Delhi, since March 2011. He worked as the official translator to H. H. the Dalai Lama since 2005. In 2003, the Office of H. H. the Dalai Lama sent him to Cambridge University, England, for proficiency English studies. He was a Visiting Fellow at Girton College, University of Cambridge. He is an erudite scholar of Buddhist philosophy, epistemology and logic.

Debajyoti Gangopadhyay studied pure physics at the Universities of Calcutta and Baroda (M. S. University). He is Professor in Annada College, Vinoba Bhave University, Hazaribagh. He has been engaged during the last 10 years to figure out some 'meaning' overlap between the foundational issues in physics and philosophy in its Eastern and Western versions.

Alex Hankey After being a senior scholar at Rugby School, he won a scholarship to Trinity College at the University of Cambridge, where he gained a triple First in the

Natural Sciences Tripos and led university teams at two sports. After a year at Stanford Linear Accelerator Center, he joined universities established by His Holiness Maharishi Mahesh yogi, for whose organization he worked for about 30 years. Now he is associated with Swami Vivekananda Yoga University at Bangalore.

Bijayananda Kar retired as Professor and Head of the Department of Philosophy at Utkal University, Bhubaneswar. He was senior fellow of the Indian Council of Philosophical Research and also of the Indian Institute of Advanced Study, Shimla. He is an author of several books and research papers.

Radhey Shyam Kaushal, M.Sc. (Phys., AMU), Ph.D. (Phys., IIT/K), Ph.D. (Phil., DU) and an Alexander von Humboldt Fellow, is presently working as Guest Faculty in the Department of Physics and Astrophysics, University of Delhi, after superannuation in 2009. The author of five books and the co-author of a sixth, Dr. Kaushal have published more than 110 research papers and has more than 1100 citations in the fields of theoretical nuclear and particle physics, classical and quantum mechanics, dynamical systems and in philosophy of science.

Rana Purushottam Kumar Singh started his academic career as an Assistant Professor in the Department of Buddhist Studies, University of Delhi, after completing M.Phil. and Ph.D. from the same university. In 2010, he was appointed Assistant Professor in the Department of Pali, Nava Nalanda Mahavihara, Nalanda, Bihar, a deemed university under the Ministry of Culture, Government of India. Dr. Rana has done field work at different archaeological sites in Thailand, Cambodia, Vietnam, the Philippines, Pakistan, Malaysia, Sri Lanka, China, etc. in search of tangible as well as intangible Buddhist heritage. In 2015, he was sent to Medan and Jakarta as expert on Buddhism during the Buddha Carika Exhibition organized by the Ministry of Culture, Government of India.

Geo Lyong Lee majored in Indian philosophy and yoga. He obtained M.Phil. in 1990 from the Radhakrishnan Institute for Advanced Studies in Philosophy, Madras University. He also received Ph.D. degree in 1995 from Delhi University. He is Professor and Dean of Graduate School of Integrative Medicine, Sun Moon University, Korea. He was a Vice-Chancellor of the Seoul University of Buddhism, Korea, and a Visiting Professor at Madras University between 2005 and 2006 to teach Korean language and Korean Buddhist thought. He is a Principal of Korea Yoga Academy Leeashram established by him in 2006.

Rajeshwar Mukherjee is the research officer at the Philosophico-Literary Research Department, Kaivalyadhama, which is a research centre recognized by the Pune University and supported by the Ministry of Human Resource Development, Government of India. He did his Ph.D. on the topic ‘The World Order and Consciousness: Synthesis of Mathematical Physics and Vedanta Philosophy’. His research interest, in general, comprises science of consciousness, quantum interpretations, modern physics vis-a-vis Indian philosophy, Vedanta, yoga, Buddhism, etc. He collaborates with the NRCVEE-Indian Institute of Technology Delhi in a major research project. He has more than 15 national and international level research publications to his credit. He is the editor of three books.

Amalkumar Mukhopadhyay obtained an M.Sc. in pure mathematics following a good honours degree in mathematics from the University of Calcutta. Having uniformly brilliant academic career, he learnt Sanskrit grammar and Siddhanta Jyotish and obtained a Ph.D. in history of mathematics in 1950. He also studied statistics in evening, diploma course in Indian Statistical Institute, Baranagar, Kolkata, and also law. He retired from a Group A post under the Government of West Bengal.

Bhakti Vijnana Muni has obtained Ph.D. in chemical engineering from IIT Kharagpur with a thesis on nanofiltration, an advanced oxidation process. He has received education in the Vedantic tradition of Sri Chaitanya Saraswat Math, Nabadwipa, India, and Bhakti Vedanta Institute, Princeton, USA. He is sannyasi and is engaged in the scientific presentation of the Vedantic Wisdom. Now he is serving as the President of Sri Chaitanya Saraswat Institute.

Shreekala Nair is Professor and Head of the Department of Philosophy at the Sree Sankaracharya University of Sanskrit, Kalady, Kerala. She has been Visiting Professor in several universities and research institutes in India and abroad and associated with many academic bodies there. She has several publications to her credit.

Surendra Singh Pokharna did B.Sc., M.Sc. and Ph.D. in physics from the Udaipur University, Rajasthan, India. His Ph.D. thesis was on theoretical study of liquid helium. He also did postdoctoral in biophysics from the Maharishi Indian Research Academy in collaboration with the Indian Institute of Science, Bangalore, and postgraduate diploma in operations research from the Operational Research Society of India. He worked as an assistant professor of physics from 1979 to 1986 at Sukhadia University, Rajasthan. He was a Senior Scientist at Space Applications Center, Indian Space Research Organization, from 1986 to 2004. At ISRO, he worked on the applications of remote sensing technology and modelling techniques to study the properties of land resources in general and agriculture in particular. He was a chief operating officer of Juriscape Legal Research Pvt. Ltd., a unit of Hi-Tech Outsourcing Services from 2005 to 2011. Presently, he is acting as a consultant at the same company and also associated with the Bhagwan Mahaveer International Research Center (BMIRC) of Jain Vishwa Bharti Institute (JVBI), Ladnun, Rajasthan, and pursuing multidisciplinary research on consciousness, quantum physics, science and Jainism in particular.

Tabish Qureshi is professor of physics at the Centre for Theoretical Physics, Jamia Millia Islamia, New Delhi. His research interests are foundations of quantum mechanics, quantum entanglement and quantum information.

S. S. Rama Rao Pappu got his B.A. from Panjab University in 1961, M.A. from the University of Delhi in 1963 and Ph.D. from Southern Illinois University, USA, in 1968. He taught philosophy in Miami University, Oxford, Ohio, USA, for 45 years retiring in 2013. The National Endowment for the Humanities, USA, appointed him as Visiting Philosopher to Wittenberg University, USA. He is currently professor of philosophy at Miami University, USA, and honorary professor, GITAM University, Visakhapatnam.

S. Rammohan holds a Ph.D. degree, besides six master's degrees in subjects like philosophy, strategic studies, economics, M.B.A. Sanskrit, Tamil and Spanish. He was a member of Civil Service (IRAS) and retired with the rank of Special Secretary to Government of India. He is the Editor of *Mountain Path* in English and *Ramanodayam* in Tamil.

Anand Rangarajan studied electrical engineering at the Indian Institute of Technology Madras, getting his B. Tech in 1984. He then pursued graduate studies at the University of Southern California, Los Angeles, and obtained a Ph.D. in 1991 focusing on the area of computer vision. After his graduate work, he joined Yale University, first as a postdoctoral associate and later as tenure tracks faculty in diagnostic radiology (focused on medical image analysis). In the year 2000, he moved to the Computer Science Department at the University of Florida, Gainesville, where he now works in machine learning, computer vision and the science of consciousness.

Devulapali V. Rao had a brilliant academic record at Andhra University, India, where he got the B.Sc. (Honours), M.Sc. and D.Sc. degrees in physics and also taught for 2 years. He spent 2 years each at Duke and Harvard Universities as a Postdoctoral Fellow. He has been teaching at the University of Massachusetts, Boston, since 1968 where he is currently distinguished professor of physics. He also holds Adjunct Professor Position at Amherst and Lowell campuses where he guides Ph.D. students. He was elected as a Fellow of the American Physical Society, Division of Laser Science in 2010.

Sisir Roy is a theoretical physicist now working as T.V.Raman Pai Chair Visiting Professor, National Institute of Advanced Studies, IISc Campus, Bangalore. Previously he worked as professor at Physics and Applied Mathematics Unit, Indian Statistical Institute, Kolkata, during 1993–2014. Professor Roy did his postdoctoral in Institut Henri Poincaré, Paris, France. He worked as distinguished visiting professor in many US and European universities. He is member of editorial boards of various international journals and editor of e-book series.

Bhakti Niskama Shanta graduated in mechanical engineering from Utkal University in the year 2000. He obtained his master's degree from the Department of Mechanical Engineering of the Indian Institute of Technology, Guwahati, on fluid and thermal science, in the year 2003. He obtained his Ph.D. on coastal hydrodynamics, from the Department of Ocean Engineering of Indian Institute of Technology, Kharagpur, in the year 2008. He worked as an invited Scientist in Korea Ocean Research and Development Institute from May 2007 to May 2008. He is now serving as Secretary in Sri Chaitanya Saraswat Institute.

Rashmi M. Shetkar studied clinical and applied psychology and cognitive neuroscience and is a trained management graduate, with skills in computational programming and software engineering. She is a doctoral scholar at Swami Vivekananda University and an exchange visiting research scholar at the University of Florida, USA. She has a diverse experience in industry, academics and research. From the carrier spanning from last 14 years, she has worked in ERP-SAP implementation for D-Link India Ltd. and IBM, for 4 years. She was soon absorbed by her alma mater at Goa Institute of Management as an Assistant Professor, after finishing her MBA, and is now a doctoral scholar, neuroscientist and vedantist at the Swami Vivekananda University, Bangalore.

Bal Ram Singh, Ph.D., has been a professor since 1990 at the UMass Dartmouth (until 2014) and Institute of Advanced Sciences (current), conducting research on botulinum and tetanus neurotoxins and also on yoga, mind and consciousness. He is the President of the Institute of Advanced Sciences. He has been visiting professor at Georgetown University, Harvard Medical School, National Yang-Ming University (Taiwan) and Jawaharlal Nehru University (India).

Kashyap Vasavada was a research associate with the NASA at the Goddard Space Flight Center, Greenbelt, Maryland, USA, from 1964 to 1966. After that, he became an assistant professor of physics at the University of Connecticut at Storrs, Connecticut, USA, from 1966 to 1970. From 1970 to 1974, he was an associate professor of physics at the Indiana University-Purdue University Indianapolis, Indiana, USA. There, he became professor of physics in 1974, retired in 2003 and has been an emeritus professor at the same institution since 2003. He was a visiting professor at Cornell University, Ithaca, New York, USA, in 1985–1986 and a visiting scientist at the University of California, Berkeley, California, USA, and Stanford University, Palo Alto, California, USA, in 1976. In addition, he has held visiting positions at a number of physics laboratories. He has published a number of scientific articles in refereed international scientific journals.

Table of Diacritical Marks

Key to Transliteration (Using Diacritical Marks)

अ	इ	उ	ऋ
a	i	u	ṛ

आ	ई	उ	ए	ऐ	ओ	औ	अं	अःa
ā	ī	ū	e	ai	o	au	am	aḥ

Classified Consonants

क	ख	ग	घ	ङ
k	kh	g	gh	ṅ
च	छ	ज	झ	ञ
c	ch	j	jh	ñ
ट	ठ	ड़	ढ	ण
ṭ	ṭh	ḍ	ḍh	ṇ
त	थ	द	घ	न
t	th	d	dh	n

प	फ	ब	भ	म	म् (intermidate)
p	ph	b	bh	m	ṃ

Unclassified Consonants

य	र	ल	व	श	ष	स	ह	क्ष	त्र	ज्ञ
Y	**r**	**l**	**v**	Ś	**ṣ**	**s**	**h**	**kṣa**	**tra**	**jña**

Anusvaraḥ-(·)	⇨	**ṁ**
Visargaḥ-(:)	⇨	**ḥ**

Part I
Theory of Quantum Reality

Chapter 1
Quantum Reality, Spiritual Concepts, and Modern Optics Experiments

Devulapali V. Rao and Lalitha D. Rao

Abstract Truth is One; wise men call it by various names. Both religious/spiritual philosophers and modern scientists have the same goal – pursuit of the ultimate truth. Science looks at "objective reality" independent of the human observer, whereas old Indian mystics adopted a holistic approach involving the observer and the observed – isolating the two is arbitrary. In terms of basic concepts, quantum theory contains some parallels to Hindu scriptures – the fuzzy and nebulous world of atoms sharpens into the world of reality only when an observation is made, similar to the Hindu concept of Aham Brahmasmi (I am Brahman). The well-known Schrodinger's cat is a thought experiment illustrating this bizarre concept. The famous physicist John Wheeler's thought experiment about the observer creating even retroactive reality is spectacularly confirmed by recent experiments on helium atoms scattered by laser light. Abstract concepts in Hindu scriptures are getting translated to real-world objective demonstrations in modern science and technology. For example, the Sanskrit sloka "*Om Pūrṇamadaḥ, Pūrṇamidam, Pūrṇāt Pūrṇamudacyate, Pūrṇasya Pūrṇamādāya, Pūrṇamev āvaśiṣyate*" can be illustrated by a hologram (Pūrṇa or Whole) recorded by coherent light generated by lasers. The *śloka* implies: That (pure universal consciousness) is full (perfect); this (manifest universe of matter, individual consciousness) is full. This fullness has been projected from that fullness; what remains is fullness. When the hologram is illuminated by light, the whole event is displayed as virtual reality, close to the concept of *Māyā*. One can cut the hologram into any number of small pieces; each piece contains the whole information and displays the event sharply.

Keywords Quantum and reality · Quantum mechanics · Hologram · *Śūnyāta* · *Purṇata* · *Brahman*

D. V. Rao (✉) · L. D. Rao
Physics Department, University of Massachusetts, Boston, MA, USA
e-mail: raod@umb.edu

S. R. Bhatt (ed.), *Quantum Reality and Theory of Śūnya*,
https://doi.org/10.1007/978-981-13-1957-0_1

1 Introduction

The two revolutionary theories of relativity and quantum mechanics provide the framework for all modern physics. In terms of some of the basic concepts, they can be compared to the Hindu scriptures Bhagavadgītā and Yogavāśiṣtha. Max Planck suggested in 1900 that heat is radiated from any hot body as small discrete packets or quanta. Einstein considered particles of light called photons. On the other hand, physicists were already familiar with the wave nature of electromagnetic radiation such as light and heat. A few years later, experiments indicated that particles of matter such as electrons also display wavelike behavior. In the microscopic atomic world, physicists encountered the bizarre behavior of wave-particle duality. One cannot clearly say whether any basic elementary particle, photon, electron, proton, or neutron, is actually a wave or a particle. Heisenberg's uncertainty principle is also a basic tenet of quantum mechanics. It states that at a given time, the position and momentum (velocity times mass) cannot both be determined precisely; if one is known exactly, then there is more uncertainty in the other parameter to compensate so that the product of the two cannot be below a fixed value equal to Planck's constant h. One can experimentally determine the value of this constant in the lab. This quantum uncertainty is not due to lack of sophisticated instrumentation but is intrinsic in the very nature of things. Quantum tunneling is a well-known phenomenon where moving particles suddenly disappear at one place and appear at another place. They can cross barriers, swerve around corners, and exist in many places at once. There is experimental evidence for this peculiar behavior. In fact it has many applications in technology, for example, in transistors. In the macroscopic world, this is similar to a Yogi confined to a locked room appearing elsewhere transcending all barriers.

According to Niels Bohr, one of the founders of quantum mechanics, the fuzzy and nebulous world of the atom sharpens into concrete reality only when an observation is made. In the absence of an observation, the atom is a ghost. It only materializes when you look for it and you can decide what to look for! Further the reality that the observation sharpens into focus cannot be separated from the observer and his choice of measurement strategy. If all this seems paradoxical or mind-boggling to accept, Einstein would have agreed with you. The classic remark "anyone who is not shocked by quantum theory means he has not understood it" is attributed to Niels Bohr (Fig. 1.1).

The well-known Schrödinger's cat is an interesting thought experiment illustrating the weirdness of the quantum mechanical concepts. The cat is in a sealed box which contains a bottle of cyanide poison. There is also a radioactive source in the box which radiates particles at random. When a particle is emitted, it triggers a mechanism to break the poison bottle and the cat dies. This is an absolutely random process. According to the tenets of quantum mechanics, the cat should exist in both the possible states, live or dead, until an observer opens the door and looks for it.

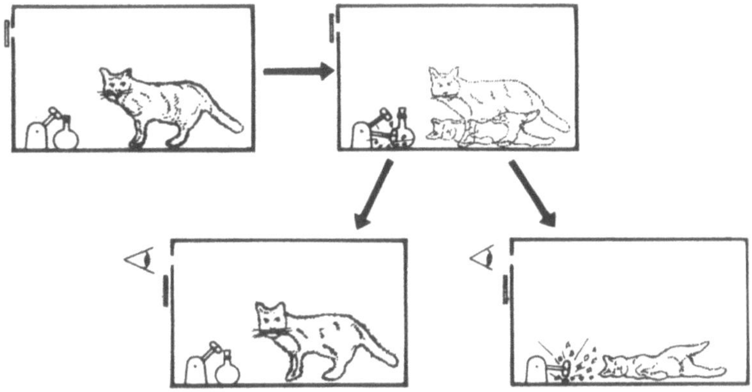

Fig. 1.1 Schrodinger's cat

Fig. 1.2 Wave interference – whenever a crest coincides with a trough, the water surface is flattened (decrease in amplitude). A crest coinciding with a crest yields a double crest (increase in amplitude)

The study of hybrid states of the quantum world, "entangled states," is a hot topic of current interest in basic research and also for potential applications in quantum computation and communications.

Let us now look at a few simple experiments in optics which a college student carries out in introductory physics course. Everybody is familiar with waves in a pond (Fig. 1.2), when a stone is dropped and the interference of two waves causes enhanced crests and troughs.

In the classical interference experiment in the lab (Fig. 1.3), a laser beam illuminates the two narrow vertical slits A and B, and the pattern is observed on screen C. The experiment can also be done using a mercury lamp as a source with the light confined to a pinhole going through a narrow band color filter, but with a laser

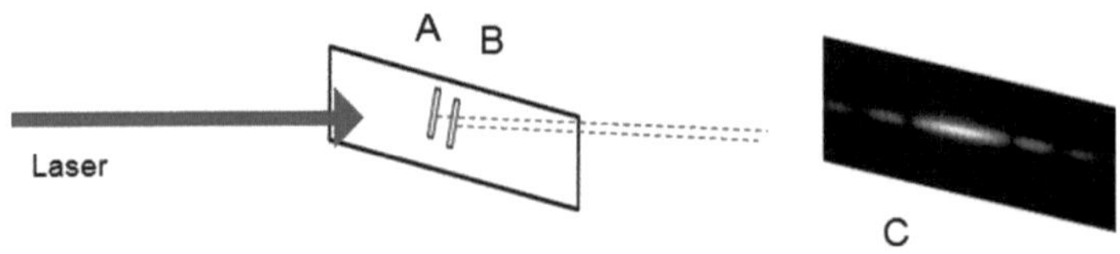

Fig. 1.3 Experimental arrangement for double-slit interference. A, B, narrow vertical slits; C, screen showing the interference pattern

Fig. 1.4 Delayed choice experiment. A, B, photodetectors; V, venetian blind

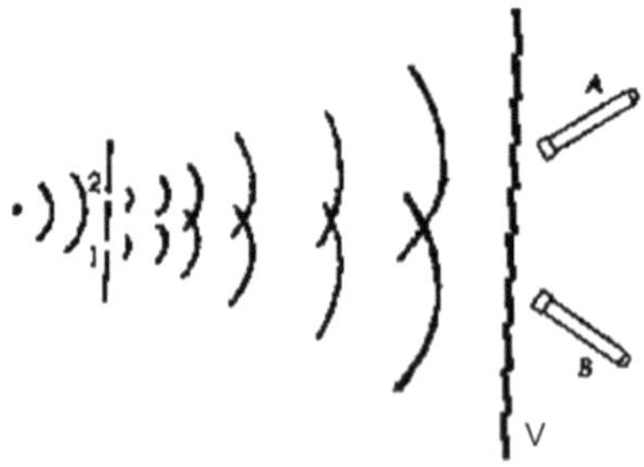

which is highly monochromatic (sharp in color), bright, and highly directional, the experiment is easy and the picture is sharp. The pattern observed contains a series of bright and dark bands called interference fringes. Depending on the light from the two sources arriving at the screen in-step or out-of-step at the location of the screen, we see a bright or dark band, thus demonstrating the wave nature of the light.

This is not the whole story. When the intensity of the laser is turned down so low that only one photon at a time is radiated, given long enough time, the same interference pattern is observed. The weirdness is obvious, any particular photon can pass through only one hole, but interference requires two overlapping waves one from each slit. Experiments with other quantum particles like atoms, electrons, or neutrons yield the same kind of interference pattern. In terms of the photon model, this looks bizarre. A wave can go through both the slits and recombine generating interference, but how can a particle go through both slits, sometimes expressed by saying that the photon went through both the slits – it was in two places at the same time! Bohr's interpretation goes like this. Think of the case when the photon goes through slit A as one possible world A and when it goes through B, world B. Both these worlds A and B have to be presently superimposed. We cannot say that the world of our experience is either A or B but is a genuine hybrid of the two. A significant factor to be mentioned is that the interference pattern is observed only if no attempt is made to determine which slit the photon has gone through. If this is done in the experimental arrangement, no interference is observed. This is illustrated in Fig. 1.4 for the delayed choice experiment where the screen is replaced with a venetian blind and two detectors A and B behind.

When the blind is closed, the detectors are not operative, and there is no way to know which slit the photon has gone through, and the usual interference pattern is observed. When the blind is open, the detectors can see the path of the photon, and

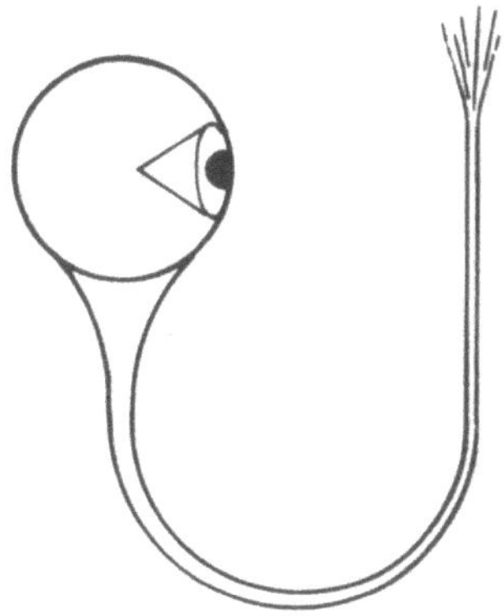

Fig. 1.5 Universe as a self-observing system. The tail represents early stages of the universe promoted to concrete reality through later observation by consciousness which itself depends on that reality

the observer knows whether it is from A or B, and so no interference is observed indicating particle behavior. Thus the observer decides the outcome of the experiment! If he wants to see the photon particles, he opens the blind. If he wants to see wave behavior, he closes the blind. The most interesting case is the experimenter can delay the choice of wave or particle, right up to the moment the photon arrives at the venetian blind thus delaying the decision until after that world comes into existence. The precise nature of reality has to await the participation of a conscious observer. Mind can be made responsible for the retroactive creation of the reality – even a reality that existed before people – retroactive causation. This bizarre possibility is illustrated by the famous physicist John Wheeler as shown in the Fig. 1.5.

Recently physicists at the Australian National University, A.G. Manning et al., actually conducted "Wheeler's delayed choice gedanken experiment," seemingly impossible when it was proposed in 1978, with atoms scattered by laser light successfully [1].

As of now, the quest for finding God through science according to some cosmologists[1] and physicist turned theologians has come to a dead end. Some of them are taking a critical look at eastern mysticism: "Reality is One. The practice of mysticism consists in finding ways to experience this unity directly. The One has variously been called the Good, God, the Cosmos, the Mind, the Void, or (perhaps most neutrally) the Absolute. No door in the labyrinthine castle of science opens directly onto the Absolute. But if one understands the maze well enough, it is possible to jump out of the system and experience the Absolute for oneself...But, ultimately, mystical knowledge is attained all at once or not at all. There is no gradual path...."

Abstract concepts in Hindu scriptures are getting translated to real-world objective demonstrations due to recent advances in science and technology. The spiritual scientist looks at the full circle with no beginning and no end as representing the

---

[1]For an in-depth study of these aspects, the reader is referred to books by Paul Davies:

The mind of God (ISBN 0-671-52806-8)
God and the new physics (0-671-79718-2)
Cosmic jackpot (0-618-59226-1)

concepts of both *pūrṇa* (fullness) and *Śūnya* (void). *Śūnya* also represents the absence of mind and fullness of spirit. The very term *adhyatma* means that which has no beginning and no end. According to their viewpoint, science is like the half circle C which begins at one end and ends at the other with a gap between the two points which is invisible and incomprehensible like dark matter.

The Invocatory *śloka* from Īśopaniṣhad [2]:

> *Om pūrṇamadaḥ pūrṇamidam pūrṇāt pūrṇamudacyate. Pūrṇasya pūrṇamādāya pūrṇamev āvaśiṣyate.*

It means, "That is full, this is full, from full arises full, when this full is taken out from that full, the full still remains full." The full circle with no beginning and no end represents spirituality. The concept is also familiar to mathematicians talking about infinity subtracted from infinity is still infinity. This abstract concept can be spectacularly demonstrated with the hologram of modern optics made possible with lasers.

The normal two-dimensional photograph is a record of light intensity scattered off the object when it is illuminated by light. A stereoscopic picture is a three-dimensional view of the object synthesized from two views of the object recorded in two mutually perpendicular directions by the human brain. The stereoscopic picture provides a three-dimensional view of the object, but it is still one view of the scene. On the other hand, the hologram is a record of the whole event – one cannot perceive any difference between the actual object and its holographic display. Many airports display holographic image of a hostess welcoming passengers. We cannot discriminate between this virtual reality and the person. This comes close to the Hindu concept of Māyā. Here is a brief description of the process involved recording holograms. Light represented by a wave contains two parameters – amplitude which represents the height of the crest or the depth of the trough and phase which represents the state of vibration at a given time on the amplitude versus time plot. In nature everything has a rhythm. Ordinary light is "incoherent," and only one parameter, intensity which is an averaged value of the square of the amplitude, can be recorded. Phase is jumbled and is not accessible. With the advent of "coherent" light with lasers since the 1960s, both the parameters can be recorded, and so information about the scene is available for the first time, and hence it is called a "hologram," the whole. Light from the laser source illuminates the object. A small fraction of the same light obtained from a beam splitter as shown in Fig. 1.6 serves as a reference beam. The interference pattern of the light scattered off the object and the reference beam is recorded on a photo sensitive plate.

The "coded" pattern, looking like bright and dark smudges on the plate, when illuminated with laser light or even ordinary light generates the whole event in virtual reality. Looking at the generated view, we may not be able to distinguish between this and the actual scene. There is another significant feature of the hologram. When you cut it into pieces, each piece contains all the information and displays the whole scene. It can be cut into any number of tiny bits, yet each bit contains the whole and is capable of displaying the whole event. It may not be as bright due to lack of sufficient intensity. The famous quantum physicist David Bohm

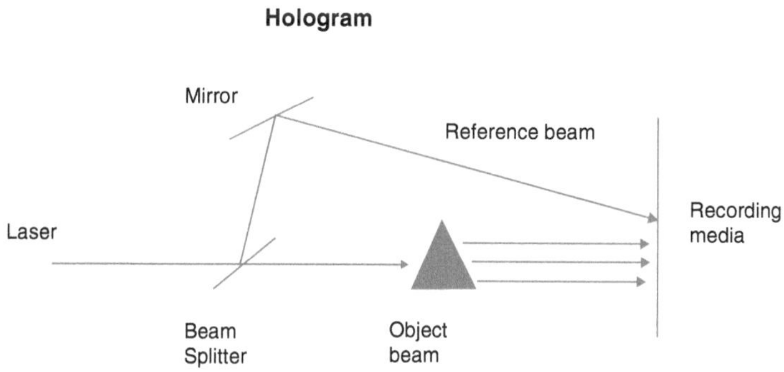

Fig. 1.6 Schematic of recording a hologram

extended this principle to the universe at large. Our world is constructed on the same holographic principles. The world is an indivisible whole and each human being represents the whole. The universe is a super hologram in which individual brains are actually indivisible portions of the whole and everything is interconnected. The well-known neurophysiologist of Stanford University Karl Pribram firmly believes in the holographic nature of reality and that the human brain uses holographic principles to mathematically convert the frequencies of the signals received through the senses into the inner world of our perceptions. Holography offers an astounding capacity for information storage achieved by simply changing the angle between object and reference beams in the same region of space of a few cubic centimeters. Pribram's theory explains how the human brain can store so much memory in so little space. The human brain has the capacity to memorize something of the order of ten billion bits of information, about the same amount contained in five sets of *Encyclopedia Britannica*. The most mind-boggling aspect [3] of Bohm and Pribram's view is that the universe itself may be a giant hologram, quite literally a kind of image or construct created, at least in part, by the human mind. We are of a universe in which individual brains are actually indivisible portions of the greater hologram and everything is infinitely interconnected. In their opinion this remarkable new way of looking at the universe explains not only many of the unsolved puzzles of physics but even religious and mystical experiences such as feelings of cosmic unity and miraculous healings.

A recent popular article in the Sunday Review, February 20, 2015, "The Reality of Quantum Weirdness," compares the situation with Akira Kurosawa's famous film "Rashomon." A samurai was murdered; each person supposedly involved presents a contradictory version. This movie provokes a deeper inquiry: Is there a true story, or is our belief in a definite observer-independent reality an illusion, reminding us of Einstein's famous question "Do you really believe the moon is there only when you look at it?" Frenkel concludes the article suggesting that we regard the paradoxes of quantum physics as a metaphor for the unknown infinite possibilities of our own existence elegantly expressed in the Vedas "As is the atom, so is the universe; as is the microcosm, so is the cosmic body; as is the human mind, so is the cosmic mind."

Christopher Fuchs [4] of the University of Massachusetts Boston describes physics as "a dynamic interplay between story telling and equation writing. Neither one stands alone, not even at the end of the day." The theory of quantum Bayesianism or QBism developed by Fuchs and his collaborators Caves and Schack is successful in solving quantum paradoxes, but the Bayesian probabilities used involve subjective degrees of belief about the system. Though quantum theory is very successful in explaining all experimental observations and facilitated many technological advances resulting in laptops, smartphones, and iPads, after about a century of debate, physicists and philosophers are still arguing about the nature of quantum reality [5]. One of the founders of modern quantum theory Wolfgang Pauli commented on the wrong-headed hypothesis of the "detached observer" [6]. His interactions with the famous psychologist Carl Gustav Jung provide a landmark forum of science and philosophy dealing with the psychophysical [6].

It may not be appropriate to compare the subjective element of the quantum world with religious or spiritual mysticism. The thoughts of mystics have no bounds of space-time, and their experiences are unique and indescribable in words. The void of Buddha or Hindu scriptures is not the same as the void of quantum physics where all kinds of elementary particles pop up and vanish. The Heart Sūtra of Mahāyāna Buddhism[2] implies that form is emptiness and emptiness is form. The Absolute in reality is formless, without name or attribute. It refers to a manifestation of the Absolute which is the totality of forms of all kinds both physical and mental. It is only through the intuitive wisdom of *Śūnyatā* that one is ferried across to the shore of liberation. The concept of no difference between *Śūnya* and *Pūrṇa* is implied in the Mahāyāna Buddhist Nāgārjuna's *śloka* from the Mādhyamika-Kārika:

Śūnyamiti na vaktavyam, aśūnyamiti vā bhavet
Ubhayam nobhayam ceti, prajñaptyartham tu kathyate.

Thomas Merton[3] experienced the fullness of his Christian faith in the fullness of the Hindu Upaniṣads noted in his spiritual diary dated December 24, 1971:

A being lost in my source, a being lost in my fulfillment. And in this very loss, I am.

Jesus is this mystery of Advaita in which I can longer recognize myself separately. Lost as much is space (*ākāśa*) of the heart as in the shining, the radiance that empties me.

And I am the Fullness, *pūrṇam*, precisely in this letting-go of myself everywhere. . . .

And my *pūrṇam* is precisely this emptiness of all self. The kenosis (self-emptying) of Christ.

The paradox of Void and Fullness is discussed at length by Paddy Meagher[4] from the perspective of Christian sacred writings in her article – The Pleroma of God, Jesus Christ and the Christian.

[2]Samdhong Rinpoche Śūnya and Purna Ref. 2. Pages 19–32.

[3]Thomas Merton Ref. 2. Pages 7–8.

[4]Paddy Meagher The Pleroma of God, Jesus Christ and the Christian Ref. 2. Pages 49–82.

The Greek word pleroma which means "that which has been filled" or "that which fills" or "with which a thing is filled" is used to express plenitude/fullness/divine perfection of God.

The *śloka* from Īśopaniṣad [2] "Om pūrṇamadah..." already quoted in this article implies the interpretation that when the infinity of the *Ātman* is deducted from the infinity of *Brahman*, the remainder is still infinite. The English version of the *śloka* goes like this:

Om that or the Reality is the embodiment of fullness. This (visible) world is also of the nature of fullness. From the Reality, the embodiment of fullness, the world which is infinite, emanates. (Thereafter) the fullness of the nature of the Reality remains as it was before even after the emergence of the world which is fullness embodied.

The word *Pūrṇa* connotes the nature of the Ultimate Reality, which is concisely expressed as *This* and *That*. This is the world we live in and know. There is something beyond This which is called That. It looks like we can never be there (physicists concede that as of now a single objective reality is an illusion) which may be the reason why spiritual mystics called it That. Heisenberg's principle of uncertainty perhaps implies the same concept.

References

1. Manning AG, Khakimov RI, Dall RG, Truscott AG. Wheeler's delayed-choice gedanken experiment with a single atom. Nat Phys. 2015. https://doi.org/10.1038/nphys3343.
2. Brata D, Sharma S. Concept of fullness or Purnatva in Indian philosophy. In: Baumer B, Dupuche JR, editors. Void and fullness in the Buddhist, Hindu and Christian traditions. New Delhi: D.K. Printworld; 2005.
3. Talbot M. The holographic universe. New York: Harper Collins; 1991.
4. Gefter A. A private view of quantum reality. http://www.quantamagazine.org/20150604-quantum-bayesianism-qbism/.
5. Friedman A. The nature of quantum reality. http://www.pbs.org/nova/blogsysics/2015/05/. See also Herbert N Quantum Reality, New York: Doubleday; 1985.
6. Von Baeyer HC, Fierz M. His character and his worldview. In: Atmanspacher H, Fuchs CA, editors. The Pauli-Jung conjecture and its impact today. Exeter: Imprint Academic; 2014.

Chapter 2
Can A Quantum Field Theory Ontology Help Resolve the Problem of Consciousness?

Anand Rangarajan

Abstract The hard problem of consciousness arises in most incarnations of present-day physicalism. Why should certain physical processes necessarily be accompanied by experience? We begin with the assumption that experience cannot exist without being accompanied by a subject of experience (SoE). Strawson has elaborately defended the notion of a thin subject—an SoE which exhibits a phenomenal unity with different types of content (sensations, thoughts, etc.) occurring during its temporal existence. Next, following Stoljar, we invoke our ignorance of the true physical as the reason for the explanatory gap between present-day physical processes (events, properties) and experience. We are therefore permitted to conceive of thin subjects as related to the physical via a new, yet to be elaborated, relation. While this is difficult to conceive under most varieties of classical physics, we argue that this may not be the case under certain quantum field theory ontologies. We suggest that the relation binding an SoE to the physical is akin to the relation between a particle and (quantum) field. In quantum field theory, a particle is conceived as a coherent excitation of a field. Under the right set of circumstances, a particle coalesces out of a field and dissipates. We suggest that an SoE can be conceived as akin to a particle which coalesces out of physical fields, persists for a brief period of time, and then dissipates in a manner similar to the phenomenology of a thin subject. While it is odd at first glance to conceive of subjects of experience as akin to particles, the spatial and temporal unity exhibited by the particle as opposed to fields and the expectation that selfons are new kinds of particles paves the way for cementing this notion.

Keywords Classical physics · Quantum field theory · Consciousness · Physicalism · Physical processes · Subject of experience · Thin subject · Particle

---

A. Rangarajan (✉)
Department of Computer and Information Science and Engineering, University of Florida, Gainesville, FL, USA
e-mail: anand@cise.ufl.edu

S. R. Bhatt (ed.), *Quantum Reality and Theory of Śūnya*,
https://doi.org/10.1007/978-981-13-1957-0_2

1 The Hard Problem of Consciousness

In its simplest formulation, the hard problem of consciousness is stated as follows: "Why is anything accompanied by experience?" To elaborate using naturalistic language, why are some physical processes and events associated with an interior—a qualitative space which is felt from the inside? The presence of experience forces us to acknowledge that nature has an interior, private, and qualitative space and that there is no obvious a priori reason for this to exist. In recent times Chalmers [5] often gets the credit, and rightly so, for putting the hard problem in its starkest terms. The hard problem of consciousness does not imply a Cartesian subject or soul as the rest of this paper should make clear. Further, and this is even more relevant to an Indian audience, no a priori transcendental idealist stance can dissolve the hard problem: the key issue is interiority pure and simple, and this needs to be explained. For a physicalist the problem is "why are physical processes accompanied by experience?," and likewise for a transcendental idealist, the problem is "why is Brahman accompanied by interiority?," though the latter is not something that one hears very often.

The physicalist response in the past few decades has mainly been *emergence* [9]. My colleagues, for example, constantly seek to reassure themselves by asserting that under the right set of conditions and at the appropriate level of complexity, consciousness pops out from brain activity. The conditions are rarely spelled out: "We don't know enough about the brain" is the refrain and the complexity measures (and necessary thresholds) never formulated. Two other defense mechanisms stand out. If it is pointed out that brain mechanisms are actually supervenient on a more basic level of physics, the charge is one of reductionism. If new relations are postulated to bridge the explanatory gap between brain processes and experience, the charge is one of sui generis interactionism. With these two moves, further philosophy is discouraged and tacit commitments to naïve identity theories left unchallenged and in their place—with the complexity problem of emergence left unaddressed. Henceforth, we use the term *taboo physicalism* for this approach cluster since it encompasses emergentist and eliminativist approaches alike while seeking to corral the mind-body problem for its exclusive preserve. Emergence (and eliminativism [7]) should be carefully separated from *holism* (as will become clear). The various emergentist approaches usually get started by pointing out the existence of lively self-organizing phenomena which adhere to the second law of thermodynamics in letter while violating its spirit. The mind can then be seen as "popping out" from the brain at a suitable level of complexity. In sharp contrast, holism asserts that compositionality is not mere combination and that mind could be a holistic property of matter. Holism goes back to Aristotle in the west and seems to be popular in Indian circles, but it is fair to say that mind-body holism has not seen a full-fledged defense in recent years. For these reasons, we separate *holistic physicalism* from taboo physicalism introduced earlier.

Idealism has seen better days in western philosophy. With roots stretching back several hundred years [30], one would have been led to expect a strong response from the idealist camp by taking a cue from Bishop Berkeley and adding German and Indian idealism to the mix. And, arguably there has not been a strong response from Eastern (and Indian) idealism either. As we have argued above, the onus is on the transcendental idealist to explain why there "seems" to be spatiotemporally bounded interiority (experience) in the universe or in *līlā*. Further, and this is again just based on anecdotal observation, recently it has become somewhat fashionable to assert that the universe is a simulation and this grounds idealist intuition. First, without additional underpinnings, the simulated reality hypothesis is still physicalist (with a quantum computer [21] or equivalent generating the simulation). Second, the question of why such a simulation is accompanied by experience still needs explanation. In other words, the explanatory gap persists in this setting as well despite the virtual reality twist. In recent years, a surprisingly strong response has come from *panpsychism* [29, 33, 30]—the ancient doctrine that qualia extend all the way down in nature and savaged by the critique that this meant that "rocks have feelings." In hindsight, this robust response is not surprising. Since complexity thresholds for the emergence of experience are hard (if not impossible) to come by, the alternative that consciousness gradually fades as one goes down in scale was always viable. In other words, interiors are complex and gradually become more and more simple as one goes to smaller scales. Complexity per se does not lead to consciousness. Instead consciousness gradually becomes more complex on an evolutionary time scale. Panpsychism faces the combination problem [26]: How do proto-experiences combine to form the full-fledged experiences of which we are familiar? Or to put it in a different way, more complex experiences do not seem to be built up from simpler experiences in the same way that, say, polypeptide chains are made up of atoms. This is an active area in philosophical circles at present. Panpsychism has to explain why consciousness is spatiotemporally localized and is a very rare occurrence in nature—unless our intuitions in this matter are seriously misguided. It may turn out that a sophisticated panpsychism will eventually emerge which is neither pan nor "psychic" but ends up assigning proto-experientiality to a small subset of nature. To summarize, while panpsychism is a genuine alternative to emergence, it faces huge problems of its own. This response to emergence also begs the question: Why can't we begin with a revamped physicalism which is not emergentist and is not a priori panpsychist? Our efforts here belong in this camp.

Another sophisticated response is neutral monism [2] stated simply as "Mind and matter are manifestations of something neutral." Here, the challenge is the explication of neutral properties, substances, etc. and how they give rise to both mind and matter. Further, there's an asymmetry between mind and matter which is not addressed in neutral monism. There is far more matter than mind in the universe. And, it's not clear why we can't begin with a physicalism which eschews matter as a starting point while accommodating experience. It is exactly the lack of a deeper physicalist theory which is the problem in most versions of neutral monism. Due to this, neutral monism often slides into a dual-aspect theory [1] wherein the interior aspect of nature is pushed all the way down with interior/exterior parallels established at various levels. Such a pan-interiority clearly has echoes in panpsychism which is why neutral monism, dual-aspect theory, and panpsychism are often discussed together. Clearly there are numerous differences between these myriad approaches. Since the present work's focus is an expanded physicalism [31] that can accommodate experience, we choose not to focus on the doctrinal differences in the above and instead move on to presenting our case.

Panpsychism's strength is that it takes consciousness as fundamental. Neutral monism goes beyond both matter and mind in search of something more fundamental. Dual-aspect theory considers nature's interiority sacrosanct. Panpsychism faces the combination problem: how do experiences combine. Neutral monism faces the fundamental problem of a lack of a deeper theory to ground the neutral "substance." Dual-aspect theory faces the problem of an a priori split between interior and exterior which is left unexplained. Finally, emergentist approaches still remain true to their anti-reductionist origins while facing the complexity problem. The expanded physicalism we envisage here has much in common with all four. The goal of this work is to suggest an expansion of physicalism [32] which, while accommodating experience, remains plausible. Indeed, we invoke ignorance of the true physical in embarking on this course. It is here that quantum field theory (QFT) ontologies are especially useful. Since the ultimate constituents of nature are still a mystery in QFT, we face the unusual situation of a physical *theory* which is effective empirically but whose physicalist underpinnings remain murky. In what follows, we examine QFT ontologies in search of ways in which experience can be accommodated. Before this, however, we need to first shore up the basic phenomenological intuition informing our physicalist expansion.

2 Subjects of Experience

If the intuition behind the hard problem is unclear to the reader, we suggest the following experiment. Pinch yourself hard, but not so hard as to hurt yourself (unless you're a masochist and are into that sort of thing). Done? As the (hopefully mild) pain recedes into the horizon, notice the immediacy, interiority, and qualitative feel (qualia) of the pain—which has an onset, a peak, and a decay. It is also spatiotemporally localized. When you pinch yourself, other people don't feel the pain (or don't appear to). For these reasons, the first phenomenological intuition guiding our approach is that experience is always associated with a subject of experience (SoE). This is hugely controversial in analytic philosophy but often taken for granted in Indian philosophy. In the Anglo-American tradition, the harsh dualism associated with Descartes of (eternal) soul and (temporal) body has largely made talk of SoEs unpalatable. Recently, however, Strawson [34] has made a yeoman effort in resurrecting the more limited concept of a "thin subject." A thin subject is an SoE which is spatially and temporally bounded, arises, and decays and has phenomenal content associated with it during its (typically) short existence. From an Eastern perspective, a thin subject is a (Mahāyānist?) middle ground between the extremes of an eternal subject (often seen in Vedānta) or selfless experiential *dharmas* (often seen in Theravada). Rather than try and correlate the thin subject with the mere self (and other nominal entities) [15], we reiterate Strawson's point that the thin subject cannot be easily denied. A thin subject of experience—exactly THAT which is conscious of reading this sentence right NOW—is a center of awareness, which is necessarily separate from the phenomenal content which arises in awareness [6]. Strawson takes this as a given and enshrines it as a thin subject of experience. The natural concern here is the doubling of mysteries: we don't understand experience, and adding an SoE to the mix seems to unnecessarily complicate matters. But, following Strawson, our strategy will be to draw an analogy between the SoE/phenomenal content/experience cluster and the object/property/process cluster. Insofar as we face serious problems with the concepts of objects, properties, and processes, we can expect to face the same in the consciousness cluster. The difference is the apparent lack of relationship between SoEs and physicalism. If an SoE can be seen as a certain type of object in an expanded physicalism, then an experience can be unpacked as a spatiotemporal property of this new type of object. To prefigure things, we plan to ground SoEs by suggesting that QFT ontologies can support a new relation in which an SoE can be seen as akin to a new kind of particle. Since QFT ontologies are still in flux, this does not entail a commitment to a new substance nor does it imply that the new particles are mere properties.

Subjects of experience do figure in western neo-Aristotelian traditions (as opposed to the analytic tradition). Recently, Lowe [22] and van Inwagen [36] have attempted to take into account SoEs in a neo-Aristotelian metaphysics. van Inwagen, in particular, poses the special composition question which we present (in modified form) below: Can an SoE be seen as a new whole but composed of existing parts such that the new whole has phenomenal properties associated with it? The difficulty with this question is that it is not clear what we mean by "existing parts" since the physicalist base may itself be unknown. The second difficulty is that "the whole is greater than the sum of the parts" implies a new conception of compositionality or holism which must be adequately defended before this project can get off the ground. Recall that we mentioned earlier the difference between holistic and taboo physicalism. van Inwagen, in posing the special composition question, is clearly hoping for a holistic theory (if not a physicalism) which we have reconfigured to try and address the problem of experience. Such an approach raises compositionality up as a new principle of wholeness (with emergence being its natural contrast). Or as a slogan, experience is a holistic and not an emergent property of the brain. The question for us then is this: Can a new compositionality relation connect SoEs to the physical? And, how is this different from being yet another emergent property?

To answer this question, we turn to the recent work of Koslicki [16]. More than any other person, Koslicki, in recent times, has rescued compositionality from becoming irrelevant. Lack of space does not permit us to do full justice to her work, so we attempt a brief summary. Koslicki has shown that a new object can be distinguished from its parts provided there are independent reasons for believing that the new object exists in the first place. For Koslicki, if the new object can be shown to be a member of a *natural kind* [20], then a weak supplementation principle can be invoked to relate the object to a set of material *and formal* parts. (The presence of formal parts in the new object clearly marks this approach as a neo-Aristotelian one.) In an exchange, rare in analytic philosophy, Koslicki engaged in a debate with an opponent who held the view that arbitrary wholes could be formed by part combination (thereby rendering the concept toothless). The upshot was that if objects can first be justified on natural kind grounds, then weak supplementation could be invoked as described above. This raises the question: What is a natural kind as opposed to an artifactual kind, and what do natural kind terms pick out in this world? Unfortunately, this is a very deep subject with the arguments mostly driven by modal logic. Instead we merely note that natural kind terms are very prevalent in biology ("a neon tetra is a *kind* of fish") and we are seeking here to apply the same common sense approach to the problem of experience (and therefore ultimately to psychology). We note the presence of the individual and the collective in the fleshing out of a natural kind. Finally, it should be clear that biological natural kind objects are clearly composites but so are motorcycles (Koslicki's example). Since we are only concerned with natural kind objects, there will be no need for discussion of the holistic properties (of the lack thereof) of artifactual objects like motorcycles. The former are our concern and not the latter since we seek to connect SoEs to natural kind objects.

We have independent reasons for believing that SoEs are natural kind objects in the (physical) world. Even the most extreme solipsist would probably (inadvertently and occasionally) grant subjecthood to another person in dialog. And, due to the fundamental role played by second-person intersubjectivity in human culture [13], the notion that SoEs form a natural kind should not be too controversial (provided one buys into the SoE concept in the first place). For at least two decades, the subdivision of consciousness studies into first, second, and third person has been in place. Consequently, the notion that the second-person "experience" acts as a bridge linking different first-person experiences can be taken for granted. Note that we are not attempting to explain intersubjectivity here. Instead, we again appeal to common sense and note that we are aware of "the other" via the second-person experience. Despite our inability to directly see into the interior of another person, we accept the existence of the other with experiences similar to ours. If this is accepted, then we have, per Koslicki, independent grounds for connecting SoEs to the physical via a new compositionality relation. The work is not yet done since the specifics of the new relation are where the real payoff will occur. And it may turn out that the new relation is not explicitly a neo-Aristotelian compositionality relation. All we really require is the grounding of SoEs as new natural kind objects which have to be related to the physical. We need to carefully expand physicalism in such a way that SoEs (which are new natural kind objects) can be accommodated [25]. And since experience is always associated with an SoE, phenomenal content can be conceptualized as spatiotemporal properties of the new SoE objects. Regardless of the ultimate status of the object/property/process cluster, SoE objects have phenomenal properties and are part of a natural process. While this expansion of physicalism could have been denied had physicalism been complete, our ignorance of the true physical combined with the murky underpinnings of QFT allows for an expansion.

3 Quantum Field Theory Ontologies and Selfons

Interpretations of quantum mechanics have been a mainstay in consciousness studies for at least 30 years [38]. The lack of a single interpretation acceptable to nearly everyone has played a role in stoking the flames. Quantum mechanics has strange features: (i) a fundamental uncertainty, (ii) wave function collapse, (iii) possibilities rather than trajectories, and (iv) entanglement, to name a few. Quantum mechanisms in the brain [12] have been suggested as playing a role in consciousness despite objections ranging from irrelevance to decoherence [10]. At this juncture, the orienting generalization for us is the simple fact that quantum mechanics (as opposed to quantum field theory) is a story regarding particles and how they (occasionally) behave like waves. We are interested in going in the opposite direction: How and when do waves (or more technically fields) behave like particles? The reason: in a field theory, particles cannot be seen as basic but instead as related to their fields. First quantization takes particles for granted and is focused on how they behave like waves in certain contexts like the two slit experiment or exhibit nonlocal correlation as in the EPR experiment. First quantization is still a mystery, undoubtedly due to Schrödinger's unusual change of variable approach [27] in a particle Lagrangian variational principle. Second quantization attempts to obtain particles from field Lagrangians and is the basis of quantum field theory. It is much less of a mystery with the basic recipe taught in QFT textbooks. In our approach, we seek to leverage our ignorance of the true physical in order to accommodate SoEs, thereby explaining experience. The essential new idea: in a nutshell, we would like to expand physicalism by conceiving of SoEs as akin to new kinds of particles related to a set of basic quantum fields.

The ontology of quantum fields is far from settled [18, 17]. A simple reason is the presence of multiple QFTs: (i) standard QFT in Hilbert and Fock spaces, (ii) algebraic QFT, and (iii) Feynman path integrals. Consequently, it is difficult to firm up metaphysical commitments. However, regardless of the availability of competing QFTs, an orienting generalization we can make is that any QFT ontology will have to deal with fields, particles, and their relations. QFT (other than Einstein's theory of gravity) is one of the most empirically successful physical theories. Indeed the standard model of particle physics [28] (which does not include gravitation) is one of the most rigorously tested QFTs. The division of matter into fermions (quarks and leptons) and bosons (Higgs, W and Z, photons, and gluons), their relations to fermionic and bosonic fields, and their interrelationships are taught in standard textbooks. Particle interactions in the presence of three forces (strong, weak, and electromagnetic) are mediated by corresponding bosons (gluon, W and Z, and photon, respectively). During the course of cosmic evolution, the quantum gravity epoch (where gravity and the other forces are unified) gives way to the grand unification epoch (where the three forces are unified but separate from gravity). This in turn leads to the inflationary epoch (in which the strong force separates from the electroweak) and then the quark epoch (where the three forces in the standard model have all separated and are distinct from gravity). The sequence of symmetry breakings which cause a unified field (except for gravity) to separate into three fields and their associated bosons and fermions is well understood. The particle-field relation plays a fundamental role in all epochs with forces between fundamental fermions mediated by bosons. Indeed, a standard move in contemporary textbooks is to present second quantization as a way in which quantum fields behave like particles. In this view, all particles arise out coherent field excitations, persist, and decay (except for the photon).

The particle-field relation is central to QFT. Fundamental issues remain as to whether particles and fields are substances, tropes, properties, or something else entirely. There are important no-go theorems in QFT such as Malament's [23, 17] which seem to exclude the possibility of spatiotemporally localized particles at all. These are typically augmented with FAPP (for all practical purposes) principles in which particles reappear with a FAPP "as if" status despite the underlying theory not allowing for their possibility. Some authors have been driven to argue that particles in QFT are supervenient on quantum fields and their appearance in empirical measurements has merely a FAPP status. A second concern regarding fundamental particles is indistinguishability with Wheeler quipping that there was only one electron in the universe. To the best of our knowledge, QFT ontologies have not progressed to the point of posing and answering the question of the existence of sharp, spatiotemporally localized, distinguishable particles. The intriguing aspect of QFT ontologies from our perspective is the ability of the universe to manifest particles from field excitations as opposed to particles being the sole foundation in nature. Whether individual particles are spatiotemporally localizable or not is secondary, likewise for indistinguishability.

We have now arrived at the point of payoff. We wish to appeal to QFT ontologies in support of setting up a new relation connecting SoEs to the physical. In this new conception, SoEs are akin to new particles—selfons—that arise out of basic fields, persist, and dissipate. Selfons mirror the phenomenology of a thin subject and are associated with phenomenal content (thoughts, feelings) which are now conceived as new spatiotemporal properties. Selfons can also be seen as objects of a new natural kind which are always accompanied by experience. Fundamental issues of indistinguishability, spatiotemporal localization, sharpness, and the viability of the object/property/process cluster will all have to be worked out within the framework of a QFT ontology: likewise with the issue of whether selfons are supervenient on a more basic QFT ontology. We seem to have answered our question in the affirmative: there does not appear to be (at present) a knockdown argument against conceiving of SoEs as selfons. The ignorance hypothesis (of the true physical) permits the construction of a plausible new physicalism which includes selfons. As long as we have independent evidence of SoEs (from phenomenology and intersubjectivity) followed by the conception of SoEs as objects of a new natural kind—selfons—there does not appear to be any barrier to their inclusion into physicalism. Clearly, this is a scientific hypothesis driven by a philosophical starting point (physicalism) and therefore could be wrong. For example, while we have asserted that selfons are akin to new particles, perhaps the QFT-selfon relation is totally different. We feel that it is important to begin with the above selfon hypothesis at this juncture for the reasons of (i) simplicity and (ii) grounding in present-day QFT.

All the standard objections apply: here's a sample. (i) Basic physics is settled with no place for selfons; (ii) this is just neutral monism with fields being the neutral entity and ordinary particles and selfons corresponding to matter and mind, respectively; (iii) there are no sharply localized particles, hence, no selfons in principle; (iv) you have gotten confused between particles and states, and your approach actually suggests that consciousness is a new state of matter; (v) this is reductionist to the core since you have reduced consciousness to (new) particles. There is no space to address all these (good) objections. Instead, we settle for a brief rejoinder to (i) above. We take it as a given that basic physics is not settled and that (despite string theory) we are still far away from the grand unification of gravity with QFT. When we describe selfons as akin to particles, we should clarify that these new "particles" could better be conceived as composites rather than fundamental. But, composites how? Note that we alluded to a weak supplementation principle of holism pioneered by Koslicki. Selfons could arise from a new holistic principle linking basic fields to new particles (as opposed to being compounds made up of existing fermions and bosons). This is one way things could go. But this way has already been pointed out by Lowe, van Inwagen, Strawson, and Koslicki. All that remains is for physicalism to follow this lead. While this is certainly a very difficult task, it is not in principle impossible. This route also bypasses historical divisions between matter and mind. We reiterate that the conception of SoEs as selfons is suggestive of a new organization principle linking basic fields to new particles (without being an emergent property of matter). Finally, this is actually an attempt to do both philosophy *and* physics with the end result open to multiple interpretations—due to the unsettled nature of QFT ontologies.

4 A Criticism of and an Opportunity for Indian Philosophy

We close by pointing out the relevance to Indian philosophy since that setting is presumably more familiar to the audience than continental or Anglo-American philosophy. A sharp criticism is first offered which will hopefully be seen as tough love rather than condemnation. This is followed by a discussion of the new opportunities presented by the hard problem of consciousness in general and QFT ontologies in particular.

4.1 Criticism

Given the fundamental (and some would say insurmountable) difficulty posed by the hard problem, one would be led to expect an appropriate response from Indian philosophy especially when the literature's obsession with the Ātman-Brahman relation for over two and a half millennia is taken into account. This is not to be found. Dialog with Indian philosophers (and this is clearly anecdotal) reveals the usual mystical and idealist leanings and a failure to take the hard problem seriously. As we have previously mentioned, even when we restrict focus to transcendental idealism, there is no sustained exploration or rapprochement with the idealisms of Berkeley, Fichte, Schopenhauer [4], etc. When one repeatedly hears the statement that consciousness is fundamental, the world is an illusion, and therefore there is no hard problem, it's clear that there is no appreciation of the limited, localized, interior, private, and ineffable nature of qualia and the need to relate them to something beyond. If the reader at this point continues to think that some form of absolute idealism resolves the hard problem, please note that the split between subjective, spatiotemporally bounded qualia and the world still needs to be explained even in an idealist metaphysics. And, despite the Indian literature's obsession with awareness (as opposed to phenomenal content), we have not seen any exploration of the triadic relationships between SoEs, phenomenal content, and the world. As we have mentioned above, recently there has been an uptick of interest in treating the universe as a simulation. Regardless of the physical basis of the simulation, in this picture, there is a clear and pressing need to relate virtuality and consciousness [14]. We have not seen much exploration in Indian philosophy of treating the world as virtual (and not merely as an illusion) and attempting to situate subjective experiences within such a world. Philosophers trained in the analytic tradition continue to express their frustration with Indian philosophy in private. The criticisms range from the following: (i) it is a cul-de-sac with no interest in other traditions; (ii) it has become dominated by different schools of Vedānta [8] leaving Nyāya, the Cārvāka materialists, and the Samkhya dualists [35] behind; (iii) despite the rise and rise of worldcentric Buddhism in the past 40 years, there's no adequate response from Indian philosophy; (iv) no attempt has been made to connect the recent interest in panpsychism with primordial Indian philosophical traditions; (v) too much material remains trapped in Sanskrit with no interest shown in translation and dissemination; (vi) Indian philosophers (and to be clear, this criticism is mainly aimed at Indian philosophers disseminating Indian philosophical views) have a tendency to keep repeating their talking points and remain unable to engage with a radically different audience; (vii) the (usually tacit) idealist metaphysics hobbles and curtails all discussion to the Jīva-Brahman relation and does not permit exploration of the Jīva-world relation.

4.2 Drawing Inspiration from Buddhism

We attempt to elaborate on (iii) above in an impressionistic manner. Prior to the 1980s, there was not much engagement between Buddhism and the west. The explosion of interest in new paradigm (mainly non-reductionist) thinking and the rise of postmodernism in the 1980s allowed Buddhism to establish a beachhead in the western academy [24]. Facile analogies between the Buddhist (Abhidharma) no-self doctrine and emergentist (and embodied) theories of consciousness [37, 39] led to further dissemination of Buddhist ideas. The same was not true of Vedānta and Indian philosophy in general (but with some exceptions [11]). The comparatively more theistic leanings of Vedānta coupled with the culture wars in the USA and Canada ensured that American Vedānta, for example, did not have as large a footprint as American Buddhism. Consequently, while Buddhism became the de facto standard bearer (for Eastern philosophy) in consciousness conferences, Vedāntic and other Indian philosophical concepts did not have a similar impact. Buddhism has faced serious problems in its transition from specific Eastern traditions to becoming a worldcentric religion and philosophy. Take, for example, the embodied self-arguments in [37]. After first deconstructing the self in standard Buddhist fashion using Abhidharma concepts, the authors back away from these arguments by deploying the Mādhyamika (Śhūnyatā) no-view dialectic on the Abhidharma. In other words, the sharp and clear no-self Abhidharma doctrine gives way to emptiness of all views, Mādhyamika approach. Despite the danger of using emptiness as a "crowbar" to justify a certain view, the authors return at the end of the book to embodiment of mind as their position. This allows the authors to (a) introduce Buddhist concepts like Śhūnyatā to a wider and secular audience while (b) promulgating a particular view—embodiment. In the two plus decades since the book's publication, Abhidharma and Śhūnyatā have entered the consciousness lexicon, but other Buddhist doctrines that build on these approaches have not. It is quite common to encounter people at consciousness conferences discussing Buddhist no-self doctrines and arguing that the emptiness doctrine should properly be viewed as mysticism but one rarely hears discussion of Buddhist tantra or Chinese Hua-yen [3], for example. Admittedly, these are impressionistic and anecdotal observations, but it should be possible to perform data analytics on this point for confirmation. The takeaway for Indian philosophy is to follow the example of Buddhism's transition from a diverse set of Eastern cultures to a more unified, worldcentric one.

4.3 Opportunity

If the above criticisms and impressions are on the mark, Indian philosophy has a huge opportunity. Despite being rich, vibrant, and much better funded philosophical traditions, continental and Anglo-American philosophy are in crisis. Their continued inability to create a satisfactory quantum field theory ontology is perhaps not an accident and may be due to a foundational mismatch between a two and a half millennia western philosophical tradition founded on atomism and modern (and postmodern) physics. With QFT underpinnings up for grabs, it is astonishing to see continued talk of the matter-mind split given the swampland foundations of matter. Too much of present-day analytic philosophy still remains obsessed with interpretations of quantum mechanics instead of delving into QFT and formulating exactly what is meant by a particle. The arrival of the new century has seen the calls for

"Everything must go" [19] at the foundations which are clearly indicative of something amiss. Indeed, just as Indian philosophy has been mired in mining the Ātman-Brahman relation, the west has been bogged down by the matter-mind split. If, as we have suggested, the matter-mind (or subject-object) split cannot be solved without shaking up the foundations of matter (via QFT ontologies and the like) and without reintroducing subjects of experience into the mix, then western philosophy may be ripe for a revolution. It is here that Indian philosophy can play a worldcentric role by (i) arguing the Jīva-world relation using all of the tools at its disposal; (ii) setting up a three-way relationship between the world, Jīvas, and Brahman; and finally (iii) bringing back the other schools of Indian philosophy [35] which have been neglected due to the univocal emphasis on Vedāntic idealism. (Western scholars do not seem to be aware of the qualified nondualism of Viśiṣtādvaita and the dualism of Dvaita, for example, and think that Advaita is the only game in town.) First, selfons offer a bridge between the Jīva and the world. The phenomenology of a selfon includes awareness of phenomenal content (thoughts, feelings, visualizations, sensations, etc.). Since selfons are objects of a natural kind, the hermeneutics (cultural and mutual co-creation) of selfons include intersubjectivity and protolanguage. Next, the structure of a selfon is based on a new compositionality relation with fundamental fields (and is foundationally different from that of matter). In contrast to a cartoon Cartesianism, the temporal nature of a selfon is that of a thin subject which lasts for only a few seconds. As opposed to emergentist theories, selfons are constituted via a new (holistic) compositionality relation. A potential slide to neutral monism is avoided by asserting an asymmetry between mind and matter and by preserving a physicalist foundation. Dual-aspect theory is transcended as selfons can interact with the world (with both being physical). The approach is explicitly physicalist without being a priori panpsychist. The triad—selfons, phenomenal content, and the physical world—serves as a bridge from physicalism to different schools of Indian philosophy, whereas such a bridge was nonexistent between materialism and Indian idealism. Finally, it is the present-day ignorance and uncertainty regarding the foundations of QFT that allow us to expand physicalism in this manner. There may be unexpected resonances between QFT ontologies and Indian philosophy that are yet to be fleshed out. Issues such as indistinguishability and no-go theorems on sharp, localized particles may point to very intriguing connections between the selfons that form a natural kind and the nonlocal self in Indian traditions. For example, if there's no way to have a sharp, spatiotemporally localized, distinguishable particle in QFT, then this may apply to selfons as well. In that case, selfons would in principle be indistinguishable, and we can follow Wheeler down the rabbit hole and assert that "there is only one SelfOn" despite the FAPP insistence on the existence of many. (We do not need to point out the relevance of the many actually being one to Indian philosophers.) While there's a tendency among many Indian scholars to assume that metaphysics is settled (undoubtedly due to the weight and age of the tradition), the unsettled nature of QFT ontologies should give them pause. Finally, given the impact Indian idealism has had on European philosophy, a revamped Indian materialism (with aid from the other Indian philosophical traditions) may be well positioned to make worldcentric contributions to the hard problem of consciousness.

References

1. Atmanspacher H. Dual-aspect monism *à la* Pauli and Jung. J Conscious Stud. 2012;19(9–10):96–120.
2. Banks EC. The realistic empiricism of Mach, James, and Russell: neutral monism reconceived. reprint edn. Cambridge University Press; 2016.
3. Cleary T. Entry into the inconceivable: an introduction to Hua-Yen Buddhism. Honolulu: University of Hawaii Press; 1983.
4. Cross S. Schopenhauer's encounter with Indian thought: representation and will and their Indian Parallels. Monographs of the Society for Asian and Comparative Philosophy. University of Hawaii Press; 2013.
5. Chalmers DJ. The conscious mind: In search of a fundamental theory. Philosophy of mind. rev edn. Oxford University Press; 1997.
6. Deikman A. 'I'=awareness. J Conscious Stud. 1996;3(4):350–6.
7. Dennett DC. Consciousness explained. 1st ed. New York: Back Bay Books; 1992.
8. Deutsch E. Advaita Vedanta: a philosophical reconstruction. Studies in the Buddhist traditions. University of Hawaii Press; 1980.
9. Freeman A (ed). The emergence of consciousness. J Conscious Stud. 2001;8. Imprint Academic, Exeter
10. Georgiev DD. Falsification of Hameroff-Penrose Orch OR model of consciousness and novel avenues for development of quantum mind theory. NeuroQuantology. 2007;5(1):145–74.
11. Griffin DR, Smith H. Primordial truth and postmodern theology. Suny series in constructive postmodern thought. Albany: New York University Press; 1990.
12. Hameroff SR. Quantum computation in microtubules? The Penrose-Hameroff "Orch OR" model of consciousness. Phil Trans R Soc A. 1998;356:1869–96.
13. Hargens S. Intersubjective musings: a response to Christian de Quincey's 'the promise of integralism'. J Conscious Stud. 2001;8(12):35–78.
14. Hoffman D. Conscious realism and the mind-body problem. Mind & Matter. 2008;6(1):87–121.
15. Jinpa T. Self, reality and reason in Tibetan philosophy: Tsongkhapa's quest for the middle-way. Routledge critical studies in Buddhism. 1st ed. London: Routledge; 2002.
16. Koslicki K. The structure of objects. reprint edn. Oxford: Oxford University Press; 2010.
17. Kuhlmann M. The ultimate constituents of the material world: in search of an ontology for fundamental physics. Philosophische analyse/philosophical analysis (Book 37). Berlin: De Gruyter; 2010.
18. Kuhlmann M, Lyre H, Wayne A, editors. Ontological aspects of quantum field theory. River Edge: World Scientific; 2002.
19. Ladyman J, Ross D. Everything must go: metaphysics naturalized. Oxford/New York: Oxford University Press; 2009.
20. LaPorte J. Natural kinds and conceptual change. Cambridge studies in philosophy and biology. 1st ed. Cambridge: Cambridge University Press; 2009.
21. Lloyd S. Programming the universe: a quantum computer scientist takes on the cosmos. reprint edn. London: Vintage; 2007.
22. Lowe EJ. Subjects of experience. Cambridge studies in philosophy. Cambridge: Cambridge University Press; 2006.
23. Malament D. In defense of dogma: why there cannot be a relativistic quantum mechanics of (localizable) particles. In Clifton R, editor. Perspectives on quantum reality: non-relativistic, relativistic, and field-theoretic, volume 57 of Western Ontario series in philosophy of sci-ence, chapter 1, pages 1–10. Springer Science and Business Media; 1996.
24. Nāgārjuna. The fundamental wisdom of the middle way: Nāgārjuna's Mūlamadhyamakakārikā. Oxford University Press; 1995. Translated and Commentary by J. L. Garfield.
25. Rangarajan A. Panpsychism and compositionality: a solution to the hard problem. In: The Science of consciousness. University of Helsinki; 2016.

26. Rosenberg G. A place for consciousness: probing the deep structure of the natural world. Philosophy of mind. Oxford: Oxford University Press; 2004.
27. Schrödinger E. Quantisierung als eigenwertproblem. Ann Phys. 1926;384(4):361–76.
28. Schwartz MD. Quantum field theory and the standard model. Cambridge: Cambridge University Press; 2013.
29. Skrbina D. Panpsychism in the west. Cambridge: MIT Press; 2007.
30. Sprigge TLS. The vindication of absolute idealism. 1st ed. Edinburgh: Edinburgh University Press; 1984.
31. Stoljar D. Ignorance and imagination: the epistemic origin of the problem of consciousness. Philosophy of mind. Oxford: Oxford University Press; 2009.
32. Stoljar D. Physicalism, volume 2 of new problems of philosophy. London/New York: Routledge; 2010.
33. Strawson G. Consciousness and its place in nature: does physicalism entail panpsychism? Imprint Academic, 2006.
34. Strawson G. Selves: an essay in revisionary metaphysics. rev edn. Oxford University Press; 2011.
35. Tigunait PR. Seven systems of Indian philosophy. Honesdale: Himalayan Institute Press; 1983.
36. van Inwagen P. Material beings. Ithaca: Cornell University Press; 1995.
37. Varela F, Thompson E, Rosch E. The embodied mind: cognitive science and human experience. rev edn. The MIT Press; 1992.
38. Wheeler JA, Zurek WH, editors. Quantum theory and measurement. Princeton Legacy Library. Princeton: Princeton University Press; 1983.
39. Wilber K. Sex, ecology, spirituality: the spirit of evolution. 1st ed. Boston: Shambhala; 1995. (page 708, note for page 497, chapter 14)

Chapter 3
Emergent Reality in Quantum from Classical Transition

Tabish Qureshi

Abstract The very fact that a quantum measurement changes the quantum state of a system in an uncontrollable way implies that the measurement does not reveal the objective reality that existed before the measurement. We argue that the nature of certain special quantum states that emerge due to decoherent interaction with the environment is such that one can *measure* the expectation value of any observable of the system in a single measurement. This can be done even when such states are a priori unknown. The possibility of measuring the expectation value of any observable, without any prior knowledge of the state, points to the objective reality of such states.

Keywords Emergent reality · Quantum measurement · Decoherent incorrection · Hermitian operators · Eigenvalue · Expectation value

1 Introduction

The very fact that a quantum measurement changes the quantum state of a system in an uncontrollable way implies that the measurement does not reveal the objective reality that existed before the measurement. We argue that the nature of certain special quantum states, which emerge due to decoherent interaction with the environment, is such that one can *measure* the expectation value of any observable of the system in a single *protective* measurement. This can be done even when such states are a priori unknown. The possibility of measuring the expectation value of any observable, without any prior knowledge of the state, points to the objective reality of such states. Thus objective reality emerges from quantum mechanics, due to decoherence.

T. Qureshi (✉)
Centre for Theoretical Physics, Jamia Millia Islamia, New Delhi, India
e-mail: tabish@ctp-jamia.res.in

S. R. Bhatt (ed.), *Quantum Reality and Theory of Śūnya*,
https://doi.org/10.1007/978-981-13-1957-0_3

The emergence of the familiar classical world from quantum mechanics has been a subject of endless debate for about a century now [1, 2]. If one believes that quantum mechanics is a fundamental theory governing the dynamics of all particles, it is only a natural inference that classical physics should emerge from it. However, the unitary nature of the Schrodinger equation appears to allow states which are never observed in our familiar classical world. Quantum mechanics allows existence of superposition of macroscopically separated states, whereas classical mechanics doesn't allow a particle to be delocalized. There have been attempts of a very diverse nature to resolve the issue, but there is no consensus on it till date.

Measurement in quantum mechanics is another aspect which is perplexing. If there is a single system, described by an unknown state $|\psi\rangle$, many believe that the state has an objective reality. However, if one wants to get some information about this supposed reality, all one can do is to perform a measurement of an operator, say $\mathcal{A}$, of one's choice. Assuming that the operator has a set of eigenstate and eigenvalues given by $\mathcal{A} \mid a_n\rangle = \alpha_n \mid a_n\rangle$, the measurement will yield one of the eigenvalues, say α_k, and the state of the system will change to $|s_k\rangle$. This, however, gives one no information about the state $|\psi\rangle$. So, a measurement on a single system does not reveal anything about the reality of the state of the system which existed before the measurement. Because of this nature of quantum mechanics, many have argued that the quantum state has no objective reality. On the other hand, the classical states that we observe in the macroscopic world do have an objective reality. Since we believe that the macroworld emerges from the microworld, the question one then asks is how does this objective reality of the macroworld emerge from the quantum microworld?

One way the emergence of classicality can be understood is by recognizing that macroscopic objects are almost impossible to shield from their environment. The interaction with the environment is unavoidable and leads to decoherence [3–5]. The decoherence approach has become very popular because it attempts to explain the emergence of classicality while staying within conventional quantum mechanics and has demonstrated predictive power. Controlled influence of environment has been successfully studied in numerous experiments.

The process of decoherence can be represented in the following way. Let the state of a system and its environment be written as

$$\mid \psi\rangle = \left(\sum_p c_p | \, p\rangle\right) \mid E^0\rangle, \tag{3.1}$$

where the state of the system is expanded in terms of certain basic states $|p\rangle$ which we shall call {\em pointer states}, and $|E^0\rangle$is the state of the environment. The meaning of pointer states will be clear in the following discussion. Assume a Hamiltonian and time evolution of the form [6]

$$\mathcal{H} = \sum_p | p\rangle\langle p | \otimes \mathcal{H}^{(p)} \tag{3.2}$$

$$\mathcal{U}_t = \sum_p | p\rangle\langle p | \otimes \mathcal{U}_t^{(p)} \tag{3.3}$$

where $\mathcal{U}_t^{(p)} = e^{-i\mathcal{H}^{(p)}t/\hbar}$ and the $\mathcal{H}^{(p)}$ are certain unspecified Hermitian operators involving the environment. The state (3.1), with the above-specified time evolution, evolves into

$$| \psi(t)\rangle = \sum_p c_p | p\rangle \mathcal{U}_t^{(p)} | E^0\rangle. \tag{3.4}$$

If one wants to look only at the system and forget about the environment, it is useful to write the reduced density matrix, which is obtained by writing the density matrix for (3.4), and trace over the states of the environment, which has the form

$$\rho_r(t) = \sum_{p,p'} c_{p'}^* c_p | p\rangle\langle p' | \langle E^0 | \mathcal{U}_t^{(p')\dagger} \mathcal{U}_t^{(p)} | E^0\rangle. \tag{3.5}$$

Under the kind of time evolution specified above, the diagonal components of the density matrix, in the basis $|p\rangle$, remain unchanged, while off-diagonal elements are reduced by a factor $\left\langle E^0|\mathcal{U}_t^{(p')\dagger}|\mathcal{U}_t^{(p)}|E^0\right\rangle \leq 1$. The time dependence of the suppressing terms will, in general, depend on the specific model of the environment and its interaction with the system, but for a wide variety of models, the suppressing terms have been found to rapidly decay over short time scales [4, 7–10].

Over a time scale, called *decoherence time scale*, the off-diagonal terms in (3.5) disappear for all practical purposes, and one is left with a diagonal density matrix of the system

$$\rho_r = \sum_p |c_p|^2 | p\rangle\langle p | . \tag{3.6}$$

The off-diagonal elements which are associated with quantum superpositions are no longer present. Thus the system *appears to* behave classically. This special set of states is selected by the interaction with the environment to emerge as classical states. In the literature they have come to be known as pointer states, and this process of environment-induced selection is called *einselection* [11, 12]. The Hamiltonian in (3.2) is a generic form which leads to the system ending up in diagonal density matrix in the pointer states basis. In general, a microscopic Hamiltonian is needed to find out what states form the pointer basis. What should be the pointer states for a particular system is a question that is not easy to answer. This question has been answered only for a few cases. For example, the pointer states of a simple harmonic oscillator are believed to be coherent states [13], and those for a free particle have been exactly shown to be minimum uncertainty Gaussian states [14]. There are some indications that Gaussian states emerge as pointer states for a particle in Stern-Gerlach experiment [15]. In the limit of weakest interaction with the environment, energy eigenstates have been shown to emerge as pointer states [16].

A question one might ask is whether such pointer states exist out there on their own or do they need an observer to bring them into existence. If decoherence is indeed the mechanism for the emergence of classicality from quantum mechanics, the pointer states should have an objective reality. Ollivier, Poulin, and Zurek have addressed this issue from the point of view that pointer states leave their imprint on the environment, which can be read out by an observer without disturbing the state [17]. Here we take a different approach and ask if the pointer states themselves have some properties which make them robust enough to allow getting information from them without disturbing them. According to the standard quantum mechanics lore, one cannot get any information about an unknown quantum state. The process of measurement necessarily destroys the original quantum state, unless of course, in the trivial case, it is already in an eigenstate of the observable being measured. In the following we will show that pointer states are special in the sense that they allow a *measurement* of the expectation value of any observable without destroying the state.

Let us now explore the process of measurement on a system which is undergoing decoherence. We assume that the system of interest is interacting with the environment. Let us represent the pointer states of the system by $\{|p\rangle\}$. In addition, the system is assumed to be interacting with an apparatus. So, the system is expected to undergo decoherence because of its interaction with the environment and also cause a shift in the state of the apparatus, which essentially constitutes the process of measurement. The Hamiltonian of the system, apparatus, and the environment is given by

$$\mathcal{H} = \mathcal{H}_A + g\mathcal{Q}_S\mathcal{Q}_A + \sum_p |p\rangle\langle p| \otimes\mathcal{H}^{(p)}, \tag{3.7}$$

where $\mathcal{H}_A$ is the Hamiltonian of the apparatus and the second term on the right-hand side represents the interaction of the apparatus with the system, with a strength parameterized by g. Operators $\mathcal{Q}_S$ and $\mathcal{Q}_A$ are the operators of the system and the apparatus, respectively, through which the interaction takes place. The "free" Hamiltonian of the system is ignored because the measurement interaction is supposed to act for a short time and dominate the time evolution during that period.

The last term represents the interaction between the system and the environment and is taken to be of a form which leads to the states $\{|p\rangle\}$ emerging as pointer states, as demonstrated in the preceding discussion. This term may not necessarily represent the true interaction of the system with the environment but may be an *effective interaction*. If one believes that pointer states do emerge out of decoherence, the effective interaction must have this form.

We assume that the apparatus is so constructed that $\mathcal{Q}_A$ commutes with the free Hamiltonian of the apparatus, i.e., $[\mathcal{Q}_A, \mathcal{H}_A] = 0$, so that we can have eigenstates$|a_i\rangle$ such that $\mathcal{Q}_A \mid a_i\rangle = a_i \mid a_i\rangle$ and $\mathcal{H}_A \mid a_i\rangle = E_i^a \mid a_i\rangle$.

The initial state of the system, apparatus, and environment is assumed to be

$$\mid \Psi_0\rangle = \Big(\sum_p c_p \mid p\rangle\Big) \mid E^0\rangle \mid \phi_a\rangle, \tag{3.8}$$

where the first term represents the unknown state of the system, written in terms of the pointer state basis, and $|\phi_a\rangle$ is the initial state of the apparatus. It should be emphasized that at this point, neither the initial state of the system nor the pointer basis is known. Here $|E_0\rangle$ is the initial state of the environment.

For the measurement process, we let the apparatus interact with the system for a time T. In addition to this, the system is continually interacting with the environment. The combined state of the system, apparatus, and environment, after a time T, is given by

$$\mid \Psi_T\rangle = e^{-\frac{iT}{\hbar}\left[\mathcal{H}_A + g\mathcal{Q}_S\mathcal{Q}_A + \sum_{p'} |p'\rangle\langle p'|\mathcal{H}^{(p')}\right]} \Big(\sum_p c_p \mid p\rangle\Big) \mid E^0\rangle \mid \phi_a\rangle. \tag{3.9}$$

We introduce a complete set of states $\sum_{\mathcal{P}''} \mid \mathcal{P}''\rangle\langle\mathcal{P}''\mid$ before the initial state,$\mid \mathcal{P}\rangle$ being the exact eigenstates of the full Hamiltonian given by (3.7). Doing that, the above can be rewritten as

$$\mid \Psi_T\rangle = \sum_{\mathcal{P}''} e^{-\frac{iT}{\hbar}\left[\langle H^A\rangle + g\langle\mathcal{P}''|\mathcal{Q}_S|\mathcal{P}''\rangle\mathcal{Q}_A + \langle\mathcal{P}''|(\sum_p |p\rangle\langle p| \otimes \mathcal{H}^{(p)})|\mathcal{P}''\rangle\right]}$$
$$\mid \mathcal{P}''\rangle\langle\mathcal{P}''\mid \Big(\sum_p c_p \mid p\rangle\Big) \mid E^0\rangle \mid \phi_a\rangle$$

The above cannot be solved in general. However, if the interaction between the system and the apparatus is *weak and adiabatic*, one can treat the interaction as a perturbation, and the exact eigenstates $\mid \mathcal{P}\rangle$, in first-order perturbation theory, are unperturbed and equal to $|p\rangle$ times the eigenstates of $\mathcal{H}^{(p)}$. In the first-order perturbation theory, the above can be approximated by

$$\mid \Psi_T\rangle = \sum_p e^{-\frac{iT}{\hbar}\left[H^A + g\langle p|\mathcal{Q}_S|p\rangle\mathcal{Q}_A + \mathcal{H}^{(p)}\right]} c_p \mid p\rangle \mid E^0\rangle \mid \phi_a\rangle. \tag{3.10}$$

Since $\mathcal{H}^{(p)}$ commutes with the other terms in the exponent, the above can be simplified to

$$| \Psi_T\rangle = \sum_p e^{-\frac{iT}{\hbar}[\mathcal{H}_A + g\langle p|\mathcal{Q}_S|p\rangle \mathcal{Q}_A]} c_p \mid p\rangle \mid \phi_a\rangle e^{-\frac{iT}{\hbar}\mathcal{H}^{(p)}} \mid E^0\rangle. \tag{3.11}$$

The fact that $\mathcal{H}_A$ commutes with $\mathcal{Q}_A$ further allows us to separate the first exponent into two parts:

$$| \Psi_T\rangle = \sum_p e^{-\frac{iT}{\hbar}\mathcal{H}_A} c_p \mid p\rangle e^{-\frac{igT}{\hbar}\langle p|\mathcal{Q}_S|p\rangle \mathcal{Q}_A} \mid \phi_a\rangle \mathcal{U}_T^{(p)} \mid E^0\rangle. \tag{3.12}$$

Let us now assume that the apparatus is prepared in an initial state which is a wave packet of eigenstates $|r\rangle$ of an operator R_A such that $[R_A, \mathcal{Q}_A] = i\hbar$. The wave packet may be centered at (say) r_0. It is straightforward to see what the effect of $e^{-\frac{igT}{\hbar}\langle p|\mathcal{Q}_S|p\rangle \mathcal{Q}_A}$ on $|\phi_a(r_0)\rangle$ will be. It will simply translate the packet by an amount $gT\langle p|\mathcal{Q}_S|p\rangle$, thus yielding an apparatus state which is shifted by an amount proportional to the expectation value of the system observable $\mathcal{Q}_S$

$$| \Psi_T\rangle = \sum_p e^{-\frac{iT}{\hbar}\mathcal{H}_A} c_p \mid p\rangle \mid \phi_a\Big(r_0 - gT\langle \mathcal{Q}_S\rangle_p\Big)\rangle \mathcal{U}_T^{(p)} \mid E^0\rangle, \tag{3.13}$$

where $\langle \mathcal{Q}_S\rangle_p = \langle p|\mathcal{Q}_S|p\rangle$.

This is equivalent to, what is known in the literature as, *protective measurements* [18–24]. Protective measurements are generally restricted to the stationary states of the system. However, the states that emerge out of decoherence also appear to be stationary.

This is the central result of this work. It can be interpreted in more ways than one. In the conventional treatment of decoherence, one believes that one state from the pointer states basis emerges as the reality, which happens with a probability. One can write a density matrix for the state represented by (3.13) and take a trace over the degrees of freedom of the environment. Doing that we find

$$\rho_r(p, p') = \sum_{p,p'} c_p c_{p'}^* \mid p\rangle\langle p' \mid e^{-\frac{iT}{\hbar}\mathcal{H}_A} \mid \phi_a(r_0 - gT\langle \mathcal{Q}_S\rangle_p)\rangle\langle \phi_a^*(r_0 - gT\langle \mathcal{Q}_S\rangle_{p'}) \mid$$
$$\times e^{\frac{iT}{\hbar}\mathrm{H}_A}\langle E^0 \mid \mathrm{U}_T^{(p')\dagger}\mathcal{U}_T^{(p)} \mid E^0\rangle. \tag{3.14}$$

For realistic decoherence, the factor $\langle E^0 \mid \mathrm{U}_T^{(p')\dagger}\mathrm{U}_T^{(p)} \mid E^0\rangle$ is expected to decay to close to zero over a very short time scale. One is then left with a density matrix which is approximately diagonal in the pointer basis

$$\rho_r(p,p') \approx \sum_p | c_p |^2 | p\rangle\langle p$$
$$| e^{-\frac{iT}{\hbar}\mathrm{H}_A} | \phi_a(r_0 - gT\langle \mathcal{Q}_S\rangle_p)\rangle\langle\phi_a(r_0 - gT\langle \mathcal{Q}_S\rangle_p) | e^{\frac{iT}{\hbar}\mathcal{H}_A} \tag{3.15}$$

The above implies that a particular pointer state $|p\rangle$ emerges as a reality with a probability $|c_p|^2$, while the apparatus state gets shifted by an amount proportional to the expectation value of the observable of the system being measured ($\mathcal{Q}_S$) in that particular state. Thus we end up measuring the expectation value of $\mathcal{Q}_S$ in the state which emerged out of decoherence, which we had no knowledge of. Note that the measurement process doesn't disturb the state – the only natural disturbance is due to decoherence.

Another way to interpret the result would be the following. Let us denote the states of the environment appearing in the final expression as $| E_p\rangle = \mathcal{U}_T^{(p)} | E^0\rangle$. Beyond the decoherence time scale, the states $|E_p\rangle$corresponding to different p are expected to be nearly orthogonal to each other. The combined state of the system, apparatus, and environment has the form

$$| \Psi_T\rangle = \sum_p e^{-\frac{iT}{\hbar}\mathcal{H}_A} c_p | p\rangle | \phi_a\Big(r_0 - gT\langle \mathcal{Q}_S\rangle_p\Big)\rangle | E_p\rangle. \tag{3.16}$$

Looking at the sum over the states $|p\rangle$, one might wonder that the state appears entangled. However, from the point of view of decoherence, when the results gets correlated with certain orthogonal states of the environment, the measurement is considered complete. The terms corresponding to different $|p\rangle$ may be considered as independent *branches* which are robust (no possibility of recoherence), each representing an independent classical reality [4]. For example, in kth branch the measured expectation values of the observable will correspond to only one state$|k\rangle$. A kind of modified many-worlds interpretation might be needed here, if one doesn't believe in a real collapse of the quantum state [25].

All this is possible because of the very nature of pointer states, namely, because of their being entangled with certain orthogonal states of the environment. In all this analysis, at no stage did we need any information about the states $|p\rangle$ – their being pointer states is enough to provide the possibility of measuring the expectation value of any observable. This analysis has demonstrated that it is possible to *measure* the expectation value of any observable of a system if it has undergone decoherence and pointer states have emerged as its quantum states. This definitely points to the objective reality of pointer states.

The result (3.16) has another very interesting meaning, which can have some practical use. In a conventional measurement, the state of the apparatus gets shifted by an amount proportional to an *eigenvalue* of the observable being measured. Our result says that if the system being measured is undergoing decoherence, then the shift in the apparatus state will be proportional, not to the *eigenvalue* but to the *expectation value* of the observable being measured.

In conclusion, we have shown that the quantum states that emerge out of interaction with the environment allow the *measurement* of the expectation value of any observable of choice, with a single protective measurement. The measurement process does not disturb an emerging pointer state. Just the expectation value of the arbitrarily chosen observable is picked out by the measuring apparatus. Since we are able to measure the expectation value of any observable of our choice in a pointer state which is emerging because of decoherence, it is sufficient reason to assign them an objective reality.

References

1. Wheeler JA, Zurek WH, editors. Quantum theory and measurement. Princeton: Princeton University Press; 1983. p. 182–213.
2. Auletta G. Foundations and interpretation of quantum theory. Singapore: World Scientific; 2000.
3. Zeh HD, On the interpretation of measurement in quantum theory. Found Phys. 1970;1:69; Joos E, Zeh HD, The emergence of classical properties through interaction with the environment. Z Phys. 1985;D 59:223.
4. Giulini D, Joos E, Kiefer C, Kupsch J, Stamatescu I-O, Zeh HD, editors. Decoherence and the appearance of a classical world in quantum theory. Berlin/London: Springer; 1996.
5. Zurek WH. Decoherence and the transition from quantum to classical. Phys Today. 1991;44 (10):36.
6. Gogolin C. Environment-induced super selection without pointer states. Phys Rev. 2010; A81:051127.
7. Breuer H-P, Petruccione F. The theory of open quantum systems. Oxford: Oxford University Press; 2002.
8. Zurek WH. Decoherence, einselection, and the quantum origins of the classical. Rev Mod Phys. 2003;75:715.
9. Zurek WH. Environment-induced superselection rules. Phys Rev. 1982;D26:1862.
10. Hornberger K. Introduction to decoherence theory. Lect Notes Phys. 2009;768:221.
11. Zurek WH. Environment-induced superselection rules. Phys Rev. 1982;D26:1862–80.
12. Zurek WH. Decoherence, einselection, and the quantum origins of the classical. Rev Mod Phys. 2003;75:715–75.
13. Zurek WH, Habib S, Paz JP. Coherent states via decoherence. Phys Rev Lett. 1993;70:1187.
14. Eisert J. Exact decoherence to pointer states in free open quantum systems is universal. Phys Rev Lett. 2004;92:210401.
15. Qureshi T. Decoherence, time scales and pointer states. Physica A. 2012;391:2286–90.
16. Paz JP, Zurek WH. Quantum limit of decoherence: environment induced superselection of energy eigenstates. Phys Rev Lett. 1999;82:5181.
17. Ollivier H, Poulin D, Zurek WH. Objective properties from subjective quantum states: environment as a witness. Phys Rev Lett. 2004;93:220401.
18. Aharonov Y, Vaidman L. Measurement of the Schrodinger wave of a single particle. Phys Lett. 1993;A178:38.
19. Aharonov Y, Anandan J, Vaidman L. Meaning of the wave function. Phys Rev. 1993; A47:4616.
20. Vaidman L. Protective measurements. In: Greenberger D, Hentschel K, Weinert F, editors. Compendium of quantum physics: concepts, experiments, history and philosophy. Berlin/ Heidelberg: Springer; 2009.

21. Hari Dass ND, Qureshi T. Critique of protective measurements. Phys Rev. 1999; A59:2590–601.
22. Qureshi T, Hari Dass ND. Protective measurements: probing single quantum systems. Curr Sci. 2015;109:2023.
23. Schlosshauer M. State disturbance and pointer shift in protective quantum measurements. Phys Rev. 2014;A90:052106.
24. Gao S, editor. Protective measurement and quantum reality. Cambridge: Cambridge University Press; 2015.
25. Schlosshauer M. Decoherence, the measurement problem, and interpretations of quantum mechanics. Rev Mod Phys. 2004;76:1267.

Part II
Concept of Śūnya

Chapter 4
Śūnya and *Pūrṇa*

S. S. Rama Rao Pappu

Abstract Being and Nothingness, Fullness and Void, Plenitude and Vacuity, and One and the Zero are some of the categories metaphysicians use to describe Ultimate Reality. In Vedāntic Philosophy, *Brahman* is the Ultimate Reality which is Pūrṇa, Being, Fullness, and the One. In Mādhyamika Buddhism, Ultimate Reality is *Śūnya*, Nothingness and Void. Scholarly studies generally maintain that (a) *Pūrṇa* and *Śūnya* are ultimately the same or (b) they are different. Without taking sides on these positions, I would like to point the crisscrossing of the concepts and arguments used in both of these schools in their conceptions of Ultimate Reality. (e.g., both *Brahman* and *Śūnya* are infinite, unknowable, and silence, Brahman is *advaita*, and *Śūnya* is *advaya*).

It seems to me whether one accepts *Śūnya* or *Pūrṇa* as Ultimate Reality is dependent on whether one is a tough-minded or tender-minded philosopher. Consider the following popular example. Suppose we have before us a glass half-filled with water. How do we describe the glass and the water? We can say "the glass is half empty" or "the glass is half full." Tough-minded philosophers describe the glass as "half empty," and tender-minded philosophers describe it as "half full."

Keywords Being · Nothingness · Fullness · Void · Plenitude · Vacuity · Zero · *Advaita* · *Advaya*

1 Introduction

Being and Nothingness, Fullness and Void, Plenitude and Vacuity, and One and the Zero are some of the categories metaphysicians use to describe Ultimate Reality. In Advaita Vedānta, *Brahman* is the Ultimate Reality, which is Being, Fullness, Plenitude, the One, etc. Śūnya is a foundational concept in Buddhism, and in

S. S. Rama Rao Pappu (✉)
Department of Philosophy, Miami University, Oxford, OH, USA
e-mail: pappuss@miamioh.edu

S. R. Bhatt (ed.), *Quantum Reality and Theory of Śūnya*,
https://doi.org/10.1007/978-981-13-1957-0_4

Mādhyamika Buddhism, Śūnya is the ultimate reality which is Nothingness, Vacuity, Zero, and Emptiness. In this paper I shall discuss Śūnya and *Brahman* and how they crisscross into each other.

The Abhidharma tradition of early Buddhism denies the existence of objects and persons and reduces them to *dharmas* and *skandhas*. In the famous *Questions of King Milinda*, when King Milinda says that he came in a chariot to visit Nāgasena, the monk asks the King what a chariot is. Is the chariot, the axle, the wheels, the yoke, the spokes, the wheels, the chariot body, individually, or collectively? Or is there something outside the parts? The king says the parts are not the chariot nor is there anything outside of it. Nāgasena therefore concludes that chariot is a mere name, a convenient designation. The parts of the chariot are also further analyzed into *dharmas*. *Dharmas* are like atoms, combinations of which constitute the material objects. Earth, water, fire, etc. are some such *dharmas*. There is no consensus on how many types of *dharmas* there are.

The same logic which dissolves objects into its constituent *dharmas* is applied also to persons. Who is Nāgasena? Is Nāgasena the skin, the flesh, the bones, the kidneys, etc.? No. These are parts of Nāgasena's body. Is Nāgasena then his psychological characteristics – sensations, perceptions, predispositions, and consciousness combined? No. It is then concluded that "Nāgasena" is a mere name. The Abhidharma tradition thus establishes the concept of "No-Self." What we call Self is nothing but the five *skandhas* – *nāma-rūpa* (name and form), *Vedanā* (sensation), *samjñā* (perception), *samskāra* (mental formation, dispositions), and *vijñāna* (consciousness).

Because of the doctrine of *anitya* (impermanence) and *pratītya-samutpāda* (dependent origination), the *dharmas* and the *skandhas* are not static realities but "flowing" transient states. It should however be noted that in the early Abhidharma tradition, the *dharmas* and the *skandhas* are irreducible, ultimate, and transient realities. The *dharmas* have an essential nature (*svabhāva*) to them.

Composite objects like chairs do not have a *svabhāva* to them; but the *dharmas* which constitute the chair have an essential nature (*svabhāva*) to them. In the Abhidharma tradition, śūnya or emptiness refers to the composites not to the *dharmas* with which composite objects are made up of.

It was Nāgārjuna and the Mādhyamika Buddhists who carry forward Śūnya or Emptiness to apply not only to objects and persons, but also to *dharmas* and *skandhas* by showing nothing has a *svabhāva* or essential nature to it. So not only objects and persons are empty, but also the *dharmas* and *skandhas* are also empty, thus paving the way to make the world śūnya or emptiness. Nāgārjuna refutes the assumption that anything has a *svabhāva* or essential nature to it. If things have *svabhāva* or essential nature, they cannot be brought about by any causal process nor can they change into something different. But given the doctrine of *anitya* and *pratītyasamutpāda*, things change and change incessantly. For Nāgārjuna, therefore, not only objects and persons are śūnya, but the *dharmas* and *skandhas* which constitute them are also śūnya.

Nāgārjuna using his dialectical skills develops Śūnya as the Ultimate Reality in the following manner. First, to know the world is to "pigeonhole" the world into "categories" (like Existence-Nonexistence, One and Many, Substance-Attribute, etc.). Nāgārjuna examines some categories like cause and effect, motion and rest, Self and No-Self, substance and qualities, etc. and shows that they are contradictory and therefore empty. Why we cannot have knowledge of the world without using the categories is the following. Suppose I say "This is a chair." My knowledge that this is a chair is possible, because I use the categories of Existence-Nonexistence (everything either exists or does not exist; the chair exists), cause and effect (everything is either a cause or an effect; there is a cause of the chair, the wood, etc.), motion and rest (everything is either in motion or at rest; the chair is static, not in motion), and substance-quality (everything is a substance, something which underlies all the qualities). In the case of the chair, it is the chairness, which is the essence of the chair, and its color, texture, its weight, etc. are its qualities. Suppose we do not have these categories; can we say anything about an object we see? Can we say anything about the world? Obviously not.

2 The Next Step in Nāgārjuna's Dialectic Is to Formulate a Criterion of Reality

For Nāgārjuna, that which is real must be Non-relative and Noncontradictory. Using his dialectical method, he subjects some important categories with which we understand the world, like cause-effect and motion-rest, and shows they are relative and contradictory. It is important to note that if any concept or proposition is contradictory, the object or state of affairs cannot be real. The concept "square-circle" is contradictory; therefore square circles are not real. The proposition "John is a married bachelor" is contradictory, and therefore we do not need to inquire whether John is indeed a married bachelor.

In his *Mūlamādhyamikakārikā*, Nāgārjuna examines the categories of cause-effect, rest and motion, Ātman (self)-Anatman (no-self), etc. and subjects each of the categories to four alternative positions and shows the contradictions involved in each of these positions. Take, for example, the category of cause-effect. The four alternative positions he examines are as follows: (1) Effect is a self-expression of the cause. But there is no point in self-duplication. (2) Cause and effect are independent of each other. If they are independent of each other, where is the relation? Anything can cause anything else. (3) Effect is identical with but different from the cause. Here we have to show what it is that continues from cause to effect and what is new. (4) Effect is a chance occurrence from the cause. This is to give up causation altogether.

Having shown each of these alternative positions on the relationship of cause and effect leads to contradictions, he draws the conclusion that cause and effect relationships cannot be rationally explained. They are just conceptual devices used to mislead rather than enlighten us. He examines other categories like motion and rest, etc. in his treatise.

Having shown how the categories we use to understand the world are relative and self-contradictory, he turns his attention to the *concepts* which we use to describe reality. The object "chair" which we see before us is different from the concept chair which we use to describe the chair. Needless to say, concepts are in the mind, they are thought processes. The question now is how can thoughts reach reality and establish the reality of things and objects. How can my concept "chair" reach the object "chair" out there and certify that I am indeed looking at the chair? *Nāgārjuna* rejects the classic "correspondence theory of truth." He draws a wedge between thought and reality saying there is no way that thoughts, which are mental, reach reality, which is physical. The Vijñānavādin idealists agree with Nāgārjuna that there are no things; there are only ideas. Nāgārjuna however will dismiss the Vijñānavādin idealism and say that his relativity criterion applies to ideas as much to things. Not only the proposition that "The seed acorn is the cause of the Oak tree" is false, because cause-effect relationships are relative and contradictory but also "the *concept* of an oak is the cause of the *concept* of an Oak tree" is also false, because relativity applies as much to the conceptual world as to the real world. He even interprets the Buddhist doctrine of *pratītyasamutpāda* (Dependent Origination) stating that "depending on the concept cause, the concept effect arises"; depending on the concept "acorn," the concept "oak" arises.

Therefore, if the worlds of objects are śūnya, and the concepts with which we think of the world are also śūnya, how can we ever know Ultimate Reality? His answer is we cannot know Ultimate Reality. Nāgārjuna does not deny that we plant acorns to raise oaks, we see fires and smoke associated with fire, etc. He says that this is empirical reality, or conventional truth (*samvṛttisatya*), where reality is built around conventions and agreements among a linguistic group. Śūnya as *paramārthasatya* cannot be known.

Śūnya then is the Ultimate Reality for Nāgārjuna. It is the Transcendental Truth (*paramārthasatya*). What is "Śūnya"? Śūnya or Emptiness is, after all, a concept. Because all concepts are relative, depending on one concept, another concept arises; the concept śūnya is relative to the concept Aśūnya, and therefore it is also empty. So, Śūnya as Emptiness is itself empty, because it is relative to Aśūnya.

The philosophically interesting thing to note is that śūnya of śūnya or emptiness of emptiness is not like the emptiness of objects, emptiness of *dharmas*, or emptiness of self. The *dharmas* are empty means that there are no *dharmas*; the self is śūnya means there is no self. But when we say that Śūnya is Śūnya, it does not mean there is something called "Emptiness" which is Emptiness. That is why Nāgārjuna says ultimate reality is not Śūnya or Emptiness. It is neither śūnya nor aśūnya nor both nor neither. In other words, he makes it impossible for us to *know* Śūnya, to *speak* of Śūnya, and to even *think* of Śūnya. In other words, Nāgārjuna takes away Śūnya from any philosophical or rational mode of thinking and understanding. Like Wittgenstein he would say "whereof one cannot speak, thereof one must be silent." Śūnya is Silence, silence not as an attribute of śūnya but an admonition that we should not speak anything of Śūnya. But what is not rationally possible to know may be possible to know in a nonrational or non-conceptual way. That may be possible through *prajñā* or meditation. That is why Nāgārjuna says that there is the least difference between *nirvāṇa* and *samsāra*. The world looked at through categories and concepts is *samsāra* and gives us the *samvṛttisatya*, and the same world looked at without categories, without concepts, and without a relativistic outlook is *nirvāṇa*.

Let me now pass from Śūnya or Emptiness in Buddhism to *Brahman* which is Fullness and the One in Vedānta.

As early as the Ṛg Veda, in the Hymn of Creation,[1] Ultimate Reality is conceived as *tadekam* (That One), and in Hymn[2] it is said that everything that exists is One which has taken many forms (*Ekaivosahsarvamidamvibhāti, ekamvaidam vi babhuvasarvam*). Again Ṛg Veda[3] states That which is One, the sages call it by many names like Agni, Yama, and Mātarisvan (*Ekam sad viprābahudhāvadanti, agnimyamam, mātarisvānamāhuḥ*). And one of the *Mahāvākyas* which occur in the *Chāndogya Upaniṣhad*[4] states *Ekam brahmadvitiya nāsti; neha nānāsti kiñcan* (*Brahman* is one, without a second, there is no multiplicity) at all. *Brahman* is not only The One, but it is also "pūrṇa." The famous Upaniṣadic hymn "*Om pūrṇamadaḥ, pūrnamidam, pūrṇātpūrṇamudacyate, pūrṇasyapūrṇamādāya, pūrṇamevā vaśiṣyate*" (That is Fullness, this is the Fullness; from Fullness Fullness comes; take away the Fullness from the Fullness, Fullness remains.)[5] Sometimes, "pūrṇa" is translated as "infinity" which means "That is infinite, this is infinite; from the Infinite Infinite comes; take away Infinite from the Infinite, Infinite remains."

In Vedānta literature, three kinds of arguments are given to prove the Oneness of *Brahman* as the Ultimate Reality. These arguments are somewhat similar to those given in Buddhism. Buddhism takes empirical objects like chariot, chair, etc. and reduces these objects to the five aggregates, and later on the Mādhyamikas show that these aggregates are *śūnya* or emptiness. In the *Chāndogya Upaniṣhad*, the teacher asks Śvetaketu to bring a fig and divide it further and further till he sees nothing. Unlike the Buddhists who argue that the fig when divided leads to Nothing, the Upaniṣadic teacher says "Verily, my dear, that finest essence which you do not perceive – verily, my dear, from that finest essence this great Nyagrodha (sacred fig) tree thus arises."[6] Vedāntic literature abounds with discussions on the nature of the Self. Like Buddhism, Vedānta denies that the Self or "I" is identical with external objects, with one's body, with one's senses, or even with one's mind. According to Vedānta, the "I" is of the nature of pure consciousness. "I" as "pure consciousness" exists, because "There is no consciousness" is a contradiction. Pure Consciousness is that which cannot be doubted or denied. Having established that the nature of the Self (*Ātman*) is consciousness, Vedānta identifies *Ātman* (individual self) with *Brahman* (Universal Self). *Brahman* as Pure Consciousness is One, because there is nothing real other than Pure Consciousness. It is infinite, because beyond

---

[1]X:129

[2]VIII:58.2

[3]I.164.46

[4]VI:2:1

[5]Brh V .1.1

[6]Vi.12.1-3

consciousness there is more Consciousness. Consciousness is also (by definition) Knowledge. So, *Brahman* is characterized as One (*ekam*), Real (*sat*), Infinite (*anantam*), and Knowledge (*jñānam*). Later *Vedāntins* describe *Brahman* not as One but as non-dual *advaita* to bring the uniqueness and indescribability of *Brahman*.

Let us now discuss where Emptiness (śūnya) and Fullness (*Brahman*) crisscross with each other. As ultimate categories, both śūnya and *Brahman* are infinite and unknowable. The famous Upaniṣadic thought (*pūrṇamadaḥ pūrṇamidam*, etc.) may as well be paraphrased into the *śūnya* language: "*śūnyamadaḥ, śūnyamidam, śūnyātśūnyamudacyate*, etc." "That is Emptiness, this is Emptiness, from emptiness there emptiness arises you add Emptiness to Emptiness, there is Emptiness; subtract Emptiness from Emptiness, there is Emptiness." As infinities, both *Śūnya* and *Brahman* are unknowable. Because knowledge presupposes a knower and the known, and if everything is *Śūnya* or *Brahman*, the knower and the known are also *Śūnya* or *Brahman*. If *Śūnya* were to be a knowable category, it is relative to *Aśūnya*, and therefore *Śūnya* will not be the Ultimate Truth. Therefore the Mādhyamikas conceive *Śūnya* as "*advaya*," not two. *Śūnya* also cannot be said to be a (mathematical) zero, because zero is relative to One. If it is not two, is it one, zero, or three or what? By stating that *Śūnya* is "*advaya*," the Mādhyamikas want to preserve the uniqueness of *Śūnya*. The only characterization we can give of *Śūnya* is that it is "*advaya*," not two.

The Vedāntins also like to preserve the uniqueness of *Brahman*. Instead of characterizing *Brahman* as the One, they describe *Brahman* as non-dual (*advaita*). To say that *Brahman* is One is to place *Brahman* in a relativistic position, because the mathematical concept One is related to Zero, on the one hand, and two on the other. We live in a world of plurality and relativity, and therefore we cannot know what One is without a second. Suppose there is only One. Can we know that it is One, in the same way that if the whole world is blue, can we know that it is blue? Just as Buddhists say that any object or idea they come across is "empty," the Vedāntins also state, using the "*neti-vāda*," that every characterization we give of *Brahman* is "*neti, neti*" "not this, not this."

Like *Nāgārjuna* who says that the best way to know Śūnya is Silence, the Vedāntins also say we should be silent when it comes to the knowability of *Brahman*. In the *Chāndogya Upaniṣhad*, it is said[7]: Where one sees nothing else, hears nothing else, that is plenum. In the Munḍaka Upaniṣad,[8] it is said that Silence is Ātman. Śankara, in his commentary on the Brahma Sūtras, mentions that when Bhaskalin asks Bhadva about the nature of *Brahman*, he replies that "*Brahman* is Silence."

Given the unknowability of both *Śūnya* and *Brahman* and the Buddhists and Vedāntins use similar expressions, *advaya* and *advaita*, to characterize the ultimate reality, whose conception is better than the other? It depends! Consider the following

[7]vii.24.1
[8]II.2.6

popular example. Suppose we take a glass filled with half water and half empty. We can describe the glass as "the glass is half empty" or "the glass is half full." Tough-minded philosophers tend to describe the glass as "half empty," and tender-minded philosophers describe it as "half full"! The Buddhist argument is From Emptiness only Emptiness comes. The Vedāntins agree that from emptiness only emptiness comes. But Reality for them is not Emptiness, because something exists for certain (i.e., My Self, the Ātman). Even if the world is *Māyā*, *Māyā* must have a substratum. The substratum of *Māyā* cannot be *Māyā* but something real. Therefore, Ultimate Reality is not Emptiness. The *Chāndogya Upaniṣhad* states[9] "How from Non-Being could Being be produced? On the contrary, in the beginning this world was just Being, one only without a second."

Buddha did not answer some metaphysical questions, the *avyakrtas* (e.g., whether the world is eternal or non-eternal or both or neither, etc.), and kept silent. *Nāgārjuna* admonishes us to keep Silence when talking about *Śūnya*. That *Brahman* is Silence is a popular theme in *Vedāntic* literature too. In other contexts again, the Vedāntins use the via *negative* method ("*neti-neti-vāda*" not this, not this), to point to the uniqueness of *Brahman*.

[9]Text

Chapter 5
Quantum Reality and the Theory of *Śūnya*

Amalkumar Mukhopadhyay

Abstract According to the traditional Indian philosophy as reflected in the *Upaniṣad*, all objects or incidents relating to the material or phenomenal world being mutable are regarded as unreal (*asat*) as opposed to the only Real (*sat*) termed the One in the *Ṛgveda*, *Brahman* in the *Upaniṣad*. According to *Nāgārjuna*'s philosophy as presented in the *Mahāprajñāpāramitā Śāstra*, all elements, physical as well as mental, are impermanent, and so, they are non-substantial (*śūnya*) and not unconditioned. The ultimate reality, or the eternal *dharmalakṣaṇa*, does not come from anywhere nor does it go anywhere; in it there is neither birth nor extinction and therefore reminds us the Absolute of the *Ṛgveda* or *atman* of *Upanisad*. This Absolute, together with the all the animates and inanimates in the material or the phenomenal world, gives rise to the concept of the single whole termed *Brahman*. This is in essence the doctrine termed the *Viśiṣṭa Advaitavāda* expounded by *Rāmānuja* (c. twelfth century A.D.) in his *Śrībhāya* based on *Brhamasūtra*'s commentary written by *Bodhāyana*. This unified entity is the 'Quantum Reality' from Indian metaphysical standpoint. This is supported by the *Nāsadīyasūkta* in the Ṛ*gveda*, followed by the *Śrīsūkta* in the Ṛ*gveda*, in all ancient *Upaniṣad*, and also by Lord *Buddha* in the *Aggaññita Sutta*. *Brahman* has been described as *Śūnya*, the space unlimited. This idea has later been reflected as *Śūnya Brahma* or *Śūnya Puruṣa* in the *Jagannātha* or in the *Mahimā* cult in *Odisa*. Starting with concept of voidness mentioned as '*vyoma*' and '*tucchaya*' in the *Ṛgveda* and the roundness of the Universe as mentioned in *Śatapathabrāhmaṇa* (middle of the seventh century B. C.), ancient India adopted a small circle as a symbol to denote void.

Keywords Veda · *Upaniṣad* · Viśiṣṭa Advaitavāda · Srībhāṣya · *Mahaprajnaparamitā Śāstra* · Brahman · Śūnya Puruṣa

---

A. Mukhopadhyay (✉)
Department of Ancient Indian Mathematics & Astronomy, Sri Ramakrishna Centre of Research in Indology, Kolkata, India
e-mail: bns@scsiscs.org

S. R. Bhatt (ed.), *Quantum Reality and Theory of Śūnya*,
https://doi.org/10.1007/978-981-13-1957-0_5

1 Introduction

The paper actually consists of two interrelated topics, namely, the Quantum Reality and the theory of śūnya. The first part is on the 'Quantum Reality' which means 'the aggregate of reality (bhūmnaḥ)' and its implications as detailed in the qualified non-dualism (Viśiṣṭa Advaitavāda). It was explained in essence by Śrī Rāmakṛṣṇa. The explanation has been elaborated with the help of a few quotations from the Srībhāṣya of Brahma sūtras by Rāmānuja along with the Nāsadīyasūkta, the Sṛṣṭisūkta, both from the ṚṚgveda, concerned verses from the Upaniṣads, the Gītā, etc., supported by elucidations and (as and where suitable) by Śrī Rāmakṛṣṇa and Swami Vivekananda. The reflections of Rabindranath Thakur expressed with his poetic elegance has not only thrown light on the origin of the Universe following the prophetic vision of ancient Indian sages but also take a reader to the Ṛgvedic age with feelings of the Absolute's omnipresence amidst the fragrance of eternal purity. The second part, i.e. the 'Theory of Śūnya', throws light on:

(i) The metaphysical concept of Śūnya describes Brahman as the Supreme space, Brahman and *agni* being synonymous (Ṛgveda). The round-shaped World being the body of agni (Śatapathabrāhmaṇa) and the whole World (*parimaṇḍala* – circular/spherical) had to be meditated upon as the replica of agni vis-à-vis the Supreme space (the *Brahman*) by those eager for knowing Brahman (*Chāndogya Upaniṣad*) and thus a small circle became a symbol for what is required to be known
(ii) The introduction of the concept of Śūnya as a numeral in Mathematic (Piṅgala's Chanda sūtra, circa 400 B.C.E./200 B.C.E.) with some sort of symbolic notation (might be a 'dot', the earliest symbol)
(iii) The development of the concept of zero in terms of Mathematical operations by Brahmagupta (seventh century C.E.)
(iv) The incorporation of the round symbol in Mathematics for zero (probably prior to the eight century C.E. by scholars as the round symbol was made public in the eighth century)
(v) The extension of the concept of śūnya by Bhāskarācārya (b. 1114 C.E) in the light of the philosophical doctrines of Kaṇāda, Nāgārjuna, etc. from the primary status of 'void' to the superfine concept of 'infinitesimal' apt to find out an application of what is known today as a differential

According to lexicon (Oxford Dictionary), the word 'quantum' used in physics as a noun means a discrete frequency of radiation corresponding to a single photon or to a transition of an atom between energy states; but when used as an adjective to qualify a 'jump' in physics, the word means an abrupt transition of an electron/atom from one state to another, whereas the meaning of the word 'reality', when considered from the angle of philosophy, of course, Western, signifies an absolute existence of objects or of phenomena in the phenomenal or material world, independent of human decisions or conventions, but must necessarily within the limits of human perception. The word 'quantum' comes from the Latin word 'quantus', meaning 'how much'. To explain the joint term 'Quantum Reality', some scholars may resort to the philosophical attitude of Copenhagen's interpretation of quantum physics. This attitude arises [1] from the duality between the real world (i.e. phenomenal or material world) and our mind, i.e. between the classical concept (the probability of occurrence of natural phenomena as experienced by human beings) and why the classical concepts do not follow the structure of the real world. Precisely spoken, Copenhagen's interpretation means that our description of nature is determined by and related to the inherent structures of human perception. The word 'perception' (originated from the Latin word perception/*perceptio*) is the organization, identification and interpretation of sensory information in order to represent and understand the environment.[1] All perception involves signals arising out of physical or chemical stimulation of the sense organs, the perceptual systems of the brain. This system leads an individual to experience the material world around him as stable, though the sensory information is often typically incomplete and rapidly varying.[2] Due to an inaccurate input of the sensory organs to the brain, illusion occurs, and so long as one is within the veil of illusion, identification of what is actual is hardly possible. Allegorically speaking, it is either a snake or a rope but never both simultaneously. The material or the phenomenal world including all objects in it appears as stable, though subjected to death and decay or, in other words, impermanent due to mutability. We therefore conclude that Copenhagen's interpretation of the Quantum theory is not aptly qualified to explain 'Quantum Reality' in relation to the world we live in. According to the traditional Indian philosophy, all objects or incidents relating to the material or phenomenal world being mutable are regarded as unreal (*asat*) as opposed to the only Real (*sat)* termed the One or the Absolute in Ṛgveda[(A)] and Brahman (*√bṛṁha + man or √bṛha + man*, literally meaning growth or expansion) of the Upaniṣads. According to Nāgārjuna's philosophy [2] as presented in the Mahāprjñāpāramitā Śāstra, all elements, physical as well as mental, are impermanent, and so, they are non-substantial (*śūnya*), and being non-substantial, they are not real, not unconditioned. The ultimate reality (relating to worldly objects), called the *svabhāva-śūnyatā dharma* or the eternal *dharmalakṣaṇa*, does not come from anywhere nor does it go anywhere; in it there is neither birth nor extinction and is free from all determinations.[3] We remember that in Kaṭhopaniṣad (verse I. ii. 18 -vide S-1 in the Appendix) and also in Gītā (verse II.20- vide S-2, Appendix), *ātman* (the self/soul) has been described as devoid of birth and death, with no origin, originating from nothing, eternal with no decay and not injured even the body is destroyed. Almost a similar concept has the word *pudgala* used by scholars following Jaina religion [3] in the sense that *pudgala* is not destroyed by water and fire (vide S-3, Appendix). Here it is to be noted that the Absolute (or the One) of the Ṛgveda has been described as *ātman* or Brahman in Upaniṣads. It is worth

[1]Wikipedia free encyclopedia.

[2]Wikipedia free encyclopedia.

[3]Venkata 1996, pp. 270–271.

mentioning here that there is a gulf of difference between Buddhism or Jainism and Vedic religious concept. Atheism is the central theme of both.

Buddhism and Jainism, while Vedic religion inculcates monotheism. This Absolute, together with all the animate and inanimate objects in the material world, gives rise to the concept of the single whole termed *Brahman.* This is, in essence, the doctrine called the Viśiṣṭa Advaitavāda or qualified non-dualism. It is the philosophy of the Śrī Sampradāya, one of the most popular schools of Vedānta (or Upaniṣad) ideology existing in India long before the times of Rāmānuja (c. twelfth century A. D.) [4] who is now considered as the main proponent of the philosophy. According to Ramanuja, as he said in his Śrībhāsya of Brahmasūtra, his commentary (*bhāṣya*) on the *sūtras* was based on the elaborations (or *vṛtti*) of the philosophy written by Bodhāyana [5]. In the theory of the origin of the Universe as described in the Nāsadīyasūkta in the Ṛgveda[4] (verses 1–7, 10.129), the revelation of the One followed by the evolution of the seed-like primordial state of the world attests the same union of the Absolute with the microcosmic unreality since the beginning of the evolution of the World. After a certain period of time, following the evolution of the cosmic seed (the primordial stage of the Universe) grows or expands, as hinted at the Sṛṣṭisūkta and also in the Puruṣasūkta in the ṚṚgveda. In Upaniṣads, this One termed *Brahman* has been described as the space unlimited (the *parama vyoman* – Taittirīya Upaniṣad, II.i.1).' The eyes cannot behold Him, neither voice nor mind can reach Him', says Kenopaniṣad (verse I.3). 'Whatsoever is moving in the Universe, is to be regarded as covered by the Lord' (verse-1, Īśopaniṣad). The concept of *Śūnya* as the space unlimited as is available in the concept of *Śūnya Brahma* or *Śūnya Puruṣa* in the Jagannātha or in the Mahimā cult in Orissa[5] (each pleading for non-dualism in the way of Śaṅkarācārya with Jagannātha in the place of Brahman) is but a reflection of the idea '*Brahman* is all, the Supreme Space, nothing is superior to the Supreme'. The philosophy inherent to the word *Śūnya*, deep-rooted as it is in Indian ethos, is so intimately mingled with our Vedic religious concept that the word does not simply mean the materialistic void as in vogue in the Western style. Our present discussion will throw light on (*i*) how the metaphysical concept of '*vyoma*' and '*tucchaya*' ushered in the thought of ancient Indian sages during the times of ṚṚgveda gradually assumed definite material or rather physical status with an abstract approach in the Vaiśeṣika Philosophy of Kaṇāda in the age of Mahābhārata and later transcended to the larger field of Mahāyānic Buddhistic Philosophy of *Śūnyatā* as put forward by Nāgārjuna in the beginning of the second century A.D.; (*ii*) how the Universe expands, as hinted at the Sṛṣṭisūkta and also in the Puruṣasūkta in the Ṛgveda; (*iii*) how the concept of the circular shape of the universe described in the Śatapatha Brāhmaṇa together with the metaphysical concept of *Brahman* (*sarvaṁ tatparamaṃ śūnyaṃ*) gave rise to the circular shape of the symbol of

---

[4]Winternitz M's, *Geschichte der Indischen Litteratur*, Vol. I, 1920 (dealing with the Vedic Literature) (Eng Translation by Mrs. S. Ketkar, p. 258, C.U., 1927.

[5]Rekharani Khuntia, Similarities Between Jagannath Cult and Mahima Cult: A Comparative Study, Oḍisha Review, 2011.

voidness later taken for the symbol of zero along with the recognition of the concept of zero as a numeral; and (*iv*) how the synthesis of the Indian traditional philosophical ideas of the concept of zero along with its concretized definitions compiled in Amarakośa by Amarasiṁha of the fifth century A.D. paved a very broad way before ancient Indian mathematicians like Āryabhaṭa- I (fifth century A.D.), Brahmagupta (seventh century A.D.) and Bhāskarācāya (twelfth century A.D.) to equip them with mathematical tools to renovate the traditional concept of zero for interpreting and devising rules of algebra and astronomy in a way erstwhile unknown in the world. A bit elaboration will hint at a better understanding of this point. Āryabhaṭa I used small circular symbols (the symbol was known by the name *kha* meaning 'sky') to denote absence of a numeral in his place-value notation either in conformity with the rule '*adarśana lopaḥ* I.i.60'), i.e. non-visibility due to non-existence advanced by Paṇini in the fourth century B.C.[6] in his *Aṣṭādhyāyī* (Sanskrit Grammar), or in accordance with *abhāva* theory due to Kaṇāda (though in practice, a dot was most probably in use to represent 'zero' in his days; but as a single 'dot' lacks in prominency, he devised his own method of representing numerals by Sanskrit alphabets) in the place-value notation of numbers using small circular symbol to denote the absence of a numeral at a place in the notation). Both Brahmagupta and Bhāskarācārya adopted this *abhāva* theory; but Brahmagupta advanced a step further than his predecessors. 'Śūnyaṃ Ākāśaṃ', said he and incorporated Kaṇāda's idea of omnipresence of *Ākāśa* among all worldly materials in operations with numbers in the sense that a number remains unchanged when zero is added to or subtracted from it. In addition to this, Bhāskarācārya went further and translated Nāgārjuna's view of '*Śūnyatā*' (that '*Śūnyatā* is the final stage to which everyone tends to; but none can reach the final stage exactly') in the language of mathematics by considering the case when a variable tends to nullity.

2 'Quantum Reality' from Indian Metaphysical Standpoint

This has been developed as the Viśiṣṭa Advaitavāda – actually a modified form of non-dualistic ideal of the philosophy, called the 'qualified non-dualism'. It has been mentioned before the commentary known as Śrībhāṣya of Brahmasūtra due to Rāmānuja is now regarded as the most authentic source of this doctrine. In consideration of the limited scope of the present article, let us glance at the first three aphorisms as these will point at what type of topic has been included in the entire work, and also let us follow what Lord Śrī Rāmakṛṣṇa has clarified as the sum and substance of the entire doctrine with an appropriate example.

[6]Bhattacharya Sachchidananda, ibid, p.712.

3 The First Three Aphorisms in Śrībhāṣya of Brahmasūtra [6]

The very first aphorism (Athāto brahmajijñāsā, I.1.1 –S-4, *i*) of Śrībhāṣya points at the main theme that after acquiring knowledge of the work portion of the Vedas and knowing that this work-centric knowledge brings forth only temporary results far from what is regarded as the eternal and infinite, this question (how to know *Brahman*?) arises in the mind of a devotee.

In the commentary of this aphorism, Rāmānuja says, 'the innate meaning of the word *Brahman* is the supreme person *(Puruṣottama)*, free from all sorts of imperfections with endless (anavadhika) and innumerable (asaṅkheya) auspicious qualities' (vide S-4, ii, Appendix).

The next aphorism (Janmādyasya yataḥ, I.1.2) (videS-5, *i,* Appendix) has been clarified as 'That supreme Person who is the ruler of all; Who is against all evils, honest in determination, has infinite auspicious qualities like knowledge, bliss etc., Who is omniscient, omnipotent, supreme merciful, Who causes the creation, existence and annihilation of this world with its manifold wonderful arrangements beyond the bounds of thought, Who comprises within Himself all from *Brahmā* to blades of grass, destined to experience the fruits scheduled for the land and the period of time they belong to, is this *Brahman* (vide S-5, *ii*, Appendix). (Though the word Brahman is neuter gender in Sanskrit Grammar, the word has been taken here for a masculine one. This justifies the word *puṁsaḥ*.)

The third aphorism (Śāstrayonitvāt ॥ I.1.3.) clarifies (vide S-6, Appendix) that the scriptures are the source (of the right knowledge) with respect to *Brahman.*

Here by the scriptures, as Swami Vivekananda explained, [7] are meant the religious scriptures, (namely, the Brahmasūtra of Vyāsa often termed Vyāsasūtra, the Upaniṣads and the Gītā). These are the three *prasthānas*; i.e. these three scriptures are like the three legs of a tripod stand on which rests the philosophy of Viśiṣṭa Advaitavāda.

3.1 Lord Srī Rāmakṛṣṇa's Clarification of Viśiṣṭa Advaitavāda (Qualified Non-dualism)

The sum and substance of the doctrine has been clarified by Lord Srī Rāmakṛṣṇa with a nice example in the following way: 'In essence, Viśiṣṭa Advaitavāda [8] is that the *Brahman* or the Absolute is qualified by the universe and its living beings. These three – *Brahman*, the world and the living beings – together constitute the One, the Absolute'. Srī Rāmakṛṣṇa has clarified the philosophy by the following example[8]:

'Take the instance of a bel fruit (in Sanskrit it is called *Bilva – Aegle marmelos*, wood apple). A man wanted to know the weight of the fruit. He separated the shell, the kernel (i.e. the soft portion within the hard shell) and the seeds; but can a man know the weight of (the fruit) only by weighing the kernel? He must weigh the kernel, the shell and the seeds together. At first it appears that the real thing in the fruit is the kernel, neither the seeds nor the shell. Then by reasoning, you find that the shell, the kernel and the seeds all belong to the fruit; the shell and the seeds belong to the same fruit as the kernel belongs to. Likewise, in spiritual discrimination, one must first reason, following the method 'Not this, not this' (*Neti Neti*): God is not the Universe, God is not the living beings. Brahman alone is Real and all else are unreal. Then one realizes, as with the bel fruit, that the reality from which we derive the notion of Brahman is the very Reality from which evolves the idea of living beings and of the Universe. The *nitya* (perpetuity) and the *Līlā* (activity) are the two aspects of the One and same Reality; therefore, according to Rāmānuja, *Brahman* is qualified by the Universe and the living beings. This is the theory of qualified non-dualism'. It may be mentioned here that, as an analogy to explain the method *Neti Neti* (Not this, not this) mentioned above, Srī Rāmakṛṣṇa describes,[7] 'The master of a house is asleep in a dark room and someone is groping to touch the master amidst the darkness prevailing there. He touches the couch and says "No, it is not he". He touches the window and says, "No, it is not he". He touches the door and says, "No, it is not he". This is known in the Vedānta as the process of *Neti Neti.* At last his hand touches the master's body and he exclaims, "Here he is" *(iha, iha)*. In other words, he is now conscious of the "existence" of the master'.

3.2 Brahman, *the Absolute Pervades the Entire World and Is Known by Manifold Names*

All pervasiveness of the Absolute, the only Real with reference to scriptures: According to the Vedānta Deśika,[8] Viśiṣṭa Advaitavāda is defined as *aśeṣa cit-acit prakāraṃ brhamaikameva tattvaṃ*, i.e. as the doctrine describing the union of innumerable varieties of animate and inanimate types of objects with the One as the *Brahman.* According to Muṇḍakopaniṣad, verse I.i.7[9] (vide S-7, Appendix), the *Universe* emerges from the Imperishable (*Brahman*) just as a spider generates and also withdraws threads of its web (itself remaining within the web), as the herbs (and trees) grow on the earth and as hair issues out (on the body) of a male. Much more explicit is the idea that *Brahman* is in front, behind, to the right, to the left, extended above and below; the world is also this *Brahman* (verse II. ii. 11 (vide S-8, Appendix) of Muṇḍakopaniṣad).[10]

The above discussion when considered in the light of these verses connotes what *Chāndogya Upaniṣad* [9] mentions in the line 'All this is *Brahman*' (verse III. 14.1) (S-9, Appendix).

---

[7]The Gospel of SriRamakrishna, Vol. I, p114, 2002, Sri Ramakrishna Math, Maylapore, Madras.
[8]Viśiṣṭa Advaitavāda, Wikipedia free encyclopedia.
[9]Muṇḍaka Upaniṣad with commentary of Śaṅkarācārya, p. 14, 1988, Advaita Ashrama, Kolkata.
[10]Muṇḍaka Upaniṣad with commentary of Śaṅkarācārya, p. 65.

3.3 Manifold Names of the One, the Absolute

In spite of the fact that monotheism is the watch word of the Vedic religion, the One or the Absolute of Ṛgveda is often found to have been described by different names in a variety of religious scripts in the Ṛgvedic as well as in the post-Ṛgvedic age. To do away with the least possibility of confusion, ṚṚgveda univocally declares (mantra – I.164.46) [10] that this One is described by the wise in many ways like *agni*, *yama* and *vāyu (mātariśvān* – air) (vide S-10, Appendix). In Skandopaniṣad[11] (vide S-11, Appendix), the One has been described as the deity of knowledge, prescribing the names *Mahādeva*, *Hari* and also *Parameśvara.* In Yogavaśiṣṭha Utpatti Prakaraṇa [11], the names *Ṛtaṃ*, *Ātmā*, *Satyaṃ*, *Paraṁ Brahma*, etc. have been mentioned with the explanation that these names were in use by great sages (vide S- 12, Appendix) to denote the Absolute.

4 The Union of the Absolute (*Brahman*) with the Unreal Even at the Time of Evolution of the World: Swami Vivekananda's Views

(i) Ṛgvedic concept of the origin of the Universe described in the Nāsadīya Sūkta
(ii) The concept of expansion of the Universe as depicted in the Sṛṣṭisūkta — these two forming the ancient Indian conjecture of 'big bang': Supporting verses from Upaniṣad
(iii) Srī Rāmakṛṣṇa's views

4.1 Swami Vivekananda Views the Unity Amidst the Variety of Cosmic Forces as Variations of a Single Unit Energy Springing from **Prāṇa** *Meaning* **Spandana** *(i.e. Vibration)*

In the introduction it has been mentioned that Absolute remained united with the microcosmic unreality in the beginning of the evolution of the World. In Ṛgveda, the Nāsadīya Sūkta elaborates this point with an unparalleled poetic luster. It will be seen afterwards in the sṛṣṭi sūkta of the same Ṛgveda that the evolution of the Universe occurs in a cyclic order. Before throwing light on the said Ṛgvedic hymns, let us look into how Swami Vivekananda explained[12] the ancient Indian idea of cosmology in the light of the Vedas, specifically speaking, the Ṛgveda and

[11] Stavakusumāñjali (Skandopaniṣad, HariharaBrahmastotra), p. 113, *Udvodhan*, Kolkata, 1986.
[12] Swami Vivekananda, ibid, pp. 345–347.

the Upaniṣads. According to Swami Vivekananda, there is unity amidst the variety of forces (existing in the Universe, apprehensible or not), and they are variations of a single unit energy (of whatever name we may choose for it). This unit springs from what is called *Prāṇa* meaning *spandana*, i.e. vibration. At the end of a cycle, when the Universe shall have to resolve back to its primal state from which it sprang, and there it remains for a time quiescent, ready to spring forth again, the *Prāṇas* are resolved back into the primal *Prāṇa* – almost motionless, *ānīdavātaṁ* (vibrating without vibrations) (the word *ānīdavātaṁ* has been mentioned in the verse 2). 'The forces permeate all matter; they all dissolve into *Ākāśa*, from which they again come out; this *Ākāśa* is the primal matter. This Ākāśa vibrates under the action of *Prāṇa*', when the next *sṛṣṭi* comes up; as the vibration becomes quicker, the *Ākāśa* is lashed into all these wave forms which we call suns and moons and systems.' According to him, 'yadidaṁ kiñcit jagat sarvaṁ prāṇa ejati niḥsṛtaṃ (First line of the second verse, sixth Chapter, Kaṭhopaniṣad) – everything in this Universe has been projected, Prāṇa vibrating', and he continued, 'you must mark the word Ejati, because it comes from Eja – to vibrate. Niḥsṛtaṃ – projected (according to M. Monier-Williams, the word means "unsheathed–come out of a covering"), *yadidaṃ kiñca* – whatever in this Universe'.

4.2 The Nāsadīya Sūkta of Ṛgveda (10th Maṇḍala, 129 Sūkta, Verses 1 to 7) Embodies How the Seed-Like Primal Stage of the Universe (asat) *Evolved Amidst Utter Darkness and Within an Indiscriminate Covering of Water After* Mahāpralaya *(Cataclysm), When There Was Nothing to Save the Absolute* (Sat)

In spite of the fact that Swami Vivekananda explained elaborately the ancient Indian idea of cosmology in the light of the ṚṚgveda and the *Upaniṣads*, the present writer feels convinced that if the quotation of the Nāsadīya Sūkta be not given here (along with detailed meaning of the verses), the readers will be deprived of the poetic elegance and the distant vision of our forefathers:

Nāsadīya Sūkta (Hymns: Wise sayings about that which is not unreal) (S-13)[13]: Origin of the Universe:

nāsadāsīnno sadāsīttadānīṁ nāsīdrajo no vyomā paro yat—
kimārīvaḥ kuha kasya śarmannabhaḥ kimāsīd gahanaṁ gabhīraṃ—— 1.
namṛtyurāsīdamṛtaṁ na tarhi na rātryā anha āsīt praketaḥ—
ānīdavātaṁ svadhayā tadekaṁ tasmāddhānyanna paraḥ kiñcanāsa—— 2.
tama āsīttamasā gūḍhamagre'praketaṁ salilaṁ sarvamā idaṁ
tucchayenābhvapihitaṁ yadāsīttapasastanmahinā jāyataikaṁ—— 3.

---

[13]Ṛgveda Saṁhitā, ibid, pp. 737–738, Kolkata, 2007.

kāmastadgre samavartatādhi manaso retaḥ prathamaṁ yadāsīt—
satobanhumasati niravindan hṛdi pratiṣyā kavayo manīṣā—— 4.
tiraścīno vitato raśmireṣāmadhaḥ svidāsīt—
retodhā āsanmahimān āsan svadhā avastāt prayatiḥ parastāt—— 5.
ko addhā veda ka iha pravocat kuta ājātā kuta iyaṁ visṛṣṭiḥ—
arvagdevā asya visarjanenāthā ko veda yata āvabhūva—— 6.
iyaṁ visṛṣṭiryata āvabhūva yadi vā dadhe yadi vā na—
yo asyādhakṣaḥ parame vyomantso aṅga veda yadi vā na veda—— 7.

Translation: At that time i.e., after the dissolution /destruction of the world (*mahāpralaya*) (at the end of a *kalpa*) there was neither existence nor non-existence, no dust (the Earth), no sky above that, which was to be covered? who (*kuha*) was protected *(śarman)* by the sky (*nabhaḥ*)? Was then deep, inaccessible (water)? 1

Neither death nor immortality was there, neither was then the concept (*praketa*) of a day or of a night, but for the One with the intent power *(svadhayā)* (self-vibrating energy in the One remaining without manifestation) breathing of One's own unaided by air (*ānīdavātṁ* ⇐*ānīt* + *avātaṁ*, ānīt being the past tense, third person, singular of the root √an, meaning 'breathed'; *avātaṁ* = without air). There was indeed no other superior (*paraḥ*) to the Supreme One (*ha*) (*tasmāt ha anyat na paraḥ*). 2

(At first there was) darkness concealed in darkness, and indiscriminate water flooding everywhere (*idaṁ sarvamā apraketaṁ salilaṁ*, *apraketaṁ* indiscriminate, *gūḍhaṁ* concealed); when there was voidness covered in voidness *(tucchayena ābhū apihitaṁ ābhū*, exist; *apihita*, covered/concealed; *tucchaya*, emptiness or voidness), there revealed the One through austerities of sages (of the universe in the former cycle of creation) (*tapaḥ tat mahinā*). 3

As the seed of creation was present (*samavartatādhi* – existent) from the beginning, there arose in the cosmic mind (of the One) the desire for (creation). The wise men judged in themselves to conclude that the Real is attached to unrealities (*satobanhumasati*). 4

Did the reins (*raśmi*) of link (between the desire and the seed of creation) spread out *(vitata)* transversely (*tiraścīnaḥ*) or below or above these? *(svid āsīt*? -was there?). There evolved mighty power with seeds of creation (*retodhā* – the One holding the seed of creation – *dhātā, Brahmā*), worldly pleasures *(svadhā)* remain below *(avastāt)* (inferior) and devotion with effort (*prayati)* above (*parastāt)*. 5

Who does know this (evolution) certainly (*addhā*), who can properly mention (*pravocat)* wherefrom this creation was originated? The gods evolved after (*arvāk* – after) this creation, who knows whence did it evolve? 6

The One who caused this creation may or may not hold (or protect) *(dadhe* – holds/ protects) it. The One, who has been controlling this (creation) remaining Himself in the highest space, may or may not desire to know that. 7

(Comment: Almost the same description as in the Nāsadīya Sūkta is available in Taittirīya *Brāhma a* with the addition of the first line of the verse 10.31.7 of Ṛgveda.)

Analysis of the Verses

From the verses quoted above with the meanings thereof, it is clear that:

- Following the dissolution of one Universe, the evolution of the next Universe occurs (through austerities of sages belonging to the former cycle of creation – *tapaḥ tat mahimnā*).
- The seed of the Creation of the Universe, unreal (mutable) though it was, remained associated with the Real or the Absolute which was vibrating (as if breathing without air), meaning thereby that sort of energy was there with the Real *(ānīdavātaṁ svadhayā tadekaṁ)*.
- The concept of gods was developed after the Creation.
- The One, who alone is the Real, is indifferent to whatever may become of this world.

4.3 Lord Srī Rāmakṛṣṇa Identifies the Cosmic Energy from the Moment of Evolution onwards as Śakti *and All Later Stages of Expansion of the Universe as* Līlā *(Actions) of the Same* Śakti

According to Srī Rāmakṛṣṇa, the Absolute, when not in action, is called *Brahman*, and when It becomes active either in creation or in destruction, the Absolute is known as *Kalī* or *Śakti*.[14] After the dissolution of the world (*mahāpralaya*), *Brahman* revealed through austerities of sages *(tapaḥ tat mahinā)* (of the universe in the former cycle of creation) with the desire for Creation and when Creation was initiated by It. It became known by the name *Kalī* or *Śakti.* In fact, starting from state of the microcosmic energy manifested in the vibration of the One at the outset to the manifestation of macrocosmic energy at all later stages of expansion of the Universe are the evidences of *Līlā* (actions) of the same *Śakti*, the active role of *Brahman.*

[14]*The Gospel of Sri Ramakrishna*, Vol. I, p. 134–135., ibid.

4.4 *That the Expanding* **Ṛtaṁ** *and* **Satyaṁ** *(Both Meaning* **Brahman***) Is United with the World with the Gradual Addition of Several Items and Has Been Hinted at the Following Sṛṣṭisūkta (Set of Three Verses 190 Sūkta in the 10th Maṇḍala)*[15] *(Vide s-14) for the Verses with Their Detailed Meaning)*

A close analysis of the verses shows:

(i) The seed-like primal stage of the Universe (*asat*) in association with the Absolute, started expanding and gradually reached the present stage equipped with the Sun, the Moon, the layers of atmospheric air, the oceans, etc.
(ii) This evolution of the creation (in modern terminology it is called 'big bang') follows the former creation, meaning thereby that the creation of the Universe occurs in a cyclic or in an oscillatory way.

Remarks

1. It is to be noted that the first line of the first verse speaks of the expansion of *Ṛtaṁ* and *Satyaṁ,* and not *Brahman* though both the words have been used to denote *paraṁ Brahman* in Yogavaśiṣṭha Utpatti Prakaraṇa (vide Section 2.4). Here the words denote seed-like primordial stage of the Universe in association with the absolute as has been mentioned in the line 'the Real is attached to unrealities (*satobanhumasati*)' (vide the fourth verse in the Nāsadīya Sūkta). Therefore, the words denote the Aggregate or the Quantum Reality at the very out set of the creation of the World which has been regarded as Brahman in the verse II. ii. 11 of Muṇḍakopaniṣad (vide Section 2.4). That is why in the Vaiśvānara vidyā (*Chāndogya Upaniṣad*, V.xii.1), the king Aśvapati is seen to teach the five sages (approaching him for *Brahma vidyā*) that Vaiśvānara, i.e. *Agni*, has the universe as His body. (Here *Agni* is the same as Brahman, as has been mentioned in the verse I.164.46 in ṚṚgveda (vide Section 2.4). So the expansion of *Ṛtaṁ* and *Satyaṁ* implies the expansion of the Universe.
2. Taittirīya *Upaniṣad* describes this expansion rather in details in the verse II. i. 1.[16]

Discussion

Mahāpralaya (devastating destruction of the universe) (according to Indian metaphysics, caused by *Śakti* or the mother *Kālī,* the active form of *Brahman*) is called today the 'big crunch'. As Śrī Rāmakṛṣṇa has said,[17] 'One cannot think of *Brahman* without *Śakti* or of *Śakti* without *Brahman*' (as the literal meaning of the word '*Brahman*' is 'expansion' and that expansion (also contraction) are the manifestation of the action of some 'force' or energy termed *Śakti* in Sanskrit). Śrī Rāmakṛṣṇa

[15]ṚṚgveda Saṁhitā, ibid. p. 764.
[16]Taittirīya Upaniṣad, Eight Upaniṣads, Vol. I, p. 303, Advaita Ashrama, Kolkata, 2002.
[17]*The Gospel of Sri Ramakrishna*, Vol. I pp.134–135, *ibid.*

further says, 'One cannot think of the Absolute without the Relative or of the Realtive without the Absolute'. The Primordial power is ever at play. She is creating, preserving and destroying in play, as it were. This power, called *Kālī* (or *Śakti*), is verily *Brahman* and *Brahman* is verily *Kālī*. It is one and the same reality. When we think of It as inactive, that is to say, not engaged in the acts of creation, preservation or destruction, we call It *Brahman*. But when engaged in any of these activities, *Brahman* is called *Kālī* or *Śakti*. The Reality is one and the same; the difference is in the name and form. Śrī Rāmakṛṣṇa allegorically explains, 'after the destruction of the universe, my Divine mother, the Embodiment of Brahman gathers together the seeds for the next creation. After the Creation, the Primal power dwells in the universe itself. She brings forth this phenomenal world and then pervades it. In the Vedas (the) Creator along with the creation has been described as the spider and its web. The spider brings the web out of itself and then remains in it (for the concerned quotation, vide Muṇḍakopaniṣad, I.i.7). It may also be mentioned here that according to the theory of 'big bang' (expansion of the Universe from the primaeval 'cosmic egg'-like initial stage) accepted by the modern cosmologists,[18] there arose the question of deciding the fate of the Universe. The density parameter Ω (defined as the ratio of the average matter density of the Universe to the critical value of the density) has a vital part to play in determining the shape of the Universe. If $\Omega > 1$, then the geometry of space is closed like the surface of a sphere. In such a case, the gravity eventually stops the expansion of the Universe, after which it starts to contract until the entire Universe collapses to a single point, a final singularity termed 'big crunch' opposite to 'big bang'. Thus, a 'big bang' will occur after the 'big crunch' of the just preceding Universe, creating a cyclic model which is also known as an oscillatory Universe. According to the Indian ideal as mentioned in the Śatapatha Brāhmaṇa (2000 B.C.), the Universe is *parimaṇḍala* (circular/round) (vide page 11), and so, the conjecture of the Indian sages rightly speaks of this oscillatory theory of the Universe, and this has been clarified by Swami Vivekananda in the lines 'at the end of a cycle, when the Universe shall have to resolve back to its primal state from which it sprang, and there it remains for a time quiescent, ready to spring forth again' (vide page 6). According to Aggañña sutta (meaning, the foremost aphorism), Lord Buddha also pointed out this to his two prospective disciples, though he neither acknowledged his indebtedness to Ṛgvedic source nor to Indian Nyāya-śāstra in the sense that whatever expands must necessarily contract afterwards.

[18]Wikipedia free encyclopedia.

5 That *Brahman* and the Universe Consisting of All Animate and Inanimate Objects in It Are Always United and the Union as a Whole Is Also Taken for Brahman in Conformity with the Doctrine of Viśiṣṭa Advaitavāda

There are multiple citations in the Upaniṣads and the *Gītā* (described as authority of Viśiṣṭa Advaitavāda) to show that all animate and inanimate objects in the world (from the very moment of its evolution) and the Absolute together form the aggregate of reality (or in other words, the Quantum Reality) known as *Brahman* in accordance with the traditional metaphysical theory in vogue in India since the ages of ṚṚgveda. The following form a pair of specimens:

In Kaṭhopaniṣad,[19] *Brahman* has been allegorically described as an inverted *pippal* (in Sanskrit, *aśvattha*, commonly called pipal or pipul) tree (canto III 1.) (vide S- 15, Appendix), existing eternally, having no beginning with the roots spread high up and branches containing leaves downwards. The World is contained in It wherefrom nothing can slip off. The same analogy between *Brahman* and a pippal tree (*aśvattha*) is available in the verse 1, Chapter 16, of *Gītā* [12].

In both the verses, the word *aśvattha* means that it does not remain beyond the next day (*śvaḥ*) and *mūla* denotes the main cause; *ūrdhva* stands for 'the best'. The self-revealed *Brahman* is expressed by the roots spreading high up *(ūrdhva mūlaṃ)*. The implication is that even after the dissolution of the world, *Brahman*, the root cause (of evolution of the world) will exist. The word *avākśākhaṃ* (having the same meaning as *adhaḥśākhaṃ*) represents all animate and inanimate objects including *Brahmā*. In spite of the fact that pippal tree is subject to death and decay, the ancient sages (*ṛṣis*) have called it immortal because the tree has been taken to have come out of the cause (the roots) representing unrevealed *Brahman*. The whole tree is termed *Brahman* tree (*Brahmavṛkṣa)*.

According to Muṇḍakopaniṣad, verse I.i.7,[20] like a spider forming its web remains within it, similarly *Brahman*, the Creator is inseparably united with the World (vide S-7, Appendix).

6 *Brahman* and *Śūnya* vis-a-vis Ākāśa: This Article Will Throw Light on the Special Aspect of Indian Metaphysics Which Describes *Brahman* as *Śūnya*, *Parama Vyoman*

In Sanskrit, there are the terms *Ākāśa*, *vyoman* or *śūnya*. To each of them, there corresponds a term in English vocabulary; but none of the English words carry with it the inner philosophical flavour attached to the original words. As, for example,

[19]Eight Upaniṣads (Kaṭhopaniṣad), Vol.I, p. 213, *ibid.*

[20]Muṇḍaka Upaniṣad with commentary of Śaṅkarācārya, p. 84, 1988, Advaita Ashrama, Kolkata.

some translated '*Ākāśa*' as the hypothetical medium 'ether' (not the pleasant smelling, highly inflammable volatile organic liquid diethyl ether), according to M. Monier-Williams' *Sanskrit-English Dictionary*, the additional philosophical meaning of the word *Ākāśa* is 'the subtle and ethereal fluid (supposed to fill and pervade the universe and to be the peculiar vehicle of light and of sound)'; similar is the case with the word *śūnya.* It may be mentioned here that in the system of word numerals (arranged as in the place-value notation) developed and perfected in India in the early centuries of the Christian era [13], the words *śūnya*, *kha*, *gagana*, *Ākāśa*, *nabha*, *pūrṇa*, *ananta*, *viyat*, etc. were taken for synonymous.

It has been mentioned before that *Brahman* has been described as the all-pervading space. The Taittirīya Upaniṣad, describes Him as 'the supreme space (*parama vy*oman)' (II. i. 1),[21] and Kenopaniṣad (I.3)[22] explain, 'the eyes cannot behold Him, neither voice nor mind can reach Him. He is the Absolute and the world is inextricably contained in Him'. This will be clear if one considers the reply of Yajñavalka to the question of Gārgī — "That which is the Heaven above, the Earth below and Which is in between them and Which they say that It is inseparable (otaprota- interwoven lengthwise and cross-wise) from the sky"'[23] (Bṛhadāraṇyaka Upaniṣad, III. viii. 7,) (vide S-17, Appendix). Again, Tejabindūpaniṣad (verse 11)[24] describes Him as the Supreme Space and nothing is superior to the Supreme. (vide S-18, Appendix). All these remind us the *Śūnya saṁhitā* of the poet Acyutānanda Dās (sixteenth century A.D.) of Oḍissa (belonging to the popularly known Pañcaśākhās – the five branches formed by five poets Balarām Das, Jagannath Dās, Acyutānanda Dās, Yaśovanta Dās and Ananta Dās)[7] who described both Kṛṣṇa and Jagannātha as the *Śūnyapuruṣa* where the word *Puruṣa* has been used in the sense of Upaniṣad to mean the same *Brahman* as has been described in the Puruṣasūkta of ṚṚgveda as *Puruṣa*. The entire idea of Śūnya saṁhitā seems to have been originated from Śaṅkarācārya's concept of non-dualism in an air of *bhakti* envisaged by Rāmānuja in his Viśiṣṭa Advaitavāda as the way to worship of *Brahman* in the form of Kṛṣṇa or Jagannātha in lieu of Hari (or Śrī Viṣṇu). The same ideology has been reflected in the writings of Bhīma Bhoi, a prominent figure heralding the Mahimā–system of belief [14]. Bhīma Bhoi described Jagannātha as *ālekha Prabhu* (unmanifested Lord), as *Jagatara tāta* (the Father of the Universe), sometimes as *Viṣṇu* and also as *Nārāyana.* In the first chapter of the Sārasvata Gītā, the poet Balarām Das, one of the exponents of the Pañcaśākhā ideology declared Jagannātha as *Śūnya Brahman.* Some may endeavour in vain to find out a link between the Mahāyānic theory of Buddhism expounded by Nāgārjuna (flourished between 50 A.D. and 120 A.D.) and the concept of Śūnyapuruṣa or *Śūnya Brahman* culminated in the Mahimā cult and Jagannātha cult in Oḍissa. The error is at the root

[21]*Eight Upaniṣads* (Taittirīyaupaniṣad), Vol. I, p. 303.

[22]Eight Upaniṣads, (Kenopaniṣad), p. 48, ibid.

[23]*Bṛhadāraṇyaka Upaniṣad with Commentary of Śaṅkarācārya*, p. 359, 2004.

[24]Minor Upanishads (Tejabindupanisad), Text, Eng. Tr., Swami Madhavananda, p. 32 Advaita Ashrama, Kolkata, 2002.

of the misconception in the sense that Buddhism pleads for Non-theism, while both the Mahimā cult and the Jagannātha cult are purely of mono-theistic ideology. History reveals that the influence of Buddhism was radically replaced by Vedānta-based ideology when Śaṅkarācārya (ca. eighth century A.D.)[25] visited Purī and prayed to Jagannāthadeva with the prayer, '*Jagannāthasvāmī nayanapathagāmī bhavatu me*' [Oh Lord Jagannātha be reaveled before me]. According to some, Śaṅkarācārya's *Māyāvāda* had Buddhistic origin, though at the time of evolution, the cosmic egg-like form of the world has been termed as (*asat*) in the line '*sato bandhumasati niravindan*' of the Nāsadīya sūkta, and this seems to form the origin of the *Māyāvāda* in Śaṅkarācārya's philosophy as 'Māyā', the illusion, is also unreal.

7 Shape of the Universe and the Symbol Prescribed for Meditation on *Brahman* vis-a-vis the Symbol of Empty Space (Also for What Is Required to Be Known, i.e. An Unknown Quantity): A Religious Concept with Metaphysical Origin

It has been pointed out that the One or the Absolute of the ṚṚgveda has been termed *Brahma*n in Upaniṣads and *Brahman* has been described as the supreme all-pervading space in Taittirīya Upaniṣad (II.i.1).[26] In connection with the measurement of the circular *gārhapatya* altar, Śatapathabrāhmaṇa (c. 2000 B.C.) [15] mentions (7.1.1.37) (vide S-19, Appendix) that *gārhapatya* altar measures (the area of) only one *vyāma*; one *vyāma* is but one *puruṣaḥ*.| *Puruṣaḥ* is Prajāpati (the lord of the creation); Prajāpati is *Agni*; He thus makes the womb of equal size to His (*Agni's*) body. It is *parimaṇḍala* (circular); for the womb is round; and moreover, the *Gārhapatya* is this world and also (= *u*) this world is *parimaṇḍala* (circular/round).

In the Vaiśvānara Vidyā (*Chāndogya Upaniṣad*, V. xii. 1),[27] the king Aśvapati, the son of Kekaya, teaches the five sages who were yearning for *brahmajñānaṃ* that Vaiśvānara (i.e. *Agni*, the other name of the Absolute, the Self) has the Universe as the body and one should meditate on the whole (Universe) and not on the parts, namely, the heaven (*div*), the Sun (*āditya*), air (*vāyu*), the sky (Ākāśa) and water (*ap*) – the objects they had been, respectively, meditating on erstwhile erroneously.

This teaching has been echoed in the aphorism 55 of Brahmasūtra, Śrībhāṣya,[28] (vide S-20, Appendix) pleading for meditating of the whole world (round shaped as is customary during *gārhapatya* sacrifice mentioned above in the Śatapathabrāhmaṇa).

---

[25] Bhattacharya Sachchidananda, ibid, p. 801.

[26] *Eight Upaniṣads* (Taittirīyaupaniṣad), Vol. I, p. 303.

[27] *Chāndogya Upaniṣad*, with commentary of Śaṅkarācārya, pp. 384–397.

[28] Brahma-sūtras Śrībhāṣya, Part-II, p. 398. ibid.

Thus, *parimaṇḍala* (circular/round) became the symbol of the whole universe prescribed for meditation, and this round symbol began to be taken for denoting empty space (also that which is to be known). It is important to note that *Chāndogya Upaniṣad* is a work of pre-Paṇini age[29] and much before the lord Buddha was born, i.e. much before the fifth century B.C.

8 Theory of *Śūnya* in Ancient India, Origin of the Word, *Abhāvaparīkṣā* Due to Kaṇāda, Nāgārjuna's Concept of *Śūnyatā* and Concept of *Ākāśa* Due to Them

In the context of what has been stated above, we therefore are much inclined at looking into the theory of *Śūnya* developed in ancient India. ṚṚgveda is the first in the world to put forward the concept of voidness (*vyoma*) or emptiness (*tucchaya*) (vide the Nāsadīyasūkta). In this connection, one has to note the verses 1.105.3[30] and 3.33.13.[48] of ṚṚgveda using the words *śūne* and *śūna*, each indicating 'devoidness'.

Now, the word *śūne* ⇐ *śūna* (√*śvi* + *kta*) derived from the root √*śvi*, though originally means 'swollen or swelled', also implies emptiness, lack, want and absence.[31] The word *śūnya (śūna + ṣṇya)* primarily means 'non-existence or lack of any material' as has been termed '*tucchaya*' in the Nāsadīyasūkta. This primary concept of non-existence or lack or *abhāva* of materialistic substance has later been labelled as a case of *vastubhūta abhāva* and generalized from materialistic standpoint in the chapter 'abhāvaparīkṣā' in the Vaiśikadarśanaṃ[32] written by Kaṇāda. The word *śūnya* in the same sense has been used in the Rāmāyaṇa (*śūnyaṃrathaṃ* verse 2.52.41), in the Mahābhārata[33] Śatapatha *Brāhmaṇa* (verse 2.3.1.9), in Manusiddhāntaḥ (verse 57, adhyāyaḥ 4).

8.1 Kaṇāda's Concept of Abhāva

Kaṇāda's generalization of the sense of lack or *abhāva* of a materialistic substance from materialistic standpoint in the chapter *abhāvaparīkṣā*[34] (vide S-21, S-22, Appendix) in his Vaiśeṣika Darśanaṃ. The chapter consists of 15 aphorisms. The sum and substance of the aphorisms has been mentioned by the commentator thus:

---

[29]Prof. Amalkumar Bandopadhyay, Paurāṇikā, Part-I, p. 227, Kolkata, 1985.

[30]Ṛgveda Saṁhitā, p.75, ibid and p. 209.

[31]M. Monier-Williams, *Sanskrit-English Dictionary*, p. 218.

[32]Sharma Candrakanta, Commentator, Vaiśeṣikadarśanaṃof Kaṇāda, Introduction, p.(i) Calcutta, 1809 Śaka year A.D.1887–1888.

[33]Mahābhārata, Ādiparva, ch. 2, verse 13.

[34]Sharma Candrakanta, Commentator, Vaiśeṣikadarśanaṃ ibid, pp. 50–54.

(The act of) carefully examining an existence or an occurrence is *pratyakṣaṃ* (perception by senses). It may be pointed out that in Nyāya, *pratyakṣaṃ* is considered as one of the four *pramāṇas*, namely, *pratyakṣa* (perception by senses), *anumāna* (inference), *upamāna* (analogy/comparison), *śabda* or *āptavacana* (verbal authority or revelation).

- Lack or deficiency of something (*abhāvaḥ*) is of two types:

 (i) Non-materialistic (*avastubhūta*) and (*ii*) materialistic (*vastubhūta*)

When a thing does not occur anywhere even in traces (just as a lotus or something like that in the sky), the lack of the thing comes under the class (*i*), and when a thing rarely occurs somewhere but not everywhere even in traces, then lack of the thing belongs to the class (*ii*). In the course of the clarification of the ninth aphorism, *abhūtaṁ nāstītyanarthāntaraṃ*, the commentator Sharma further explains that:

- 'A thing does not occur' and 'a thing is not present' bear the same meaning.
- A thing produced or existed before and later due to abandonment or for some other cause ceases to occur and is also its absence.

That which is inexistent bears the same meaning as it is not present. The notion that 'a (particular) pitcher is not in the room' does not arise if the room was formerly devoid of the pitcher. (It is a case of *vastubhūta abhāva*.)

8.2 Concept of Śūnyatā *Propounded by* Nāgārjuna

According to K. Venkata Ramanan, the primary meaning of *śūnyatā* as expounded by Nāgārjuna in his interpretation of the Prajñāpāramitā sūtras is devoidness, which is a direct reference to the truth of things, mundane and ultimate; but it refers also to the method by which śūnyatā as truth is brought to light, namely, by rejecting the imagination of ultimacy and absoluteness in regard to what is only relative and non-ultimate. *Śūnyatā* as the mundane truth is relativity and conditioned becoming; this is brought to light by rejecting the supposed ultimacy and absoluteness of particular entities and specific concepts and conceptual systems.[35] Referring to Kumarajīva, Venkata Ramanan points out that when the diverse character of mundane things is analysed and investigated to their very end, all things are seen to enter the *anutpāda-dharma*, i.e. *dharma* devoid of birth. *Anutpāda-dharma* is also called the *dharmadhātu* or *bhūta-koṭi*. The term *bhūta* means the unconditioned reality, and the *para koṭi* denotes the mysterious skill to reach the supreme end, 'the limit' and 'the apex', of 'the realization' that particular entities, concepts and conceptual systems in the World are devoid of birth and death. In *anutpāda-dharma*, the part

[35]Ramanan K. Venkata, ibid, pp. 172–173.

anutpāda refers to *nirvāṇa*.[36] *Nirvāṇa* compasses end and beginning. It is the 'way' of attaining it by not attaining'. It is not attained by attaining.[37] Radhakrishnan [16] clarified 'Nāgārjuna's *nirvāṇa* as something not to be attained. It is the ultimate goal towards which all beings move seeking fulfilment'.

8.3 Nāgārjuna's Concept of **Ākāśa** *(Space)*

It has been mentioned earlier that in the use of word numerals, which was in vogue in India from the early centuries of the Christian era, *kha*, *Ākāśa*, *nabha*, etc. were taken for synonyms of *śūnya* in the sense of voidness. According to Nāgārjuna, *Ākāśa* is a name, and not any substance. In the mundane truth, it is admitted as the necessary condition for movement, the container of all. It is capable of containing everything precisely because it is *akiñcana*, not itself anything like a specific entity; everything dwells in it. It is known as the principle of accommodation. In the ultimate truth, it is of the same nature as *nirvāṇa* – the prototype of the ultimate reality.[38]

8.4 Kaṇāda's Concept of **Ākāśa** *(Space, Ordinarily, Sky)*

In this connection we like to focus here under the following points on the relevant portion of the philosophical doctrine of *Ākāśa* put forward by Kaṇāda in the chapter Ākāśa parīkṣā[39] (vide S-23 to S-25) of his Vaiśeṣika darśanaṃ.

(a) Entrance and exit (or in other words, egress and ingress) are the two characteristics (mark/feature) of Ākāśa. It provides room (or space) and is (meant) for entrance or exit. Where there is no compartment, these (operations of entrance or exit) do not occur. Therefrom it appears that the space which this Ākāśa provides is not a place to reside.
(b) The so-called egress and ingress denote associated operations with respect to the same object. This is because coming out (of a place/compartment) is followed by an (earlier) entrance (there); but these two are not so in the case of Ākāśa. Thus, Ākāśa has no characteristic (like a materialistic abode).
(c) The two operations occur in absence of any material contact, and lack of conjunction (contact) implies absence of any action.

In connection with this and subsequent aphorisms, Professors Sarmā summarizes:

---

[36] ibid, pp. 262–263.

[37] ibid, p. 324.

[38] Ramanan K. Venkata, ibid, p. 206–207.

[39] Sharma Candrakanta, Commentator, Vaiśeṣikadarśanaṃ, ibid, pp. 50–54.

(i) Indeed this *Ākāśa* (the space) is omnipresent, eternal and is present everywhere but is not as the cause of any action. (ii) Due to the lack of logical discontinuance, it has no beginning. (iii) It is a cause for making room for space, having no material confinement or in other words, it is a space to accommodate everything, keeping the thing unchanged. (iv) Indeed this unique Ākāśa is discriminated due to its variance in its appearances. (v) It is singular.

9 The Word '*Śūnya*' Taken so Long as Void or Space Becomes a Numeral in the Chandasūtra of Piṅgala[40]

To the philosophical concept attached so long to the idea of *śūnya*, its status as a numeral was ascertained for the first time in the *History of Mathematics* by Piṅgala[D] (prior to 200 B.C.) [17], a nāga, born much before Nāgārjuna and is well-known for his Chandasūtra. It may be mentioned here that one Piṅgala is also known to be the author of the commentary of Mādhyamaika kārikā,[41] which was later translated into Chinese by Kumarjīva. Here we are concerned with the two verses (1*) dvirardhe* and (2) *rūpe śūnyaṁ* (vide S-26, Appendix) given by Piṅgala in his Chandasūtra. The first part of the verse 1 embodies the question 'In a metre of six syllables, how many recurrences of lines (*padas/caranas*) (turn of a line – the final rhythm) are possible?'. The second part of this verse and the second verse (i.e. *rūpe śūnyaṁ*) throw light on how one should proceed to get the necessary data for computation of the answer required. The method is this:

- Put the number of letters of the metre on the ground and its half is removed, i.e. the number is divided by 2. If such a division is possible (it is possible when the number of letters in the metre is even), write the number 2 at a separate place on the ground.
- As the quotient 3 (due to division of 6 by 2) cannot be halved, we enter into the next phase prescribed in the verse rūpe śūnyaṁ. Deduct 1 from the odd number (here it is 3), and place 0 (zero) below the number 2 written formerly. As the number 2 obtained this time can also be halved, write 2 bellow 0 written in the former step. Since half of 2, i.e. 1, cannot be further halved (for getting integral quotient), again 0 (zero) is put below 2 written in the former step.

Example (In this example we use the symbol '0' for *śūnya* though it is not known what type of symbol was in use at the times of Piṅgala.)

---

[40]Samādhyāyī Sītānāth (Commentator), Piṅgala ChandaḥSūtraṃ, p. 185, Calcutta, 1840 Śaka year (1918–1919 A.D.).

[41]Ramanan K. Venkata, *ibid*, p. 15.

Taking Gāyatrī metre containing six syllables, Datta and Singh[42] have explained how the following worksheet can be prepared:

Place on the ground the number 6, its half is 3 and the numeral 2 is placed separately on the ground. As half of 6, i.e. 3, cannot be halved (to get a whole number), subtract 1 from 3 to get 2, and put śūnya separately on the ground below the previously put 2. Again half it (half the numeral 2, and place 2 separately below the *śūnya* put aside before). As the half of 2, i.e. 1, cannot be halved, subtract 1 from it to get 0, and then that *śūnya* is to be placed below the numeral 2 put aside. Thus one gets the column 1 (from the top downwards)

Column 1	Column 2
2	$(2^3)^2 = 2^6$
0	$2^2 \times 2 = 2^3$
2	2^2
0	$1 \times 2 = 2$

Calculation starts from the last item of the first column and rises upwards step by step. Corresponding to the last member 0 of the first column, take unity and multiply it by 2 to get 2. Corresponding to the number 2 just higher above, square of the number 2 is to be recorded. Corresponding to the number zero in the next higher position, multiply 2^2 by 2 to get 2^3. Lastly, with respect to 2 in the uppermost position, square 2^3 to get 2^6. This gives the number of ways of arranging two things in 6 different places.

Remark That some sort of symbolic notation of *śūnya* as well as of numerals was in practice in those days is evident in the said verses. But what type of symbol for zero was in use in those days is not known. According to Datta and Singh,[43] the earliest symbol for *śūnya* (zero) was a dot, and the small circular symbol was introduced in Indian Mathematics before the eighth century A.D. In Puliśasiddhāntaḥ of pre-Śaka period [18], *śūnya* was taken for a numeral.

10 *Nāmalṅgānuśāsana*, A Sanskrit Lexicon, Puts Forward Manifold Meanings of Śūnya

Amara Siṁha,[44] the writer of the Sanskrit lexicon *Amarakos'a* (Nāmalṅgānuśāsanaṃ) during the reign of Candragupta II (375–413 A.D.), recorded in his lexicon the three meanings of *śūnya*[45] as (vide S-27, Appendix):

[42]Datta Bibhutibhusan & Singh Avadhesh Narayan, ibid, p. 76.

[43]Datta Bibhutibhusan & Singh Avadhesh Narayan, p. 82.

[44]Bhattacharya Sachchidananda, ibid, p. 39.

[45]Bhattacharya Śrīmadgurunath Vidyānidhi, editor, Amarakoṣa (Amarārthacandrikā), Kolkata, 1988.

(a) *vaśikaṁ* (void), (b) *riktaḥ* (devoid or destitute of) and (c) *tuccha* (trifling). In fact each of these three meanings were ordinarily taken for synonymous with the term *śūnya*, though strictly speaking they are not so. In the works of Āryabhaṭa-I and Brahmagupta, the word *śūnya* is found to have been used only in the senses of *vaśikaṁ* (void) and *riktaḥ* (devoid or destitute of). It was Bhāskarācārya (b.1114 A. D.) who was the first to use of the word *śūnya* in the sense of *tuccha* (trifling) in his works.

11 Introduction of the Symbol of Zero

In the work Vāsavadattā of an ancient Indian poet Subandhu (flourished at the end of the sixth century A.D.), *śūnya* was denoted by a dot (*bindu*) certifying that it has a position among numerals but without any characteristic property of being positive or negative (as if having no dimension); but his work cannot be taken for a definite proof of the introduction of a dot as the symbol for zero during his times. 'All that we can infer', say Datta and Singh, 'is that at some time before Subandhu, the dot was in use, we may go further and state that the earliest symbol for zero was a dot and not a small circle'.[46] The small circular symbol was adopted much before the fifth century B.C. (vide p. 11) to denote void or the quantity which is required to be known (but not the numeral zero), and the idea (that the small circular symbol is an object for meditation of the unknown having the significance described in the Tejabindūpaniṣad) (Brahman is the Supreme Space, and nothing is superior to the Supreme –verse 11) was confined among the ascetics aspiring to get at the level of a *Brahmajña* (the one who is said to know Brahman). In the third century A.D., it was a general practice to take this symbol for denoting an unknown quantity or absence of a quantity in arithmetical problems only, especially, involving rule of five, rule of seven, etc., but not in the sense of zero. The earliest epigraphical record of the use of zero in the form of a small circle is found in Ragholi plates of Jaivardhana II of the eighth century A.D.[47] This small circular symbol (vide the discussion below), therefore, appeared in the use of commoners for zero in the eighth century A.D.

Discussion

David Pingree opined that 'at the beginning of the common era Greeks appeared in India in large numbers, bringing with them, among other things, astronomical tables in which, in the sexagesimal fractions, the "empty" places were occupied by a circular symbol such as are found in Greek papyri of that period' [19]. But Boyer [20] points out that 'It was once assumed that the round form (of zero) stemmed

[46]Datta Bibhutibhusan & Singh Avadhesh Narayan, Part-I, *ibid*, p.81.

[47]Datta Bibhutibhusan & Singh Avadhesh Narayan, p. 82.

originally from the Greek letter omicron, initial letter of the word "ouden" or empty, but recent investigations *seem to belie such an origin.* Although the symbol for an empty position in some of the extant versions of Ptolemy's tables of chords seem to resemble an omicron, the early zero-symbols in Greek sexagesimal fractions are round forms variously embellished and differing markedly from a simple goose egg'. Boyer's statement will justify how far Pingree's opinion is tenable in the light of historical investigations. The history of interactions between the Greeks and the ancient India shows that there was a cordial relation between the Mauryan court and the courts of the Greek kings of Egypt. This is furnished in 'the thirteenth Rock Edict of Aśoka in which the five Greek rulers of Egypt are specially mentioned' [21]. All such relations came to an end in 30 B.C. with the end of the Greek regime in Egypt following the suicide of the last Greek Queen Cleopatra there.[48] If by this time ancient Indians were aware of the symbolic notation (ô or ō) for empty space, used in sexagesimal fractions in ancient Greece, then they would not find the necessity of inventing the alphabetic system of notation (n, ṇ and vowels) for writing the numeral zero (as found in the fifth century A.D. in the *Kaṭapayādi* system) and wait for several centuries to recognize the small circular symbol to denote zero. In connection with the sexagesimal fractions as mentioned by Pingree, it may be mentioned that sexagesimal division of angles and of chords was in practice in ancient India as is available in the first Saurasiddhānta written two to three centuries earlier than the beginning of the Śaka era.[49] (The Western scholars are often confused about the period of writing this Saurasiddhanta, and they erroneously take it for identical with the modern *Sūrya Siddhānta* compiled in a much later period.) Actually Pingree seems to have been unaware of the history of interactions between the ancient Greeks and ancient Indians and appears to have been groping amidst darkness in the concerned history to establish superiority of the Western world over India, disregarding the historical credit given to ancient India for her invention of the circular symbol of zero. The round symbol to denote space, i.e. *kha* (also what is required to be known), was adopted from metaphysical standpoint much before the fifth century B.C. It has been pointed out before that in all works on ancient Indian arithmetic (*Pāṭigaṇitaṃ*), mostly in problems on proportion, the rule of five, the rule of seven, etc., this round symbol was found to be in use for denoting unknown quantities. This use of the symbol in this sense can be traced back to the third century A.D. It was used in the same sense in Bakshālī arithmetic. The Arabs also under Indian influence used this symbol for unknowns in similar problems. Similar use of the same symbol for an unknown quantity is found in Europe in the Latin manuscript of some lectures of *Gottfried Wolack* of *Erfürt* in 1467–1468 A.D. (Smith and Karpinski, pp. 53–54.).[50] This round symbol denoting void/empty/unknown quantity gained the status of the numeral zero much later.

---

[48] *Encyclopaedia Britanica.*

[49] Shankar Balkrishna Dixit, *Bharatiya Jyotish Sastra* (English translation) p.396, ibid.

[50] Datta Bibhutibhusan & Singh Avadhesh Narayan, Part-I, pp.82–83.

12 Transcendence of Philosophical Views into Mathematical Ideas in the Thoughts of the Three Prominent Indian Mathematicians of Yore

(i) Āryabhaṭa I (A.D. 476)
(ii) Brahmagupta (A.D. 650)
(iii) Bhāskarācārya (b.1114 A.D.) and analysis of their approach to show to what extent they accommodated the concerned philosophy in the field of their study:

12.1 Āryabhṭa- I's Use of Small Circular Symbol to Denote Emptiness

Āryabhaṭa I used small circular symbols (termed *kha*) to denote the absence of an integer at a place (i.e. blankness) in conformity with the metaphysical meaning of space attached to the symbol in one or more of the 18 places in the place-value notation of a number, though the use of a dot for zero was most probably in vogue at that time, not the small circular symbol and that was why he had to introduce alphabetic notations of numerals [22].

12.2 Brahmagupta's Mathematical Concept of Zero

Brahmagupta was the first to put forward a mathematical definition of zero (represented by a dot and not by the circular symbol), taking zero for a numeral throughout the verses 31–36, but regarding the identity of zero, he expressed his view allegorically as *śūnyamĀkāśaṃ*; this word *Ākāśaṃ* is not merely the word representation of zero; it has on the contrary the deeper philosophical significance following Vaiśeṣika philosophy. This will be clear in the following statements (vide S-28, Appendix) consisting of the verse 31, first line of the verse 33, second line of the verse 34 and from the first line of the verse 36 in the eighteenth chapter of Brāhmasphuṭasiddhāntaḥ [23]:

The sum of two positive (numbers) is positive, the sum of two negatives is negative, the sum of a positive and a negative is their difference and the sum of a positive and a negative quantity of equal magnitude is zero.[12]

Zero, when added to a negative (number) gives a negative, to a positive gives a positive and to zero gives zero. (first line of the verse[33]).

A negative (number) less zero is the negative (number), a positive (number) less zero is the positive (number), zero is (like) the space.

Zero multiplied by a negative (number) is zero, multiplied by a a positive (number) is zero and multiplied by zero is also zero.

A negative (number) or a positive (number) is upon (i.e. divided by) zero is *tacchedaṁ* and zero divided by a negative or by a positive (number) is zero, (though this *'tacchedaṁ* was not identified).

Analysis of Brahmagupta's Concept of Operations with Zero in the Light of the Philosophical Theories He Utilized

(a) Represented algebraically, the above rules to state that, for

$$x > 0, y > 0, x + y > 0, (-x) + (-y) < 0, \;\; x - x = 0 \tag{5.1}$$

Clearly the last one (i.e., x − x = 0) of (5.1) furnishes the mathematical definition of zero and is based on the philosophy of *vastubhūta abhāva* (a thing produced or existed before and subsequently due to abandonment or for some other cause ceases to occur and is also its absence).

(b) $x + 0 = x$ and $-x + 0 = \;-x$ are but the pictures of Kaṇāda's aphorism: (omnipresence) of *Ākāśa,*

(c) $0 + 0 = 0$ and $0.0 = 0$ originate from the idea 'space is singular'.

12.3 Bhaskaracarīya's Logistics of Zero- an Analytical View

Bhāskarācārya, well-aware of all lexicological meanings of the term *śūnya* given in Amarakoṣa (*Nāmalṅgānuśāsanaṃ),* assimilated all the philosophical ideas available in the writings of scholars prior to him and enriched his mathematical theories with the help of all the philosophical theories available to him. This will now be highlighted.

The Logistics of Zero and the Analysis of His Concept

In Līlāvatī [24] verses 45 and 46 (vide S-27, Appendix) as well as in Bījagaṇitaṃ, the logistic 3 (S-29, Appendix) Bhāskara II's concept of zero demands special credit on several grounds.

Līlāvatī says:

When a number is added to cipher, the sum equals the additive, the square and other powers of cipher is cipher; a number divided by cipher is (termed*) khaharaḥ*, it should be considered in accordance with the rule of finding remainder (*śeṣavidhau*) and should be considered by multiplying it by cipher. (Verse 45)

Cipher multiplied by a number gives zero and the product when divided by cipher gives back the original number; the same (unchanged state) occurs when cipher is subtracted from or added to a number. (Verse 46)

In Bījagaṇitaṃ the corresponding rules (vide S- 30, Appendix) are:

When cipher is added to or subtracted from *(cyutaḥ)* a positive or a negative number, the number remains unchanged. Their sign is reversed *(viparyāsaṃ)* when subtracted from cipher. Cipher multiplied or divided by a number is cipher, any power or any root of cipher is also cipher; a number divided by cipher is *khahāraḥ.*

12.4 Discussion on the Concept of Zero in the Mathematical Thoughts of Bhāskarācārya and Brahmagupta: A Comparative Study in the Light of Indian Philosophical Ideas Utilized by Them

Here, save the portions *khagaṇaścintyaśca śeṣavidhau and hāraścet puṇastadā rāśiḥavikṛ ta eva jñeya* the statements in Līlāvatī and in Bījagaṇitaṃ are the same.

It has been pointed out before that the aphorisms concerning vastubhūta abhāva and *Ākāśa* of the Vaiśeṣika philosophy motivated Brahmagupta to build up his mathematical concept of zero. Bhāskarācārya accepted Brahmagupta's ideas except the case of division of a number by zero. To be more precise in this matter, he incorporated the meaning of *śūnya* (zero) given in Amarakoṣa. Clearly, Brahmagupta accepted the meaning (a) *vaśikaṃ*, i.e. void (this explains his definition, *śūnyamākśaṃ*, i.e. zero is void or empty like sky), and (b) the meaning *riktaḥ* – devoid or destitute of' (which in conformity with above-mentioned *vastubhūta abhāva* has been clearly reflected in the line *dhanarṇayorantaraṁ samaikyakhaṁ*). But Brahmagupta could not explain the operation of division of a number by zero. He left the matter stating *khoddhṛtamṛ-ṇadhanaṃ vā* '*tacchedaṃ*', i.e. a positive or a negative number divided by zero is *tacchedaṃ* without throwing any light from mathematical interpretation on the word *tacchedam.* To remove this ambiguity, Bhāskarācārya undoubtedly resorted to the two additional points of views, namely, (i) the meaning *tuccha*, i.e. trifling or negligibly small as given in Amarakoś*a*, and (ii) the principle of *Śūnyavādaḥ* of Nāgārjuna in the sense that in Nāgārjuna's philosophy, the word *śūnyatā*, though primarily means the sense of devoidness of mundane attachments, refers to the method also by which *śūnyatā* is attained and *śūnyatā* does not mean non-existence or nothingness or complete extinction; it is like *nirvāṇa*, the ultimate end to which each tends to move.

12.5 Influence of Indian Philosophy and of Amarakosa on Bhāskara II

Analysis of the Bhāskara II's concept of addition and subtraction of zero from a number, squaring, etc. of zero, introduction of his idea of infinitesimal adeptly adapting the lexical meaning of the word *śūnya/kha* and his extension of the field of mathematics to introduce a new concept known today as 'differential'— all on the background of philosophical ideas developed in India and application of concept of differential for finding *tatkalikīgati* (instantaneous motion of a planet)

12.5.1 Kaṇāda's Vaiśeṣika Philosophy – its Impact on Bhāskara II's Explanation of the Pre-existing Logistics of Zero

Bhāskara II's concept of addition and subtraction from a number, squaring, etc. of zero in the light of Kaṇāda's Vaiśeṣika philosophy:

> The logistic of Bhāskara II in the case of (i) addition of zero to and its subtraction from a number follows the omnipresence sense of Ākāśa (space), (ii) squaring, etc. of zero (vargādau khaṃ) resembles the concept of Ākāśa in the sense of its singularity like sky (ekatvecakaśasya). It has been pointed out before that in these cases both Brahmagupta and Bhāskara II followed the same track.

12.5.2 Amarakosa Compasses Bhāskara II's Concept of Infinitesimal

Introduction of his idea of infinitesimal, in accordance with the lexical meaning of the word Śūnya:

> In the interpretation of the case of division by zero, the words kha (the sky) and śūnya (taken for synonymous) have been used as a quantity (or a number) existent in reality but negligibly small befitting the meaning tuccha (trifling or negligibly small) given in Amarakoṣa with the further refinement that this negligibly small quantity is of order zero (that means 'tends to zero'). This is clear in the statement khaguṇaḥ khaṃ. The following line, however, demands special attention:
>
> khabhājitorāśiḥ khaharosyāt khaguṇaścintyaśca śeṣavidhau.

That is, a number divided by zero shall be 'khaharo', and by the rule of finding the remainder in the course of the process of division, when a number is divided by kha, both the quotient and remainder should be taken into account after multiplying by kha (1)

Here the term 'kha', as has been pointed out by Datta,[51] is an indefinitely small quantity (of course, positive) comparable to an infinitesimal of the modern practice and is in conformity with the concept of 'tuccha' of Amarakoṣa.

---

[51]Datta Bibhutibhusan & Singh Avadhesh Narayan, ibid, p. 242.

It may be noted that Bhāskarācārya in his Siddhāntaśiromaṇi designed the smallest unit of time as $(1/3000)^{\text{th}}$ part of the time required for a single winking and called it one truṭi. This has been defined in the verse 16 (S-31, Appendix) in Kālamānādhyāyaḥ of Madhyamādhikāraḥ thus:

One-thirtieth part of nimeṣaḥ of an eye (time required for a single winking of an eye) is one tatpara; one-hundredth part of *tatpara* is one *truṭi*] and one truṭi is $(1/3000)^{\text{th}}$ part of the time for a single winking.[52]

12.5.3 Introduction to a New Concept Known Today as 'Differential' by Bhāskarācārya – on the Background of Indian Philosophical Ideas

Introduction to a new concept known today as 'differential' by Bhāskarācārya — on the background of philosophical ideas developed in India:

In fact, the time was reckoned even less than one truṭiḥ, the smallest unit of time for calculating tātkālikī gati (instantaneous motion) of a planet. This will be seen in the verses 36 and 37 in the chapter 'Spaṣṭādhikāraḥ' in the Gaṇitādhyāyaḥ of his Siddhānta Śiromaṇi.[53] To explain what has been pointed out at (1) in the former section, we proceed as follows:

If δs, denote the displacement corresponding to a very small unit of time δt (designated as *kha* in former section when δt tends to nullity but does not exactly equal zero according to the idea; 'it is not attained by attaining, and 'the way of attaining it by not attaining') and if V be the instantaneous velocity, then, by the rule (*Śeṣa* vidhau) of determining remainder (*Śeṣa* – remainder out of a process of division, vidhau – by the rule), multiplying $\frac{\delta s}{\delta t}$ when $\delta t \to 0$ by kha (i.e. $\lim_{\delta t \to 0} \delta t$ by kha), and $\frac{\delta s}{\delta t}$ denotes khaharaḥ. In modern notation,

$$ds = \lim_{\delta t \to 0} \left(\frac{\delta s}{\delta t}\right) . \lim_{\delta t \to 0} \delta t$$

$$= \lim_{\delta t \to 0} \left(\frac{\delta s}{\delta t} . \delta t\right) = V \lim_{\delta t \to 0} \delta t + \varepsilon \lim_{\delta t \to 0} \delta t$$ (where ε is also of order zero, depending on δt).

It may be mentioned here that the idea $\lim_{\delta t \to 0} \delta t$ comes from Nāgārjuna's philosophy. If the symbol '0' (zero) be taken for the state of complete *Śūnyatā* (i.e. *nirvaṇa*) towards which all are moving 'it is not attained by attaining', 'the way of attaining it by not attaining', then nirvaṇa involves 'tending to nullity' and complete Śūnyatā is the limiting case.

[52]Paṇḍit Girijāprasād Dvivedī, editor and Hindi Translator, Siddhāntaśiromaṇi Gaṇitādhyāyaḥ, p., ibid.

[53]ibid, p. 178.

12.5.4 Application of the Concept of Differential for Finding Tatkalikīgati (Instantaneous Motion) of a Planet

The above concept of differential based on Nāgārjuna's philosophy of tending to nullity has been utilized by Bhāskarācārya in finding tatkalikīgati (instantaneous motion of a planet) as opposed to tatkalikagati (proportional calculation of the mean motion of a planet at a desired epoch) of Madhyamādhikārah in Sūrya Siddhānta [25]:

Bhāskara II's concept of tatkalikīgati evolved from a more advanced idea of instantaneous motion (*sūkṣma gati*) measured precisely with reference to an indefinitely small interval of time. This will be evident from the verses 36–37 of Spaṣṭādhikārah Gaṇitādhyāyah Siddhāntaśiromaṇi (vide S-32, Appendix) which state:

The difference of the true motions of planet (at the epoch of its rise or at the midday) is its true motion in that interval (verse 36).

The motion of the mean anomaly during the interval (*mṛdukendrabhukti*) multiplied by the *koṭijyā* of the *mandakendra* (i.e. by the R.cosine of the planet's initial anomaly, when R is the radius of the *kakṣāvṛtta*) and the product is divided by the radius (of the *kakṣāvṛtta*). The result is to be added to or subtracted from the mean (daily) motion of the planet according as the kendragati lies in the region starting from first point of Cancer and extending up to the first point of Capricorn or in the region starting from first point of Capricorn and extending up to the first point of Cancer (verse 37).

Bhāskarācārya's Elucidation of the Verses in his Self-Commentary (Vāsanābhāṣya) (s-33)

The difference between the longitudes of a planet at the epoch of its rise or at the noon on a certain day and that on the next day at the same epoch is the true speed of the planet in minutes. If today's longitude be less than that of yesterday, then the motion is to be regarded as retrograde within the interval. It is proper to consider that a planet moves in this way within the interval. This is indeed a rough estimate of the motion of a planet. Now is stated the precise instantaneous (motion of a planet). The longitude of the Moon less her longitude at the apsis is her kendragati. In the cases of all other planets, the difference of longitudes of the kendras (mandakendra – mean anomaly) on 2 successive days alone denotes their daily speed. Finding the R.cosine ratio of the planet's initial anomaly (when R is the radius of the kakṣāvṛtta of the planet in question), multiply that by the daily motion of the mandakendra, and divide the product by the radius (of the kakṣāvṛtta); the result is to be added to or subtracted from the planet's daily motion accordingly as the kendra lies in the region starting from first point of Cancer and extending up to the first point of Capricorn or in the region starting from first point of Capricorn and extending up to the first point of Cancer. Thus the instantaneous motion of the mandaspaṣṭagrahaḥ will be obtained.

12.5.5 Explanation of the Above Commentary in the Light of the Mathematical Concept Developed by Bhāskarācārya

Referring to Bapudeva Śāstrī's article 'Bhāskara's Knowledge of the Differential Calculus', what Bag [26] has interpreted may be explained in the light of the mathematical concept developed by Bhāskarācārya on the Indian philosophical foundation in the following way:

Let x, y, u be, respectively, the mean longitude, the longitude of the mean anomaly (mandakendra) and the true longitude of a planet at a certain epoch (sunrise or midday) on a certain day, and let $x^{'}$, $y^{'}$, $u^{'}$ be, respectively, those on the previous day at the same epoch. Then, $x - x^{'}$, $y - y^{'}$, $u - u^{'}$ will denote, respectively, the mean motion (*madhyagati*) of the mean anomaly (*mandakendragati*) and the true motion of the planet during the period of 1 day. If a be the *antyaphalajyā* then according to Bhāskarācārya, the bhujaphala on the day on which y is the longitude of the mandakendra, the *bhujaphala* is, $\frac{a \sin y}{R}$ where R is the radius of the concentric.

Similarly, the *bhujaphala* on the previous day $= \frac{a \sin y'}{R}$, so that

$$u = x \pm \frac{a.\sin y}{R} \; \& \; u' = x' \pm \frac{a.\sin y'}{R},$$

Therefore, the true motion of the planet in the period of 1 day will be

$$u - u' = x - x' \pm \frac{a.(\sin y - \sin y')}{R} \tag{5.2}$$

Similarly, if x, y, u be, respectively, the mean longitude, the mean anomaly and the true longitude of a planet at a certain instant, say, t, and if $x + \delta x$, $y + \delta y$, $u + \delta u$ be those (in the above order) at the instant $t + \delta t$, when δt denotes a small increment in t, < one *truṭi* [= $(1/3000)^{\text{th}}$ part of a winking], then by the analogy from (5.2),

$$\delta u = \delta x + \frac{a}{R} 2.\cos\left(y + \frac{\delta y}{2}\right)\sin\frac{\delta y}{2} \tag{5.3}$$

or $\delta u = \delta x + \frac{a}{R}\cos y\delta y$; as $\frac{\delta y}{2}$ is very small, $\cos\left(y + \frac{\delta y}{2}\right)$ and $\sin\frac{\delta y}{2}$ are, respectively, taken as $\cos y$ and $\frac{\delta y}{2}$.

Now, dividing δu $\left(= \delta x + \frac{a}{R}\cos y\delta y\right)$ by δt (where δt is small enough of order zero and is taken for *kha*, or in other words, $\delta t \to 0$), if η be the remainder and u_1 be the quotient then

$$\frac{\delta u}{\delta t}\left(= \frac{\delta x}{\delta t} + \frac{a}{R}\cos y\frac{\delta y}{\delta t}\right) = u^1 + \eta, \tag{5.4}$$

where $\eta(> 0)$ evidently depends on δt, and $\frac{\delta u}{\delta t}\left(= \frac{\delta x}{\delta t} + \frac{a}{R}\cos y\frac{\delta y}{\delta t}\right) = u^1 + \eta$ is *khaharaḥ*

Then, by the rule *khaguṇaścintyaśca śeṣavidhau*,

$$\delta u\left(= \delta x + \frac{a}{R}\cos y\delta y\right) = u^1\delta t + \eta\delta t, \text{where}\, \eta \to 0\, \text{as}\, \delta t \to 0, \tag{5.5}$$

It has been pointed out before that the idea $\lim_{\delta t \to 0} \delta t$ comes from Nāgārjuna's philosophy taking the symbol 0 (zero) as the state of complete śūnytā, i.e. nirvaṇa (towards which all are moving — it is not attained by attaining, 'the way of attaining it by not attaining'), then 'nirvaṇa' means 'tending to nullity', and complete śūnytā is the limiting case. Then when$\delta t \to 0$, $\eta.\ \delta t \to 0$ and according to modern notation, we have from (5.4) $du = dx \pm \frac{a}{R}\cos y dy \Leftarrow d\left[x \pm \frac{a}{R}\sin y\right]$, where the negative sign occurs in the case of retrograde motion.

A Necessary Information

According to Bhāskarācārya, for a circle of diameter D = 1250 unit and the circumference C = 3927 unit, C/D = 3927/1250. Therefore, when the circumference $= 21{,}600'$ the diameter = 21600/1250 = 6876; whence, the radius = 3438, and one can deduce that

$2\pi^c\left(= \frac{2\times 3927}{1250}\right)^c = 360^0 = 21600'$; so, $1^c = \frac{21600\times 1250}{2\times 3927} \approx 3438$ minute.

i.e. an arc of length measuring $3438'$ of a circle of radius $3438'$ subtends an angle measuring one radian (which is approximately $3438'$) at the centre. So an angle measuring $225'$ being much less than one radian, the modern trigonometric formula $s = r.\ \theta$ (where θ radian is the angle subtended by an arc of length s at the centre of a circle of radius r) holds. It may be mentioned here that the practice of taking the smallest arc length of a circle as $225'$ was a customary matter in ancient India since the times of the Saurasiddhanta (two or three centuries before the beginning of the Śaka era).

13 Conclusion

Above we have had what scholars of yore in India used to think about the metaphysical implication of Quantum Reality, and that *śūnya* does not mean simply void; it is pūrṇa also; it is the ultimate unconditioned state wherefrom all of us have been originated and to which we are all destined to evanesce. Our discussion will be incomplete if we forget the views of the world-famous national poet Rabindranath Thākur on the metaphysical idea of cosmology and on the Viśiṣṭa Advaitavāda (or qualified non-dualism).

Views of the poet[54]:

We remember that Rabindranath Thākur has described the ancient Indian idea of cosmology in the Naivedya, verse no. 80, p. 304 (S-34, Appendix); thus:

[54]Rabindra Racaanbali, Vol. IV, Viswabharati. pp. 304 & 293, Reprint of 1402 B.S.

(Translation: an attempt by the present author):

Oh Infinite ! where You are beyond the bounds of human thought, the Highest, Inaccessible, far from the limits of vision, therefrom comes down the flow of bliss just as the Jāhnavī from the top of the Himālayas. The Highest, with whom was seen the golden trace of the World amidst primordial darkness, where there will be the annihilation of the crimson glow of the exhausted setting Sun and successive appearance of radiant new worlds like series of luminous stars will occur through ages out of the force of desire for creation in Whose heart –.

That *Brahman*, the Absolute is always in association with the entire World has been expressed by the poet in the verse no. 57, p. 293 (S- 33, Appendix) in Naivedya in the following way:

(Translation: an attempt of the present author):

It was declared in a lofty and grave voice from a shadowy hermitage, that the Absolute, the Supreme of all gods is united wholely and permanently with fire, water, trees, plants and above all, with this entire Universe. It was this India where this noble idea was originated.

The sublime metaphysical and philosophical ideas evolved in the hearts of ancient Indian sages were found to have echoed in the voice of all subsequent men of letters in India and formed the firm philosophical base on which the gorgeous structure of ancient Indian mathematics was erected.

Notes

1. All quotations (with transliteration and English translation of essentials) marked serially with the letters S-1, S-2, etc. have been given in the Appendix before Reference.
2. *The notes (marked with the letters A, B, C, etc.) related to certain names or topics in the body of the article are given at the end of the main article (before the 'Appendix', under the heading 'Notes', as, e.g. ṚṚgveda [(A)], will denote that there is the relevant note marked with the letter (A) in the section 'Notes').
3. A. According to Winternitz M's, *Geschichte der Indischen Litteratur*, Vol. I, 1920 (dealing with the Vedic Literature) (Eng Translation by Mrs S. Ketkar, 1927, p. 258), 'The available evidence merely proves, that the Vedic period extends from an unknown past, say X, to 500 B.C., none of the dates 1200–500 B.C., 1500–500 B.C., and 2000–500 B.C., which are usually assumed, being justified by facts. Only it may be added, as a result of recent researches, that 800 B.C. should probably be substituted for 500 B.C. and that the unknown date X more probably falls in the third rather than in the second millennium before Christ'[4].

B. Kaṇāda – He was a sage during the times of Mahābhārata. He was also known by the names Kāśyapa, Pippalapāda, and Ulūka[33]. Bhīṣma of Mahābhārata, when waiting for the last moment of his life at Kurukṣetra, eulogized him[34]. Now, the battle of Kurukṣetra occurred when it was the junction of the yugas Kali and Dvāpara[35] and Kali yuga started on the 18th February 3102 B.C.[55] So, Kaṇāda was a sage flourished in the beginning of the fourth millennium B.C. The name

[55]Bentley J., *A Historical View of Hindu Astronomy*, p.113, Internet Archive.

Kaṇāda was given to the author of Vaiśeṣika Philosophy most probably because he propounded that the world was formed by an aggregation of atoms (kaṇā).[56] He was called by the name kaṇabhakṣa muni by those who used to oppose his doctrine (or theory) of his Vaiśeṣika Philosophy. As Ulūpī was the name of a Nāga Princess married to Arjuna of Mahābhārata, it may be that Ulūka might have a relation with the royal Nāga family.

C. Nāgārjuna was a devoted proponent of the Mahāyāna branch of Buddhism. He was born in a south Indian Brahmin family. Nothing is known about his parentage or the name of the place his birth or his early education. We do not even know why he became a Mahāyānic Buddhist. Controversy is also there in the history of his childhood. According to Kumarajīva, who translated Nāgārjuna's works into Chinese and spent his life in preaching Mahāyānic Buddhism in China, Nāgārjuna obtained Prajñāpāramitā-sūtras from Mahānāga.[57] In the field of philosophy, Nāgārjuna flourished between 50 A.D. and 120 A.D.[58]

D. Nothing in detail is known about Piṅgala. He is the oldest author of chanda-śāstras. Varieties of identities are available of this author of chanda-śāstra, namely, (i) a sage (muni), (ii) a nāga and (iii) ācārya (a teacher). Halāyudha Bhaṭṭa (tenth century A.D.), whose famous commentary Mṛtasañjīvanī of Piṅgala's chandasūtra, describes Piṅgala as a brāhmin and as the younger brother of the famous Sanskrit Grammarian Pāṇini. Considering all historical and paurāṇik evidences, Professor Asoke Chatterjee Sastri [27] concludes that Piṅgala is to be placed in fourth century B.C. contradicting Dowson [28] who opined that Piṅgala belonged to the second century B.C. It may be mentioned here that one Piṅgala with vimalākṣa (reddish brown eyes) was known as the commentator of the Madhyamikika kārikā. This was translated in 409 A.D. into Chinese by Kumarajīva.

Appendix

S-*1* *na jāyate mriyate vā vipaścinnāyaṁ kutaścinna babhūva kaścit* |
ajo nityaṁ śāśvato'yaṁ purāṇo na hanyate hanyamāne śarīre
||*-kaṭhopaniṣad, I.2.18*

S-2 *na jāyate mriyate vā kadācinnāyaṁ bhutvā bhavitā vā na bhūyaḥ* |
ajo nityaṁ śāśvato'yaṁ purāṇo na hanyate hanyamāne śarīre —— gītā, II.20

[56]M. Monier-Williams, *Sanskrit-English Dictionary*, p. 245.

[57]Ramanan K. Venkata, ibid, p 27.

[58]ibid, p. 30.

S-3 *jalānalādibhirnāśṁ yon a yāti sa pudgalaḥ* / *25- gaṇitasārasaṁgrahaḥ*, [3], *p.3*

S-4, *(i) Aphorism-1athāto brahmajijñāsā* / *1.1.1, Brahmasūtra, Śrībhāṣya,*[6], *p.4*
Commentary adhītasāṅga-saśiraska-vedasya adhigatālpāsthiraphala-kebalakarmajñānatayā
saṁjātamokṣābhilāṣasyānanta-sthiraphala- brahma-jijñā-hyantarabhāvin /
-Brahmasūtra, Śrībhāṣya [6], *p.4*

S-4 *(ii) The explanation of the word 'brahma' following the commentary: brhmaśabdena (ca) svabhāvatonirastanikhiladoṣo'na-vadhikātiśayāsaṅkheya- kalyāṇaguṇagaṇaḥ puruṣottamo'bhidhīyate* / *Brahmasūtra, Śrībhāṣya, p.5*

S-5, *(i) Aphorism-2 janmādyasya yataḥ*‖*1.1.2, Brahmasūtra, Śrībhāṣya* [6], *p. 271*

S-5, *(ii) asya-acintyavividh-avicitraracanasyaniyatadeśakālaphalabhoga brhmādistambaparyantakṣetrajña-miśrasya jagataḥ yataḥ yasmāt sarvesvarāt nikhilaheyapratyanīkasvarūpāt satyasaṁkalpātjñānā-nandādyananta-kalyāṇguāt sarvajñāt sarvaśakteḥ paramakārūṇikāt puṁsa sṛṣṭisthitipralayāḥ, pravartante, tadbrahmeti sūtrārthaḥ*| *Brahmasūtra, Śrībhāṣya* [6], *pp. 271–272.*

S-6 *śāstraṁ yasya yoniḥ kāraṇaṁ pramāṇaṁ, tacchāstrayoni; tasya bhāvaḥ śāstrayonitvaṃ* | *tasmāt brahma-jñānakaraṇatvāt śāstrasya, tadyonitvaṃ brahmaṇaḥ* |
- Brahmasūtra, Śrībhāṣya [6], *p. 280*

S-7 *yathorṇābhiḥ sṛjate gṛhnate ca*
yatha pṛthivyāmoṣadhayoḥ saṁbhavanti /
yathā sataḥ puruṣāt keśalomāni
tathā 'kṣarāt saṁbhavatīha viśva
‖*Muṇḍakopaniṣad, I.i.7*[10]

S-8 *brahmaivedammṛtaṁ purastād brahma paścādbrahmadakṣiṇaścottareṇa* /
adhaścordhaṁ ca prasṛtaṁ brhamaivedaṁ viśvamidaṁ variṣṭhaṃ
‖*Muṇḍakopaniṣad, II.ii.11*[9]

S-9 *sarvaṁ khalvidaṁ brahma* —— *Chāndogya Upaniṣad (III. 14.1,)*[10]

S-10 *ekaṁ sadviprā bahudhā vadantyagniṁ yamaṁ*
mātariśvānamāhuḥ / *(ṚṚgveda, I.164.46*

S-11 *cijjaḍānāṁ tu yo draṣṭā so'cyuto jñānavigrahaḥ* /
sa eva mahādeva sa eva hi mahāhariḥ ‖

sa eva jyotiṣāṁ jyotiḥ sa eva parameśvaraḥ ||[12] *Skandopaniṣad*[12]

S-12 Ṛtamātmā paraṁ brahma satyamityādikā budhaiḥ /
kalpitā vyavahārārthaṁ tasya sajñā mahātmanaḥ ||*Yogavaśiṣṭha Utpatti Prakaraṇa*[13]

S-13 Nāsadīya Sūkta[14]*:*
nāsadāsīnno sadāsīttadānīṁnāsīdrajo no vyomā paro yat—
kimārīvaḥ kuha kasya śarmannabhaḥkimāsīd gahanaṁ gabhīraṃ—— 1
namṛtyurāsīdamṛtaṁna tarhi na rātryā anha āsīt praketaḥ—
ānīdavātaṁsvadhayā tadekaṁtasmāddhānyanna paraḥkiñcanāsa—— 2.
tama āsīttamasā gūḍhamagre'praketaṁ salilaṁsarvamā idaṁ
tucchayenābhvapihitaṁyadāsīttapasastanmahinā jāyataikaṁ—— 3.
kāmastadgre samavartatādhi manaso retaḥ prathamaṁ yadāsīt—
satobanhumasati niravindan hṛdi pratiṣyā kavayo manīṣā—— 4.
tiraścīno vitato raśmireṣāmadhaḥ svidāsīt—
retodhā āsanmahimān āsan svadhā avastāt prayatiḥparastāt—— 5.
ko addhā veda ka iha pravocat kuta ājātā kuta iyaṁvisṛṣṭiḥ—
arvagdevā asya visarjanenāthā ko veda yata āvabhūva—— 6.
iyaṁvisṛṣṭiryata āvabhūva yadi vā dadhe yadi vā na—
yo asyādhakṣaḥparamevyomantso aṅga veda yadi vā na veda—— 7.

S-14 Sṛṣṭisūkta[16]*: ṛtaṁ ca satyaṁcābhīddhāttapaso'dhyajāyata* /
tatorātryajāyata tataḥ samudro arṇavaḥ ||*1.*
samudrādarṇavādadhi saṁvatsaro ajāyata /
ahorātrāṇi vidadhadviśvasya miṣato vaśī ||*2.*
sūryacaṁdramasau dhātā yathā pūrvamakalpayet /
divaṁca pṛthivīṁcāṁtarikṣamathosvaḥ || *3.*

Translation *Due to the growth (or expansion) (abhīddhāt) of Ṛtaṁand Satyaṁ (both meaning Brahman), which revealed through the austerity of sages (belonging to the preceeding cycle of creation) (tapasaḥadhi ajāyata), there evolved the ocean of air and the sea. (1).*

*From oceans, there evolved whole years, days and nights and therefrom (*i.e.*from the oceans) manifested the Lord of the Universe (viśvasya vaśī, the dhātā (also called the Prajāpati), the Lord of the Universe, miṣataḥ–prakaṭībhavataḥ, manifested) (2).*

(Then) as devised earlier, the dhātā created the Sun and the Moon, days, the Earth, the sky and other higher regions (svaḥ, mahaḥ, etc.*) (3).*

S-15 ūrdhvamūlo'vākśākha eṣośvatthaḥ sanātanaḥ /
tadeva śukraṁ tad brahma tadevāmṛ tamucyate /
tasmin lokāḥ śritāḥ tadu nātyeti kaścana / *etadvai tat* ||*Kaṭhopaniṣad, canto III 1.*[12]

Translation *With the roots (spread) above and branches (with all the leaves) downwards (avāk – downwards avākśākha – that which has the branches downwards), this aśvattha (pippal, commonly called pipal or pipul) (tree) is eternal (sanātanaḥ – existing from time immemorial, having no beginning, imperishable). That alone is resplendent* (i.e. *impressive for being richly colourful – śukraṁ). That is Brahman. That alone is said to be immortal. In That is contained (śritaḥ – contained) in the world. Also (u) from That (tat) not any one (kaścana – any one) can slip off (ati + eti). This verily is that.*

S-16 ūrdhvamūlamadhaḥśākhamaśvatthaṁ prāhuravyayaṃ /
chandāṁsi yasya parṇāni yastaṁ sa vedavid—— Gītā, Chapter 15, verse 1[20]

Translation *The aśvattha with roots spreading high up and branches downwards and whose leaves stand for (worldly) desire, pleasure, is rightly described as immortal. He who knows it (the tree representing Brahman as the shelter of the whole world) knows the meaning of the Vedas.*

S-17 yadūrdhaṁ divaḥ, yadavāk pṛthivyāḥ yadantarā dyāvāpṛthivī ime, yadabhūtaṁ
ca bhavacca bhaviṣyaccetyācakṣate, Ākāśa eva tadotaṁ ca protaṁ ceti /
-Bṛhadāraṇyaka Upaniṣad, III. viii. 7)[24]

S-18 sarvaṁ tatparamaṁ śūnyaṁ na paraṁ paramātparaṃ / *Tejabindūpaniṣad (verse 11)*[14]

S-19 vyāmamātri bhavati / *vyāmamātro vai puruṣaḥ* / *puruṣaḥ prajāpatiḥ* /
prajāpatiragnirātmasammitāṁ tadyoni karoti / *parimaṇḍalā hi*
yoniraratho'yaṁ vai loko gārhapatya u va'ayaṁ lokaḥ / *–*
śatapathabrāhmaṇa, 7 / *1* / *1* / *37 manatraḥ* /

S-20 bhūmnaḥ kratuvat jyāyasvaṁ, tathā hi darśayati ||*55 – Brahmasūtra, Śrībhāṣya, p, 398.*[30]

Translation *(The meditation of) the whole (bhūmnaḥ⇐singular of the sixth case ending of bhūman, meaning the whole) is superior (jyāyasvaṁ – to be considered as superior) as in the case of sacrifice; indeed the scripture shows this.*

S-21 parīkṣitaṁbhāvānāṁpratyakṣaṃ— abhāvānāṁparīkṣyate—
dvividhohyabhāvaḥ— yatkiñcinna kadācidbhavati gaganāravindādi,
kiñcicca kadācidbhavadapi kencidrūpeṇa kutracidbhavati, na sarvadā na
rūpāntareṇa na khalvapi sarvatra—
tatrādyo'vastubhūtovastubhūtastvanyaḥ—
—abhāvaparīkṣā[36] *in the Vaiśeṣika Darśanaṃ.*

Translation *(The act of) carefully examining an existence or an occurrence is pratyakτaθ. Different types of abhāva is now analysed. A thing may never occur (even) in traces, just as a lotus or something like that in the sky, or a thing may be such as its traces may seldom occur somewhere, but not everywhere, not even always in any form whatsoever. The first kind of abhāva is avastubhūta (non-materialistic) and the latter is vastubhūta.*

S-22 *The clarification of abhūtaṁ nāstītyanarthāntaraṃ: yanna bhavati tadabhūtaṁ yannavidyate tannānāstītyanayostulyo'rthaḥ | prāgutpatte partaścātmaprahāṇānna bhavati kāryayātaṁ tadāt- manetyabhūtamiti nāstīti cocyate | --------yadyabhūtaṁ nāstītyeko'rthaḥ, nāsti ghatogehe iti prathītirnopapadyte, na hi ghatonaṁ nāstīti | - abhāvaparīkṣā*[36] *in the Vaiśeṣika Darśanaṃ*

Translation *(The statements) 'that a thing does not occur' and 'a thing is not present' bear the same meaning. A thing produced or existed before and later due to abandonment or for some other cause, ceases to occur, is also its absence... That which is inexistent bears the same meaning as it is not present. The notion that 'a (particular) pitcher is not in the room' does not arise if the room was formerly devoid of the pitcher. (It is a case of vastubhūta abhāva.)*

S-23 *niṣkramaṇaṁ praveśanamityĀkāśasya liṅgaṃ || 19. – Vaiśeṣika Darśanaṃ.* ibid.
Commentary of the above aphorism by Prof. Sarmā:: ākāśohyavakāśaṁ dadāti | tato niṣkrāmati praviśati ca | yatra cāvakāśo nāsti na tatraite bhavataḥ | tasmādanumīyate asyĀkāśaṁ yadavakāśaṁ dadātīti | asya pratyavasthānaṃ | - Ākāśa parīkṣā, Vaiśeṣika Darśanaṃ.[41]

Translation *Ākāśa provides room (or space) and is (meant) for entrance or exit. Where there is no room (compartment), these (operations of entrance or exit) do not occur. Therefrom it appears that the space which this Ākāśa provides is not a place to reside.*

S-24 *tadaliṅgamekadravyatvāt karmaṇaḥ—— 20. – Vaiśeṣika Darśanaṃ.*ibid.

Translation *As the operations (of egress and ingress) from the same object (Ākāśa is taken for one among the nine dravyas in vaiśeṣika darśanaṃ), that (Ākāśa) has no mark (or characteristic or feature).*

*Commentary of the above aphorism: niṣkramaṇaṁ praveśanañceti dvayaṁ na liṅgamĀkāśasya | kasmāt? eka dravyatvāt karmaṇḥ karmakhalvetat niṣkramaṇaṁ praveśanañcaikasmin dravye samavaiti; yanniṣkrāmati praviśati ca, na tvedakāśe'pi samavaiti; naitadĀkāśasy liṅgaṃ | - Vaiśeṣika Darśanaṃ.*ibid.

Translation *'Going out (of it)' and 'entering into (it)' are not the characteristics of Ākāśa. Why? (These are the) operations relating to the same object; and indeed* *these two operations of going out and entering into the same object (place) are associated, because coming out (of an object) is followed by an (earlier) entrance (there); but these two are not associated operations in this Ākāśa. This Ākāśa has no characteristic (like a materialistic abode).*

S-25 saṁyogādabhāvaḥ karmaṇaḥ. 22.- Vaiśeṣika Darśanaṃ.ibid

Translation *These (two) operations (of entrance and exit) occur in absence of (instrumental/material) contact (*i.e. *lack of conjunction implies absence of any action).*

Summary S-25 to S-27:

(i) *khalvākāśḥ sarvagato'pi na sarvatra karmotpattinimittaṃ* |
(ii) *vyatirekābhāvat pūrvabhāvaniyamona* |
(iii) *mūrtasaṁyogābhāvastatra heturevamākāśo'pyavakāśanimittaṁ heturbhavati* |
(iv) *khalvemĀkāśaṁ bhāvavadupādhibhedādbhidyate* |
(v) *ekatve Ākāśasya. – Vaiśeṣika Darśanaṃ,* ibid.

Translations

(i) *Indeed this Ākāśa is omnipresent, but is not everywhere as the cause of any action.*
(ii) *Due to the lack of logical discontinuance, it has no beginning.*
(iii) *It is a cause for making room for space, having no material confinement, or in other words, it is a space to accommodate everything, keeping the thing unchanged.*
(iv) *Indeed this unique Ākāśa is discriminated due to its variance in its appearances.*
(v) *It is singular.*

S-26 *dvirardhe:*
apanīte ityadhyāhāraḥyadā jijñāsite ṣaḍakṣare chandasi kati vṛttāni bhantīti? tadā taṁchandokṣarasaṁkhyāṁbhūmau sthāpayitvā tato'rdhaṁapanayet|tasmin apanīte dvau labhyate| tatastāt dvisaṁkhyāṁbhumau pṛthak prastārayet |tataḥśeṣāśtatra trayo'kṣarasaṁkhyā bhavanti|teṣāṁardhayituṁaśakyatvāt kiṁiti āha| (Piṅgala's chandasūtraṃ[42])
rūpe śūnyaṁ:
viṣamasaṁkhyāto rūpaθapanayet, tasmin apanīte śūnyaρlabhyate| tat pūrvalabdhāyā dvisaṁkhyāyā adhastāt sthāpayet|tato dvisaṁkhyā avaśiṣyate, tato'rdhe apanīte punardvisaṁkhyā labhyate|taṁśūnyādhastāt sthāpayet|tato rūpe śūnyaṁ labhyate|tad dvisaṁkhyāyā adhastāt sthāpayet| (Piṅgala's chandasūtraṃ[42]

Translation *When asked, in a metre of six syllables, how many recurrences of lines (padas/caraṇas) are possible? Then putting the number of letters of the metre on the ground, its half should be removed. After removal of that, two is obtained. Then the number two is to be placed separately on the ground. Here the remainder (after making the half of six) consists of three letters. It is said, due to incapability of getting half of that what to do?*

From the odd numeral deduct 1 and put aside the number 0 below the number 2 set aside before. The remainder 2 is halved (as this 2 is divisible by 2); its half is removed, and again 2 is to be written below the 0 (written earlier); after the removal of 1 from 2 at the preceeding stage, the remainder 1 cannot be halved, and put the number śūnya below the number 2 set aside in the former case. Hence the heading rūpe śūnyaṁ.

S-27 śūnyantu vaśikaṁ tucchariktake. 11[47]

Translation *Zero (śūnya) is also void (vaśikaṁ), devoid or destitute of (riktaḥ) and trifling (tuccha).*

S-28 dhanayordhanamṛṇmṛṇayordhanarṇayoraṁtaraṁsamaikyakhaṁ
ṛṇaikyamṛṇaṁdhanaśūnyayordhanaṁśūnyayoḥśūnyayoḥśūnyaṁ|31
*śūnyavihīnamṛṇamṛṇaṁdhanaṁdhanaṁbhavati śūnyamĀkāśaṁ|(1*st *line of 33)*
*śūnyarvayoḥkha kha śūnyarvayorvā vadhaḥśūnyaṃ| (2*nd *line of 34)*
*khodhφtamṛṇṃdhanaṃ vā tacchedaṃkhamṛṇadhanavibhaktaṃ vā| (1*st *line of 36)*
-Brāhmasphuṭasiddhāntaḥ,[52] *pp.243–244*

Translation *The sum of two positive (numbers) is positive, the sum of two negatives is negative, the sum of a positive and a negative is their difference and the sum of a positive and a negative quantity of equal magnitude is zero. (31).*

Zero when added to a negative (number) gives a negative, to a positive gives a positive and to zero gives zero. (first line of 33).
A negative (number) less zero is the negative (number), a positive (number) less zero is the positive (number)and zero is (like) the space.
Zero multiplied by a negative (number) is zero, multiplied by a a positive (number) is zero and multiplied by zero is also zero.
*A negative (number) or a positive (number) is upon (*i.e. *divided by) zero is tacchedaṁ and zero divided by a negative or by a positive (number) is zero.*

S-29 yoge khaṁkτepasamaṁ, vargādau khaṁkhabhājito rāśiḥ—
khaharaṁsyāt khaguṇaḥ khaṁ khagṇvaśca śeṣavidhau—— 45
śūnye guvake jāte khaṁkṣepasamaṁ, hāraścet punastadā rāśiḥ—
abikṛta eva jñeya, tathaiva khenonitaśca yutaḥ—— 46 – Līlāvatī [24]

Translation *When a number is added to cipher, the sum equals the additive, the square and other powers of cipher is cipher; a number divided by cipher is (termed) 'khaharaḥ', and it should be considered in accordance with the rule of finding remainder (śeṣavidhau) and should be considered by multiplying it by cipher. Cipher multiplied by a number gives zero, and the product when divided by cipher gives back the original number; the same (unchanged state) occurs when cipher is subtracted from or added to a number.*

S-30 khayoge viyoge dhanrṇa tathaiva
cyutaḥŚūnyatāstad viparyāsmeti—
vadhādau viyat khasya khaṁkhena ghāte—
khahāro bhavet khena bhaktaśca rāśiḥ—— —— 3—— :Bījagaṇitaṃ [28]

Meanings of essential words: cyutaḥ, deviated from/deprived of, here, subtracted from; Śūnyatāḥ, from zero; viparyāsaṃ, opposite in (sign); viyat, (sky) cipher; the word vadhādau has been used to mean multiplication and division of cipher by a number, product of ciphers, powers or roots of cipher.

Translation *When cipher is added to or subtracted from a positive or a negative number, the number remains unchanged. Their sign is reversed when subtracted from cipher. Cipher multiplied or divided by a number is cipher, any power or any root of cipher is also cipher; a number divided by cipher is khahāraḥ.*

S-31 yo'kṣṇornimeṣasya kharāma bhāga sa tatparastacchata bhāga uktā truṭiḥ.
— verse 16, kālamānādhyāyaḥ, madhyamādhikāraḥ, siddhāntaśiromaṇi[54].

[Meaning of essential words: akṣṇoḥ nimeṣaḥ, i.e. *of one nimeṣaḥ (time required for a single winking) of an eye, kharāma = 30, kharāmabhāgaḥ = (1/30)th part, tatparastacchatabhāgaḥ(tatparaḥ + tat + śatabhāgaḥ) = one-hundredth part of tatpara].*

Translation *One-thirtieth part of nimeṣaḥ of an eye (time required for a single winking of an eye) is one tatpara; one-hundredth part of tatpara is one truṭi, and one truṭi =* $\frac{1}{3000}$*th part of the time for a single winking* [28].

S-32 dināntara spaṣṭa khagāntaraṁ syādgatiḥ sphuṭā tatsamyāntarāle | 36
koṭiphalaghni mṛdukendrabhuktistrijyoddhṛ tā karkimṛgādikendre
tayā yutonā grahamadhyabhuktistātkālikī mandaparisphuṭā syāt | 37
-spaṣṭādhikāraḥ, gaṇitādhyāyaḥ, siddhāntaśiromaṇi,[55] *p.178*

Meanings of essential words: *khaga*, that which moves in the sky, a planet, here used to mean the mean position of a planet; *tatsamayāntarāle*, in that interval of time; *koṭiphalaghnī*, multiplying by the *koṭiphala*; *mṛdukendrabhukti*, daily motion of the *mandakendra*; *trijyodhṛtā*, divided by the radius of the *kakṣāvṛtta*; *karkimṛgādikendre*, in the *karkikendra* (the region starting from first point of Cancer and extending up to the first point of Capricorn) or in the *mṛgādikendra* (the region starting from first point of Capricorn and extending up to the first point of Cancer);

grahamadhyabhuktiḥ, mean daily motion of the planet; *tayāyutonā*, is to be increased or decreased by that; *tātkālikī*, instantaneous; *mandaparisphuṭā*, instantaneous motion of the *mandaspaṣṭagrahaḥ*.

Translation The difference of the true motions of planet (at the epoch of its rise or at the midday) is its true motion in that interval (verse 36).

The motion of the mean anomaly during the interval (mṛdukendrabhukti) multiplied by the koṭijyā of the mandakendra (i.e. by the R.cosine of the planet's initial anomaly, when R is the radius of the kakτāv tta) and the product is divided by the radius (of the kakṣāvṛtta). The result is to be added to or subtracted from the mean (daily) motion of the planet accordingly as the kendra lies in the region starting from first point of Cancer and extending up to the first point of Capricorn or in the region starting from first point of Capricorn and extending up to the first point of Cancer (verse 37).

S-33 *adyatana-śvastana-spuṭagrahayoraudakiyodinārdhajayorvātākālikayorvā yadantaraṁkalādikaṁsā sphuṭagatiḥ | adyatanācchastane nyune vakragatirjñeyā|tatsamayāntarāla iti |tasya kālasya madhye'nayā gatyā grahaścālayituṁyujyata|iyaṁkila sthūlāgatiḥ|atha sūkṣmā tātkālikālikī kathyate|tuṅgagatyunācandragatiḥkendragatiḥ|anyeṣāṁgrahāṇāṁgrahagatireva-kendragatiḥ|mṛdukendrakoṭiphalaṁkṛtvā tena kendragatirguṇyā trijyaā bhājyā labdhena karkyādikendre grahagatiryuktā kāryā|mṛ gādau tu rahitā karyā| evaṁtātkālikālikī mandaparisphuṭ ā syāt| – spaṣṭādhikāraḥ, gaṇitādhyāyaḥ, siddhāntaśiromaṇi,*[55]*, p.179*

Translation *The difference between the longitudes of a planet at the epoch of its rise or at the noon on a certain day and that on the next day at the same epoch is the true speed of the planet in minutes. If today's longitude be less than that of yesterday, then the motion is to be regarded as retrograde within the interval. It is proper to consider that a planet moves in this way within the interval. This is indeed a rough estimate of the motion of a planet. Now is stated the precise instantaneous (motion of a planet). The longitude of the Moon less her longitude at the apsis is her kendragati. In the cases of all other planets, the difference of longitudes of the kendras (mandakendra – mean anomaly) on 2 successive days alone denotes their daily speed. Finding the R.cosine ratio of the planet's initial anomaly (when R is the radius of the kakṣavṛtta of the planet in question), multiply that by the daily motion of the mandakendra, and divide the product by the radius (of the kakṣavṛtta); the result is to be added to or subtracted from the planet's daily motion accordingly as the kendra lies in the region starting from first point of Cancer and extending up to the first point of Capricorn or in the region starting from first point of Capricorn and extending up to the first point of Cancer. Thus the instantaneous motion of the mandaspaṣṭagrahaḥ will be obtained.*

S-34

Transliteration (following the mode of pronouncing of Bengali words in Gangetic West Bengal):

he ananta, jethā tumi dhāraṇā-atīt
sethā hote ānander abyakta-saṅgīt
jhariyā parhiche nāmi, adṛśya agama
himādriśikhar hote jāhnabīr sama |
jagater prātaḥkāle diyechila dekhā
ādi andhakār –mājhe, yethā raktacchabi
asta jabe jagater śrānta sandhyārabi
naba naba bhubaner jyotirbaṣparāśi
puñja puñja nihārikā jar bakṣe āsi
phiriche sṛjanbege meghkhaṇḍasama
yuge yugāntare — | Naivedya, verse no. 80, p.304[56].

S-37 *Transliteration: (following the mode of pronouncing of Bengali words in Gangetic West Bengal)*

he sakal īśvarer parama īśvar,
tapobanatarucchāye meghamandrasvar
ghoṣaṇā kariyāchila sabar upare
agnite jalete, ei biśvacarācare,
banaspati oṣadhite ek debatār
akhaṇḍa akṣay aikya | se bākya udār
ei bhārateri | Naivedya, verse no. 57, p. 293,[56]

References

1. Kober M. Copenhagen interpretation of quantum theory and the measurement problem; 2009. p. 1–9.
2. Venkata RK. Nāgārjuna philosophy as presented in Mahā-Prajñāpāramitā-Sāstra. Tokyo, Vermont: Charles E. Tuttle Co. Japan; 1966. p. 210–1.
3. Rangacharya M. Gaṇitasārasaṁgrahaḥ of Mahāvīrācārya, text with Eng. Translation. Madras: Madras Govt.Press; 1912. p. 3.
4. Bhattacharya Sachchidananda. A dictionary of Indian history. Calcutta: Calcutta University; 1972. p. 673.
5. Swami Vivekananda. Lectures from Colombo to Almora. Kolkata: Advaita Asrhama; 2005. p. 260–1.
6. Brahma sūtras. Śrībhāṣya, part-I. Kolkata: Advaita Ashrama; 2012. p. 1–86.
7. Swami Vivekananda. Lectures from Colombo to Almora. Kolkata: Advaita Ashrama; 2005. p. 263.
8. *The Gospel of Sri Ramakrishna*, Vol. II, Maylapore, Madras: Sri Ramakrishna Math; 2001. p. 733–734.
9. *Chāndogya Upaniṣad*, with commentary of Śaṅkarācārya, Kolkata: Advaita Ashrama; 1997. p. 208.
10. Ṛgveda saṁhitā, compiled by Rameschandra Datta, Kolkata: Sadesh; 2007. p. 129.
11. Ghosal BB. Mukti evaṃ tāhār sādhan. Kolkata: Udvodhan; 2004. p. 12.

12. ŚrīmadbhagvadGītā. Sanskrit commentary by Madhusūdan Sarasvatī, Bengali Transl. by BhutnathChattopadhyay. Kolkata: Naba Bharat Publishers; 1986. p. 968.
13. Datta B, Singh AN. History of Hindu mathematics, part I. Lahore: Motilal Banarasi Das; 1935. p. 53.
14. Pattanaik P. The cult of Jagannath, Orissa review; 2003
15. Datta B. The science of Śulba. Calcutta: Calcutta University; 1932. p. 10.
16. Radhakrishnan S. Indian philosophy. Bombay; 1977. p. 666.
17. Bibhutibhusan D, Narayan SA. History of Hindu mathematics, part-I. Lahore: Motilal Banarsidass; 1955. p. 75.
18. Dikshit SB. Bharatiya Jyotish Sastra (English translation). New Delhi: India Meteorological Department; 1981. p. 396.
19. Pingree D. Zero and the symbol for zero in early sexagesimal and decimal place-value systems, 'the concept of Śūnya'. New Delhi: I.G.N.C.A & I.N.S.A; 2003. p. 140.
20. Boyer CB. History of mathematics. New York: Wiley; 1968. p. 237.
21. R.C.Majumdar (G. Editor). The age of imperial unity. Bombay: Bharatiya Vidya Bhavan; 1980. p. 616.
22. Shukla K.S. (commentator). Āryabhaṭīya of Āryabhaṭa. New Delhi: Indian Notional Science Academy; 1976. p. 7.
23. Ramswarup Sharma (Chief Editor). Brāhmasputasiddhānta. New Delhi: I. I. of A. & S. R.; 1966. p. 243–5.
24. Śaṅkara & Nārāyaṇa. In: Sarma KV, editor. Līlāvatī of Bhāskarācārya with Kriyākramkarī. Hoshiarpur; 1975. p. 91.
25. Pandeya R. Hindi-translator of Sūryasiddhāntaḥ, (Gūḍhārthaprakāśikā). Varanasi; 2009. p. 45.
26. Bag AK. Mathematics in ancient and medieval India. Delhi: Chaukhambha; 1979. p. 292–3.
27. Kumar SCA. Piṅgalacchandasūtra, a study. Calcutta: Calcutta University; 1987. p. 81.
28. Dowson J. A classical dictionary of Hindu mythology and religion. 2nd ed; 1888. p. 234 - (Internet Archive).
29. Jha DC. Elucidator and Hindi translator of Bījagaṇitaṃ by Bhāskarācārya. Varanasi; 1913. pp. 13–14.

Part III
Comparative Analysis of Nature of Quantum Reality and Theory of Śūnya

Chapter 6
Two Aspects of *Śūnyatā* in Quantum Physics: Relativity of Properties and Quantum Non-separability

Michel Bitbol

Abstract The so-called paradoxes of quantum physics are easily disposed of as soon as one accepts that there are no such things as intrinsically existing particles and their intrinsic properties, but that both particles and properties are relational "observables." Accordingly, quantum physics does not offer a "description of the outer world," but rather a prescription about how to make probabilistic predictions within a participatory environment. The latter view (or rather criticism of views) looks quite radical with respect to standard Western Aristotelian ontology; but it looks natural in the context of the Indian-Buddhist concept of Pratītyasamutpāda which underpins Śūnyatā. Special attention will then be devoted to the quantum feature of non-separability, which displays remarkable similarities with Pratītyasamutpāda. Finally, the meaning of such twofold parallel between quantum physics and Śūnyatā will be discussed. This parallel will be related to the similarity of epistemological situation between knowing a world from which we are not entirely separated and knowing oneself.

Keywords Śūnyatā · Pratītyasamutpāda · Quantum *non-separability* · *Paradoxes of quantum* physics · Probabilistic predictions · Participatory environment

1 Introduction

The so-called paradoxes of quantum physics are easily disposed of as soon as a momentous philosophical step is taken. This step consists in accepting that there are no such things as intrinsically existing particles and their intrinsic properties, but that both particles and properties are relational "observables." Accordingly, quantum physics does not offer a "description of the outer world," but rather a prescription about how to make probabilistic predictions within a participatory environment. The

M. Bitbol (✉)
Archives Husserl, Ecole Normale Superieure, Paris, France
e-mail: michel.bitbol@ens.fr

S. R. Bhatt (ed.), *Quantum Reality and Theory of Śūnya*,
https://doi.org/10.1007/978-981-13-1957-0_6

latter view (or rather criticism of views) looks quite radical with respect to standard Western Aristotelian ontology; but it looks entirely natural in the context of the Indian-Buddhist concept of Pratītyasamutpāda which underpins Śūnyatā. Special attention will then be devoted to the quantum feature of non-separability or entanglement. Quantum non-separability displays remarkable similarities with Pratītyasamutpāda, in view of its lack of affinities with productive causality and its convergence with a symmetric relation of co-production. Finally, the meaning of such twofold parallel between quantum physics and Śūnyatā will be discussed. This parallel will be related to the similarity of epistemological situation between knowing a world from which we are not entirely separated and knowing oneself.

That there is a striking analogy between central features of quantum physics and the Indo-Buddhist concept of *Śūnyatā* was pointed out by many authors in the past.[1] However, most of them did not go beyond superficial similarities in the vocabulary (e.g., equating the quantum vacuum with emptiness) or superficial affinities between the alleged pictures of the world (e.g., interdependence instead of separation, flux, and impermanence instead of reifying stability, etc.). In this paper, I wish to spell out some points on which the parallel precisely works beyond mere family resemblance and to make a few assumptions about the origin of this parallel.

As a preliminary, it must be borne in mind that the relevant parallel should not concern two *metaphysical views*. It should not be question of comparing two views of the world (a Buddhist and a scientific view), and to find that, surprisingly enough, they are similar. It should not be question either of two ways (internal and external, experiential and experimental) of probing into the true essence of reality and of expressing this essence by words or mathematical formalisms which, surprisingly, seem to converge. Actually, the parallel concerns two of the most radical critiques of metaphysical views and essentialism[2] that have been proposed in the history of human thought.

Buddhism arose from a deep and basic existential concern and did not indulge in metaphysical speculation. Its aim was therapeutic rather than doctrinal. This skeptical/agnostic tendency of Buddhism was amplified in *Prāsangika Mādhyamaka*, wherein every single metaphysical view is submitted to a procedure of *reductio ad absurdum* which tends to show that this view is in fact only partial and conventional. The ultimate truth here can by no means be captured by a metaphysical picture expressed in words or symbols.

In the same way, quantum physics [3] is persistently averse to metaphysical interpretation. Despite many half-successful attempts of providing it with a so-called realist interpretation, quantum physics still relies on Bohr's and Heisenberg's initial remark according to which *no* unified *picture* of the atomic and subatomic domain can be derived from it. According to these founding fathers of

[1]See Ref. [1].

[2]This (imaginary) machine comprises a fragment of radioactive matter having one chance in two of disintegrating over the time of 1 h and a flask of poison which is released when the disintegration occurs. If the poison is released, it kills the cat[1] [2].

quantum mechanics, their theory is no "view" of the microworld, but rather a mathematical symbolism intended to predict probabilistically the outcome of experiments performed at the microscale using macroscopic devices. As we will see in Sects. 2, 3, and 4 of this paper, most "mysteries" of quantum physics (such as the Einstein-Podolsky-Rosen paradox [4] or the measurement problem [5]) are defused from the outset if one is ready to accept all the consequences of this economic construal of quantum mechanics. Recent developments in theoretical and experimental research have strengthened the anti-metaphysical reading of quantum physics. Quantum information theories, which underlie the project of quantum computers, have been shown to offer an alternative foundation to quantum physics as a whole [6]. According to them, quantum mechanics is nothing more and nothing else than a theory of *the limits of available experimental information.* It is by no means a description of the putative objects of experiments. Besides, new tests of Bell-like inequalities [7] put the very assumption of "realism" (according to which quantum mechanics is a possibly incomplete description of the *intrinsic properties* of real objects) under increasingly strong pressure.

In Sects. 5 and 6 of the present paper, special attention will be devoted to a prominent feature of quantum physics that precisely evokes the relational background of *Śūnyatā.* This remarkable feature is non-separability or entanglement. Quantum non-separability displays remarkable similarities with *Pratītyasamutpāda*, in view of its lack of similarity with productive causality, and its convergence with a symmetric relation of co-production. But, here again, instead of ascribing quantum non-separability to the world as a whole, as if it were an intrinsic feature of the universe, we will point out that non-separability itself is non-separable. Non-separability is non-separable from its instrumental conditions of manifestation (just as, according to *Mādhyamaka*, emptiness is itself empty).

At the end of the paper, in Sect. 7, the meaning of such multifarious parallel between quantum physics and *Śūnyatā* will be discussed. This parallel will be related to the similarity of epistemological situation between knowing a world from which we are not entirely separated and knowing oneself.

2 Some Quantum Paradoxes: A *Śūnyatā* Resolution

This preliminary section is devoted to showing that the majorities of the so-called weird paradoxes of quantum physics dissolve away and no longer appear "weird," once they have been approached relationally rather than absolutely. In other terms, they dissolve away as soon as one accepts that the micro entities are *empty of intrinsic being and determinations.*

There is a premise which conditions all these case studies. It concerns the evaluation of the epistemological status of quantum theories. What *are* these theories, what do they succeed in doing, and what do we have a right to ask of them? Do they offer an *explanation*, or at least a description unequaled in its precision, of subatomic processes? Have they made it possible to elucidate the *intimate mechanisms* of chemical phenomena and nuclear transformations? Have they penetrated further into the fundamental nature of things than classical theories, such as the mechanics of Galileo and Newton, thermodynamics or electromagnetism, have ever managed to do before? The great misunderstanding which runs through the debates on quantum physics stems from the fact that most of the time it is thought that this is the way things *must be*: that the three questions listed above *must* be construed as so many positive assertions. The belief of Western science in a *telos* here claims its due. However, once one accepts that, a whole cascade of disturbing conclusions burst forth. The world which is supposedly "revealed" by quantum mechanics has an appearance as unexpected and ludicrous as Alice's Wonderland recounted by Lewis Carroll. And in this case, it would seem that all that can be done is to accept the weirdness.

As it is well known, one of the first surprises is the *wave-particle duality*, this strange association between extension and point localization, between continuity and discontinuity, which is supposed to describe the "nature" of the new entities that are sometimes called "quantons" [8]. Einstein, the first scientist to have imagined entities of this sort (concerning electromagnetic radiation), was unable to hide his perplexity on this subject: "Is it possible to reconcile the quanta of energy on one hand, and the Huygens principle on the other? Appearances go against it, but God seems to have found the trick of it." It may be added that although our spontaneous intuition cannot digest this amalgam between the discontinuous and the continuous, between quanta and the "Huygens principle" for wave interference, we ourselves seem to have found a "trick" which has the reputation of being able to combine them and thus to "enter into the mind of God." The mathematical formalism of Hilbert does indeed make it possible to establish a connection both with a continuous geometry (by means of the concept of the spatial amplitude of a probability or a "wave function") and with an algebra of discontinuity (by means of the scheme of quantization). But does this formal derivation really suffice to reconcile the two contradictory concepts, or does it just amount to reject them both (setting them back-to-back) and substitute something quite different in their place?

This paradoxical situation was the first one to alert scientists about the extreme novelty of the quantum domain. Due to this and other situations, many physicists consider that there is something quite extraordinary in the occult nooks and crannies of the world, something that quantum descriptions give an oblique glimpse of, but that neither our language nor our imagination can properly grasp, and that only mathematics makes it possible to circumscribe. Nevertheless, the very same situations can also be interpreted in a diametrically opposite fashion. Once they are reconsidered in the most intellectually economical fashion, all these supposed "paradoxes" converge toward the possibility that quantum theory is nothing more than an ingenious but purely formal way of anticipating experimental information; that it does not offer an incomplete and cryptic revelation of an invisible and ineffable reality, but only a method for orienting oneself with respect to that which shows itself and is said; and that instead of penetrating further into the recondite depths of matter than any previous theories have managed to do, quantum mechanics is rather a systematic inventory of its *surface*. For the so-called wave-particle duality, paradox can be immediately *dissolved* as soon as one renounces the application to quantum mechanics of the descriptive, representationalist, "realist" conception of physical theories.

Indeed, assigning a double nature, as wave and as particle, to the objects called "quantons" is a biased, overdetermined, and prejudiced way of expressing a phenomenon which does not a priori impose *either* an ontology of waves *or* an ontology of particles. The phenomenon in question is the distribution of a large number of punctual events according to a pattern which is isomorphic to that which *would be produced* by the interference of two waves or the diffraction of a single plane wave passing through a hole. Bohr himself already criticized treating this sort of phenomenon in ontological terms when he replaced the assertion of a wave-particle *duality* by that of a *complementarity of the images* of a wave and a particle. Each image is only relevant, according to Bohr, with respect to a particular experimental context; and the contexts which render these two images appropriate are partially exclusive of each other. But that is not all. It can be shown in a quite general way that any theory capable of accounting for phenomena concerning mutually exclusive contexts predicts distributions which will have a wavelike *appearance* [9], i.e., distributions where everything happens *as though* we are dealing with waves even though there are no waves at all. In other words, far from manifesting the *absolute* wavelike properties of microscopic entities, the interference behavior of quantum phenomena could be the eloquent sign of their epistemic *relativity*; far from being an indication about the intrinsic nature and own-being of the world, the interference behavior is a sign of its *emptiness* of own-being; far from bearing witness to the deep nature of things, the pseudo-wavelike effects could well represent one of the most salient marks of the *superficial*, *interfacial* character of the phenomena that quantum mechanics makes it possible to anticipate.

A second surprising feature of quantum physics is the measurement problem. A famous way of telling the story so as to bring out its dramatic implications is the paradox of Schrödinger's cat [10].The most succinct account of this paradox plays on the contradiction between the state of the cat as *described* and as *concretely occurring*. Quantum mechanics[3] (so it is said) describes the cat subjected to Schrödinger's infernal machine as being in a *superposed state* of being both alive *and* dead. However in actual practice, it is found that the cat is found to be *either* alive *or* dead. Here, the (supposed) quantum description of the cat does not accord with what one sees of it. Dozens of solutions have been proposed to get around this difficulty. One of them consists of taking the quantum "description" literally and to suppose that each of the two terms in the superposition represents a separate "possible world": in one of these worlds, the cat really is alive (and the inhabitants of this world see the cat alive), and in the other world, the cat is dead (and the inhabitants find it dead). However the way out that is currently dominant (called

[3]This (imaginary) machine comprises a fragment of radioactive matter having one chance in two of disintegrating over the time of 1 h and a flask of poison which is released when the disintegration occurs. If the poison is released, it kills the cat.

"decoherence") amounts to refusing to confront the problem according to the standard formulation and to change the formulation in a way that is so subtle that many scientists are unaware of the sleight of hand: instead of a problem of compatibility between conjunction and disjunction, between a plurality of possibilities and the uniqueness of what actually exists, the problem that is resolved is a problem of connection between two forms of the calculation of probabilities [11].

Yet, this so-called cat's paradox can also be dissolved at once if one accepts once again to follow the lead given by Bohr and if one criticizes once again the concept of intrinsic determinations in favor of one of relational (or dependently arising) determinations. The apparent contradiction that arises in the "cat's paradox" indeed derives from the repeated use of the term "state," which actually has two quite different meanings. The superposed quantum "state" of the cat does not fit with the "state" that is manifest and observable. This apparent conflict disappears as soon as we recognize that the quantum "state," far from indicating what the cat intrinsically *is*, only makes it possible to estimate the chances one has of *seeing it* in a certain way; that far from corresponding to a "state" in the full and proper sense of the term, the quantum "state" vector is nothing other than a symbolic instrument making it possible to evaluate the *probability* of *finding* the cat in one or other of its two biological states. Indeed, no one has ever required that a probabilistic evaluation should reveal in advance the actual outcome (in the full and proper sense of the term) of the event in question; in the same way, no one should hope to reveal or to engender the actual observed state of the cat merely on the basis of the quantum probabilities. The only non-conventional aspect of quantum theory is the peculiar (nonadditive and interferential) structure of its calculation of probabilities, which is quite different from the classical calculation, because it is adapted to the contextuality of microscopic phenomena [12]. The only remaining problem thus consists of linking up (at least approximately) this nonclassical structure of probabilities with the classical additive structure which is valid for the mutually exclusive events observed in the laboratory. This technical (rather than fundamental) problem is solved, as we have already indicated, by the theories of decoherence.

A third disturbing finding was formulated for the first time by Einstein, Podolsky, and Rosen [13]. Even though the aim of these authors was to demonstrate the "incompleteness" of quantum mechanics (its incapacity to describe all the "elements of reality" attached to physical systems), what posterity has retained from their reflections is quite different [14]. The enigma which remains bears on the explanation of the strange "EPR correlations" predicted by the "entangled states" of quantum mechanics. How is one to understand the strict correlation between the values of observables measured on pairs of particles which were initially in contact but which are now situated at arbitrarily large distances from each other? Briefly, the only two explanatory frameworks which are plausible are (a) common causes and (b) reciprocal causal influence. But one and the other of these two explanatory possibilities encounters insurmountable obstacles in quantum physics. Considering that the origin of the correlations lies in common causes amounts to asserting that they are inscribed in the *properties* of the particles and that these properties were fixed ever since the initial moment when the particles were contiguous. However this option (called *local hidden variables*) is excluded by Bell's theorem [15]. The other hypothesis, that of reciprocal causal influences with an arbitrarily large speed (including larger than the speed of light), has indeed been modeled and tested experimentally in recent years [16]; but it has been refuted and must therefore be rejected in its turn. How is it possible to extricate oneself from this impasse? Two extreme options remain available. On the basis of the presupposition of "scientific realism" (according to which quantum theory describes the properties of things as they really are, including their inseparability), the only way out is to adopt an *ontological holism.* According to this doctrine, space and time are only emergent deployments of an "implicate order" [17] which is pre-spatial and pretemporal; and the two distant particles are in truth distinct manifestations of one and the same universal entity. Their correlation no longer has to be explained by any sort of *transmission*, from the past to the present or from a present here to a present over there, but simply by a statement of *identity.* At the opposite extreme, according to the most radical of the *anti*-realist options, there is simply no need to "explain" an instantaneous correlation at a distance, for the good and simple reason that the latter has no intrinsic existence. The correlation only ever sees the day *relatively* to mechanical and electromagnetic devices apt to "provide evidence for it." Now, that can only come about when the information concerning one of the correlated properties has had a sufficient time (at least the time that would be taken by a light signal) to rejoin the region of space where the information concerning the other property is available [18]. No "nonlocal influence" need be invoked in this case.

The fourth disturbing finding covers in fact a whole network of clues that point toward a conception of physical theory that is non-descriptive and nonrepresentational, but rather purely predictive and informational [19]. A large number of experiments (some of which have actually been carried out, others which are pure thought experiments) make it pretty much unthinkable that one could describe processes which are supposed to have happened before the actual act of their detection or observation; and this forces one to trust only the information drawn from such an act by basing oneself on previous knowledge of the configuration of the experimental setup as a whole. I will mention just two of these experiments: "measurements without interaction" and "delayed-choice experiments" [20]. In the first sort of experiment, information derived from an *absence* of interaction between the object and an intermediate part of the instrument has exactly the same consequences as those that would result from their actual interaction [21]. That is enough to make one think that what counts in a quantum experiment is not the detail of the hypothetical processes which may be supposed to occur between the preparation and the final detection, but rather the informational content that the whole structure of the apparatus confers on the event of detection. Indeed, in some spectacular experiments with "delayed choice," the object interacts with an elementary measuring agent (e.g., a photon), but its so-called state depends on decisions that can be made millions of years later concerning the arrangement of the device which makes it possible to collect the photon. Unless one imagines that certain influences can go backward in time (as certain physicists have been led to propose (Wheeler 1978)), it must be recognized that what may be carelessly called "the state of an object" expresses nothing other than the information made available by the observational apparatus which gives access to it *after* the moment when all the decisions concerning the device have been taken.

3 A Nonessentialist, *Śūnyatā*-Like Strategy to Make Sense of Quantum Theory

If we wish to express the lesson of these reflections in a deliberately provocative way, we might remark that quantum mechanics is better understood, in a way that avoids posing logically insoluble problems, by admitting that it reveals *rigorously nothing* about the alleged intimate nature of its objects. After all, if quantum mechanics is considered as a generalized process of evaluating probabilities, there is no more reason for it to reveal the nature of its objects than the classical theory of probabilities has of revealing the nature of objects to which it is commonly applied such as dice, or roulette tables, or fluctuations in the financial market, or the clients of an insurance company. Just like the theory of probabilities, quantum mechanics is grafted onto the outside layer of events that it aims at anticipating – without penetrating into a hypothetical "interior." Even more than the theory of probabilities, quantum mechanics rests on the surface of things, because what it anticipates are not even actual events that will come about by themselves, but merely *potential* phenomena which require a particular experimental setup in order to occur (Bohr wrote that these phenomena are *defined* by such an experimental setup). And that is not yet all. Not only does quantum theory reveal no intimate nature of things beyond the phenomena, but its success and its fruitfulness are easily explained by the fact that it incorporates in its very structure the *limits* to the exploration of phenomena. Its success and its fruitfulness come from the fact that it does not even allow any *meaning* to the belief that there might be something deeper to understand behind the superficial screen which is its own domain of validity. Heisenberg's indeterminacy relations can thus be considered as the expression of a limit to any possible knowledge of the dynamic variables of elementary particles. But these relations are at the same time a powerful tool of theoretical exploration which has made it possible to predict, among other things, the bandwidth of rays of electromagnetic emission, the lifetime of radioactive nuclei, and a number of striking effects of quantum field theory (such as virtual particles, the Casimir forces, etc.). Here, the limit to knowledge is not a matter of a provisional obstacle but determines the very form of what is to be known. Relaxing the usual struggle toward representational sensemaking turns out to be a good strategy to make sense of the efficiency of quantum mechanics.

Indeed, not only is quantum mechanics the superficial prediction of superficial phenomena, but its redoubling of superficiality is what accounts for its remarkable vocation for *universality*. If quantum theory is above all a general procedure for anticipating on a probabilistic mode the replies to experimental solicitations, or more precisely for anticipating replies which correspond to the type and the *order* of these solicitations, then it ought to be generally applicable to *any domain whatever* that is solicited. Now this does indeed turn out to be the case, which reinforces the initial "deflationist" interpretations. The recent generalization of quantum theory, which is applicable to many domains in the human sciences [22] running from decision theory to semantics by way of the psychology of perception, is a remarkable illustration of this. It does not matter who or what *responds* (human beings or things), the probabilistic structure of the responses is the same. From this restricted point of view, a set of human beings making choices which depend on the options which are presented to them, and on the order of the decisions to be taken, behave exactly like a set of electrons on which one evaluates several incompatible observables [23]. A set of speakers who have to decide on the meaning of a polysemic word, according to the propositional contexts, thus behaves exactly like a set of microscopic particles which violate the Bell inequalities [24]. There is nothing shocking about the fact that it should be so and implies strictly nothing about any community at the level of their *profound being* between electrons and humans; there is only a formal isomorphism in their situation and their "surface" reactions to being solicited.

To recapitulate, the nonessentialist conception of quantum mechanics as being doubly "superficial" (both superficially phenomenal and superficially probabilistic) makes it possible to dissolve away what are alleged to be the major paradoxes of this theory, to explain a large part of its effectiveness, and to promote its universality. As if this were not enough, one can add that this conception also maintains a remarkable degree of notional and mathematical simplicity, which contrast strongly with the ever-increasing sophistication of those ideas which aim at saving a "realist" interpretation of quantum physics. Why, under these conditions, is this conception not more widely accepted? Why does it so often find itself opposed by the indignant reactions of certain physicists who reproach it with "betraying the ideal of science," of "breaking the great dream of knowledge" [25], of being unacceptable or even "scandalous" [26]? Why, even when indignation is absent, does the exposition of the minimalist conception of quantum theory give rise to a resigned silence which manifestly expresses a profound disappointment? There is no doubt that it is because, as we have felt it coming since our introduction, we are dealing with a breach of several contracts at the level of a whole civilization. One of these is a fairly recent contract which, from the sixteenth century onwards, has instigated a collusion between the desire for a metaphysical breakthrough upheld by the clerks and the need for technological perfectionism of the craftsmen [27]. Another is a very ancient contract which has made it an obligation to seek a principle of understanding appearances in the inmost depths of things [28]. If scientific progress does not help our gaze to penetrate to the very heart of material bodies, and to *definitively guarantee* technological effectiveness by laying bare their secret, what is the point of it? If the progress of knowledge amounts merely to a kaleidoscopic deployment of the phenomenal skin of things, instead of opening up a vision of their very flesh and marrow, does it not seem in vain? It is all very well to recall that all the entities which, in the history of science, were pompously dignified at the time by the title of "realities behind appearances" have turned out to be themselves a matter of (1) *other* appearances (or phenomena) revealed by a new approach, postponing the revelation of what Goethe called the "Urphänomen" [29] to an indefinitely remote utopia, or (2) mathematical idealities which express some invariants of the phenomena reconstructed by the intelligence. This simple reminder is not enough. The "dream of reason" pursues its course; this same dream that Kant upheld at the beginning of his quest, before discarding it in his critical philosophy: the dream of managing to grasp by thought a "representation of things *as they are*" [30]. The only possibility to suspend this vain quest would be to change our cultural representations, by opening it to alternative cultural areas. The Indo-Buddhist cultural area, with its (anti-) concept of *Śūnyatā*, is one of the best possible options for this sake.

So, let us now compare the modalities and motivations of both Buddhist and quantum criticisms of metaphysics.[4] We will start our further inquiry in Sect. 4, before the era of Mādhyamaka for Buddhism and before the era of quantum mechanics in physics. This will make clear that both Mādhyamaka and quantum physics represent achievements in a long process of emancipation with respect to the naive beliefs of ordinary life. Section 5 will develop the central parallel between quantum entanglement (or non-separability) and dependent co-arising. In Sect. 6, a deliberately deflationary, non-speculative, reading of this parallel will be presented. Finally, in Sect. 7, a suggestion about the common root of the Buddhist and quantum critiques of metaphysics will be made.

4 Historical Development of the Theories of Knowledge in Buddhism and in Classical Physics

The intellectual history of both physics and Buddhism went through remarkably similar steps that can be enumerated as follows:

1. Transition from the paradigm of substance to the paradigm of causality
2. Replacement of productive causality by law-like successions
3. Transition from causal and law-like successions to co-emergence
4. Criticism of ontological construal of co-emergence and claim that co-emergence itself is relative to the cognitive act that posits it ("co-emergence co-emerges")
5. Silent return to the practices of life or agnostic return to the practices of experimental science

In Western science, the two first steps were taken by classical physics, whereas the three following steps had to wait for the birth of quantum physics. This is the

[4]HH the Dalai-Lama, *The Universe in a Single Atom*, op. cit. p. 67: "Each of these (physical) pictures is excellent in its own right and for the purpose for which it has been designed, but if we believe any of these models to be constituted by intrinsically real things, we are bound to be disappointed."

reason why this section will be devoted to the two first steps, whereas the following steps will be discussed in the next sections.

The first step from substance to causality, to begin with, was taken very early by Buddhism. The primary ambition of the concept of *pratītyasamutpāda* was existential; it was to describe (across the 12 links of dependent origination) the inexorable chain of consequences from acts to retribution or from "ignorance" (*avidyā*) to aging and death. Yet, it also had an important epistemological content, since (i) it involved a strong attack against the belief that substances and selves are permanent beings and (ii) it posited the general rule according to which every (impermanent) phenomenon arises as an effect of causes and conditions. At this stage of Buddhist thought, the use of the word "cause" evoked the concrete and familiar idea of an impelling force. Indeed, "cause" is the English translation of the Sanskrit "*Hetu*," deriving from the verbal stem "*Hi*," which essentially means "to throw ahead," "to hurl," and "to set in motion." So, one could still figure out that an effect is somehow "produced" by its causes and conditions, in the sense that causes give an impulsion that triggers the effect under certain conditions. This elementary conception of causation is precisely that which came under strong pressure with Mādhyamaka. But even in this elementary form, it performed a very important epistemological task: it showed that the apparent regularities we witness in ordinary life can be explained without assuming that *there are* enduring and intrinsically existent "things" or substances. To explain these regularities, it is enough to assume a strict causal connection between successive instantaneous events. In other terms, *causes replace substances.*

In Western science, a similar turn (though less complete) was taken during the seventeenth century, with the physics of Galileo and Descartes. But let me first sketch what was the conception of science in earlier days, from Aristotle to the middle ages. According to Aristotle, the aim of science is to show that given predicates *inherently* pertain to certain "subjects" or substances. This strong connection between each predicate and the corresponding substance is established through *reasoning*. In other terms, it is established through a chain of propositions that starts from a set of axioms (namely, from a set of accepted propositions of the same type as the one which has to be derived) and stops with the proposition ascribing the predicate to the relevant substance. Reasoning usually involves two levels of generality. At the upper level of generality, one establishes that a certain "essential" predicate is inherent to a certain species. At the lower level, this predicate is ascribed to an *individual* belonging to that species. During the middle ages, Aristotle's emphasis on substance as the foundational category of being, and on inherence of essential predicates to substances, was taken for granted, and many consequences were (often abusively) drawn from it. One consequence was the pervasive use of the concept of "substantial form": a basic, essential, characteristic of substance that persists unaltered through change.

René Descartes (the French philosopher, mathematician, and physicist of the first half of the seventeenth century) eagerly criticized the concept of "substantial form." He insisted that "Substantial forms cannot offer us a strong reason for any natural process, since even their supporters confess that they are occult and that they do not understand them" [31]. In other terms, substantial forms (namely, inherent, essential properties of substances) are fake explanations of phenomena. Instead of them, Descartes advocated mechanical explanations in which only spatial bulk and velocity are involved. A typical mechanical explanation then relied on the description of motion and collisions of extended bodies. Accordingly, Descartes developed at length the description of how "(. . .) the bodies that collide change the state of motion of one another" [32]. From then on, phenomena were explained by calculating the sudden impulses that bodies receive from other bodies that bump into them. Causes, in the most straightforward sense of thrust or impulsion, were endowed with the explanatory power which was formerly ascribed to substances. Substances were still there at this initial stage of classical physics, but they had no explanatory role by themselves; only their *relations* of mutual contact had one.

This tendency toward more emphasis on causes and less on substances was amplified by substitution of laws for productive causes.

In Buddhism, this substitution was rather obvious from the beginning. It was only amplified to its ultimate consequences by Mādhyamaka, against the residual reifications of Abhidharma. A well-known statement attributed to the Buddha himself, and quoted in many *Sūtras*, is the following[5]:

> "When this is present, that is present.
> From the arising of this, that arises.
> When this is absent, that is absent.
> On the cessation of this, that ceases."

In the former sentences, there is no question of production, generation, or alteration, only of association and correlation. This restriction to the immanence of (correlated) phenomena, as opposed to the transcendence of productive powers, has often been compared to Hume's empiricist theory of causality. Indeed, according to Hume, the notions of "power" or "productive cause" are "obscure"; experience, he writes in his *A Treatise of Human Nature*, teaches us the usual association of phenomena but not their generative causes (if any).

Now, Hume was expressing nothing else than the lesson he had gathered from his knowledge of Newtonian physics. True, at first sight, Newton's theory of gravitation represented a metaphysical regression when compared to Descartes, since it appeared to use the notion of an "occult quality" called the "attractive power" of massive bodies. No wonder that many French Cartesian philosophers of the late seventeenth century and early eighteenth century eagerly criticized Newton on this ground. But in fact, Newton carefully avoided to rely on speculative claims; even though he had developed a very elaborate metaphysics of space and material bodies during his early career, he was eager to make his mathematical physics virtually independent of it. According to him, the really important thing in mechanics is the set of *mathematical laws* that connect and unify (astronomical and earthly) phenomena. Throughout his *Philosophiae Naturalis Principia Mathematica*, he insisted that

[5]*Samyutta Nikāya*, II, p. 10, quoted by: *Encyclopedia of Buddhism*, IV, 1, p. 2, Sri Lanka Government Printing, 1979.

when he used words like "attraction" or "impulse," he did not wish to assert something about "the species of forces and their physical qualities, but their quantities and mathematical proportions."[6] As for the mathematical laws themselves, they were considered by Newton as little more than an expression of the observed regularities of successive phenomena. In his fourth "rule for the study of natural philosophy," Newton thus pointed out that "In experimental philosophy, propositions gathered from phenomena by induction should be considered either exactly or very nearly true (...)."[7] At the end of the eighteenth century (a few decades after Hume), Kant corrected this claim by pointing out that, actually, Newton's mathematical laws derive as much from the constructive power of our understanding as from pure experience. But this remark does not need to be developed any further in this context.

Our central concern is only to notice the shift of the concept of cause in physics, from "productive power" to mere regular succession. Indeed, this shift has crucial consequences, such as (here again) additional weakening of the substantialist view of the world. Moritz Schlick, a philosopher of the Vienna Circle at the beginning of the twentieth century, expressed these consequences as clearly as possible. According to him, the concept of substance (which expresses permanence of certain feature of phenomena) is *superfluous* as soon as universal mathematical laws of regular succession are formulated: "Each time one evokes substance in physics, one asserts nothing else than the existence of a specific regularity" [33].

5 Quantum Non-separability and *Pratītyasamutpāda*

We have just seen how classical physics progressively underrated the concept of substance (which only subsisted in the mathematical and non-individual form of the principles of conservation of mass, energy, charge, etc.) and simultaneously abstracted a concept of regularity law from the ordinary concept of causality. This closely parallels the Buddhist critique of the concepts of substance and productive causality.

However, nothing prevented a classical physicist from speaking (in ordinary language) *as if* the concept of individual substance were still in use, and nothing prevented him/her from providing partial explanations *as if* the concept of productive causality had not been replaced with law-like regular successions. By contrast, quantum physics brought to radical completion the critique of the concepts of individual substance and causality; so much so that it became utterly artificial to use them, even for the sake of verbal communication. This achievement of the

---

[6]I. Newton, *The Principia (Mathematical Principles of Natural Philosophy)*, University of California Press, 1999, Book I, sect. 11, Scholium, p. 588.

[7]I. Newton, *The Principia (Mathematical Principles of Natural Philosophy)*, p. 796.

critique here again closely parallels similar steps in Buddhist thought, as we will now see.

One crucial point of the quantum paradigm is the strict correlativity of attributes to one another and of entities to one another, so that the terms of a certain type of relation – let's call them its *relata* – cannot even be ascribed an existence independently of the relation that connects them (step 3 of our list in the previous section). But before we document this striking feature of quantum physics, it is worth stating its Buddhist analogue straightaway.

According to several early Buddhist texts, the conditioning which unfolds through the 12 links of dependent origination is not necessarily one-way; it is likely to be *mutual*: "The conditioning relation can be reciprocal."[8] One argument in favor of this reciprocity is the cyclic character of the chain of dependent origination. But the best way to get a flavor of what is at stake in mutual conditioning is to read the beautiful metaphor which illustrates it: "(. . .) It is just as if there stood two sheaves of reeds leaning one against the other (. . .) If I were to pull towards me one of those sheaves of reeds, the other would fall; if I were to pull towards me the other, the former would fall."[9] Here, the property "leaning against" of one sheave of reeds has no intrinsic reality, but only a relational reality: it co-arises with the corresponding property of the other sheave of reeds. Later on, Nāgārjuna and his successors of the Mādhyamaka-Prāsangika school made this identification of *Pratītyasamutpāda* with mutual co-arising even more systematic. They insisted on the equivalence of dependent origination and *Śūnyatā* (or lack of own-being), which was already stated in the *Prajñāpāramitā-sūtra*. They also indicated indirectly, through the etymology of several quasi-synonyms of *Pratītyasamutpāda*, that dependent origination is tantamount to a mode of being (or rather to an intermediate status between being and not-being), *not* to a mere causal connection between preexistent beings. One example is the word *idampratyayamātra*, which means literally "the measure (or extent) of going-with-this." This word is used by Candrakīrti in the following sentence: "things exist to the extent of their going-with-this."[10] In other terms, things have no other form of being than their going with other things or their being related to other things. Other examples are the terms *paratantra* (literally "woven-of-the-other" as opposed to *svatantra*, "woven of itself") and *paraparasiddha* (literally "established by one another"). Both words are used by Nāgārjuna to characterize dependent origination and ultimately *Śūnyatā*.[11] One can understand provisionally these words as saying that interdependent "beings" are replaced by interbeing or that the mode of being of interdependent "beings" is nothing beyond interdependence itself.

---

[8]Katthāvatthu, XV, 2, quoted by J. Macy, *Mutual Causality in Buddhism and General System Theory*, SUNY Press, 1991, p. 57.

[9]Samyutta-Nikāya, II, 114, quoted by J. Macy, *Mutual Causality in Buddhism and General System Theory*, op. cit. p. 55.

[10]Candrakīrti, *Madhyamakāvatāra*, Trad. L. de la Vallée-Poussin, Le Muséon, 1911, p. 277–278, quoted by L. Viévard, *Vacuité et compassion dans le bouddhisme madhyamaka*, De Boccard, 2002, p. 44.

[11]Nāgārjuna, *Acintyastava* (42–43), in C. Lindtner, *Master of Wisdom*, op. cit. p. 27.

Co-arising or co-relativity of properties and entities is also one of the most striking characteristics of quantum mechanics. It was taken as a central theme around 1935, in papers of Schrödinger [34] and Einstein, Podolsky, and Rosen.[12] Schrödinger's characterization of what he called the "entanglement" of the states of the microphysical systems runs thus:

> When two systems, of which we know the states by their respective (state vectors), enter into temporary interaction due to known forces between them, and when after a time of mutual influence the systems separate again, then they can no longer be described in the same way as before, viz. by endowing each of them with a (state vector) of its own. I would not call that *one* but rather *the* characteristic trait of quantum mechanics, the one that enforces its entire departure from classical lines of thought.[13]

In other terms, after an interaction has taken place between them, the two systems can no longer be ascribed two separate states (they are non-separable); their quantum states are "entangled" insofar as one can only define a single global state by combining them and ascribe this global state to the whole. Schrödinger summarized this situation by pointing out that "the best possible knowledge of a whole does not necessarily include the best knowledge of all its parts, even though they may be entirely separated."[14] Accordingly, there has been a widespread temptation to express quantum entanglement in terms of a doctrine of holism. But a more immediate and less speculative empirical consequence of quantum entanglement is the strict *correlation* of certain observable properties measured on entangled subsystems. This correlation is so strong that it can only be interpreted in terms of co-relativity of these properties. Let me give an example. A pair of spin $-1/2$ particles is prepared in a global state corresponding to an overall z-component of spin equal to 0. The individual states of these two particles are then entangled. In this case, whenever one finds that the z-component of spin of one particle is $+1/2$, a measurement of the z-component of spin of the other particle is bound to yield the value $-1/2$. Conversely, whenever one finds that the z-component of spin of one particle is $-1/2$, a measurement of the z-component of spin of the other particle is bound to yield the value $+1/2$. The crucial point is that, *before any measurement has been performed*, there is no sense in ascribing a property (or a value of the z-component of spin) to each particle taken in isolation. This means that the property "z-component of spin" has *no existence of its own* in each particle, but only *relative to* the corresponding property in the other particle. These properties of microscopic

[12]A. Einstein, B. Podolsky and N. Rosen, "Can quantum-mechanical description of physical reality be considered complete?" *Physical Review*, 47, 777–780, 1935. About the relation between the EPR argument and dependent origination, see HH the Dalai-Lama, *The Universe in a Single Atom*, op. cit. p. 64.

[13]E. Schrödinger, "Discussion of probability relations between separated systems," *Mathematical Proceedings of the Cambridge Philosophical Society*, 31, 555–563, 1935; see a comment in: H.J. Treder and H.H. von Borzeszkowski, "Interference and interaction in Schrödinger's wave mechanics," *Foundations of Physics*, 18, 77–93, 1988.

[14]E. Schrödinger, "Discussion of probability relations between separated systems," loc. cit.

particles clearly behave as the property "leaning toward" of sheaves of reeds used to illustrate the reciprocity of *pratītyasamutpāda*.

David Mermin built a complete interpretation of quantum mechanics on the basis of this remark. According to him, "Correlations have physical reality; that which they correlate does not" [35]. In other terms, in quantum physics, *there exist correlations between properties, whereas the correlated properties have no existence of their own*. However, the claim that "correlations exist" in the world out there, or, as Mermin writes, that "correlations are fundamental, irreducible and objective" [36], is itself very dubious. This will be shown below, when we come to the fourth step of our parallel between dependent origination and quantum entanglement.

But before we criticize Mermin's *relational realist* claim, we must amplify the scope of quantum entanglement by showing that not only *properties* but also *entities* are made relational by it. Let us consider again a pair of spin $-1/2$ particles with entangled spin states. In addition, let us suppose that these particles are indiscernible, namely, that all the properties they have apart from spin are identical. If the two particles are, say, electrons, they have exactly the same mass and charge. Moreover, both can be in a spatial state for which the distribution of probability of their position in space is exactly the same. One can then distinguish them neither by their masses and charges nor by their pre-measurement position; and the list of criteria of distinction stops there. In virtue of Leibniz's *Principle of Identity of Indiscernibles*, one is tempted to declare that since there is no way to distinguish these electrons, they in fact constitute one and the same object. Yet, physicists still say that they are *two* electrons, not *one* object. How can we justify their way of speaking? By invoking Quine's principle of "weak discernibility" [37]: two objects are weakly discernible if they stand in an irreflexive relation, namely, a relation that each object cannot have with itself.

A well-known illustration is provided by imagining a universe made of only two metallic spheres. The spheres have all their properties in common, which, in view of Leibniz's Principle of Identity of Indiscernibles, threatens our initial claim that there are *two* of them. However, the spheres also stand to each other in the relation "being at a distance of 10 meters from one another." This relation is clearly irreflexive because no sphere can be at a distance of 10 m from itself. It is then in virtue of such an irreflexive relation that the spheres can and should be said to be *two*, not *one*. Another, more relevant, illustration is provided by our two indiscernible but entangled electrons. They can be distinguished by no intrinsic property, not even by the putative property "having a given value of the z-component of the spin," since, as we have seen, no such property can be ascribed to one electron taken in isolation. However, they can be said to be two entities, not one, because they stand to each other in the *relation* "having an opposite value of the z-component of their spin." This relation is irreflexive, once again, because no electron can have an opposite value of the z-component of its own spin. As a consequence, not only the spin-component but also the *identity* of electrons with entangled quantum states is relational rather than intrinsic. Even their existence *qua* distinct entities is relational rather than intrinsic! In the same way as the mode of being of interdependent "beings" is nothing beyond interdependence, here the mode of being of entangled indiscernible particles is nothing beyond the irreflexive relation that arises from their entanglement.

Now, some people can still think that the co-arising, co-emergence, or strict co-relativity, which characterizes both dependent origination and quantum entanglement, is underpinned by causal influences in which the effect comes later than the cause (even though the delay between cause and effect might be imperceptibly small). Is it so? Interestingly, the answer to this question seems to be entirely negative *in the two domains we are comparing.*

To begin with, Nāgārjuna eagerly criticized the common sense features of the concept of causality and especially the idea that a cause exists before its effect. Here is one of his sharpest remarks about this point: "A cause has an effect when there is an effect, but when there is no (effect) the cause amounts to no cause."[15] Several commentaries, based on the well-known example of the seed and the sprout (the seed which has disappeared when the sprout appears and the sprout which does not exist when the seed exists) [38], have developed the meaning of this concentrated sentence. In short, there is no point calling something a cause *before* the effect in dependence of which it arises has manifested itself. But there is no cause either *after* the effect has manifested itself, since, if it has finished its work, or if it has disappeared in the process, there is no point calling it a cause *at that moment.* The basis of the argument is that no *present* experience can hold together the cause and the *subsequent* effect (except indirectly, by means of memory and interpretation, or by means of mental projection). In addition to this, the very idea of "producing" something, of having an effect "produced" by a cause, looks incorrect or artificial when what arises is *empty of own-being*: "When neither existents nor non-existents, nor existent non-existents are established, how could one propose a 'productive cause'" [39] Co-relativity and interbeing are not only distinct from causation developing in time and production of being; they make them pointless.

In quantum physics, the conclusion turns out to be very similar. A group of physicists of the University of Geneva (Switzerland) recently wondered whether the correlations that manifest between microsystems with entangled states can be ascribed to faster-than-light causal influences [40]. They pointed out that a necessary condition for such causal influences to make sense would be that causes occur earlier than their effects. Accordingly, they arranged their device in order to test the time-ordering of events by relying on relativistic effects. And they found that the outcome of the experiment is *incompatible* with the very idea of causal influences. There are correlations, but nothing can be said to *cause* them. This result shook the initial conviction of the authors of the experiment, and prompted them, here again, to adopt a holistic metaphysics: according to them, the experiment proves that the world is an inseparable whole. But, as we will now see, other (non-metaphysical) approaches of this perfectly reciprocal and non-causal kind of co-relativity are available.

---

[15]Nāgārjuna, *Sūnyatāsaptati* (6), in: C. Lindtner, *Master of Wisdom*, op. cit. p. 97.

6 Quantum Non-separability, *Pratītyasamutpāda*, and *Śūnyatā*: A Deflationary Perspective

A well-documented temptation at this stage is to claim that one has elucidated the nature of the world. In a Buddhist perspective, the temptation would amount to declare that the ultimate nature of reality *is* emptiness, that the world *is* a process involving a meshwork of dependently arisen phenomena. In physics, the temptation consists in saying that the world *is* an inseparable whole existing below the level of space and time, or that reality *is* pure structure, or that the relations *are* real, while the *relata* are not. In other terms, in both cases one tries to elaborate a new metaphysics by reifying the very tool used for the criticism of old metaphysics. But in both cases also, this attempt fails.

The Madhyamika school, to begin with, is very careful to criticize in advance any reification of emptiness or dependent origination, which can be so tempting as a replacement of the reification of entities. According to Nāgārjuna, "If there is no essence (*svabhāva*), there is no other-essence (*parabhāva*)".[16] According to Stcherbatsky's translation [41], which has pedagogical virtues, this means that if self-existence is lacking, relational existence is also lacking. In other terms, *the idea of relational existence is relative to its dialectical opposite*, namely, to the (conceptual) possibility of absolute existence. Interdependence (*Pratītyasamutpāda*) is no absolute either, but only one term in a dual system of oppositions (here, the opposition with independence). Interdependence itself is dependently arisen; *emptiness itself is empty* (*Śūnyatā is Śūnya*).

In addition to that, any system of oppositions underpinning dependent origination is in turn relative to a certain conceptual scheme or to a certain pattern of discriminative criteria. A standard example, discussed by Candrakīrti,[17] is the couple of cart and wheel. Cart is relative to wheels because without wheels it would not be a cart but only (say) a wooden box; wheels are relative to cart, because without a cart they would not be wheels but only (say) wooden circles or wooden hoops. But in addition to that, both carts and wheels are relative to their corresponding concepts and to the conceptual framework which incorporates them. For, if it were not for their concept, and the standard use of them which is made according to this concept, they would just "have" *thusness* (*tathāta*).

To recapitulate, it is clear that, in virtue of its universal applicability, *Śūnyatā* is no self-existent "thing" to be substituted to ordinary things: "If there were to be something non-empty, there would then be something called empty. However, there is nothing that is non-empty. How could there be something empty?"[18]

---

[16]Nāgārjuna, *Mūlamadhyamakakarikā*, I, 3, in J. Garfield, *The Fundamental Wisdom of the Middle Way*, op. cit.

[17]Candrakīrti, *Prasannapdā*, quoted by: C.N. Huntington & Geshé Namgyal Wangchen, *The Emptiness of Emptiness*, University of Hawaii Press, 1989, p. 50.

[18]Nāgārjuna, *Mūlamadhyamakakarikā*, XIII, 7. Translated by D. Kalupahana, Motilal Banarsidass, 1996.

A similar series of criticisms of relational being has recently developed in quantum physics. The major target of the criticism is Mermin's claim that "correlations have physical reality" or that "correlations are fundamental, irreducible and objective." Mermin explicitly considers that correlations are *local* element of reality, namely, that they are not influenced by distant correlations. In the same way, properties of particles were taken by Einstein as local elements of reality, not influenced by distant correlations. But we know that the very assumption that particles are endowed with properties construed as local elements of reality is inacceptable in virtue of the following reasoning:

(a) Such properties are connected by Bell's inequalities.
(b) Bell's inequalities are experimentally violated.
(c) Therefore, local realism is falsified.

A. Cabello [42] then wondered whether Mermin's *relational* local elements of reality can really replace Einstein's local properties. He wondered, in other terms, whether local realism could still be defended provided the predicate of "local reality" is ascribed to *relations* rather than monadic properties. But the outcome of his investigation was definitely negative. Cabello found that no quantum correlation can be isolated from the overall network of quantum correlations that constitute a physical system, irrespective of the spatial distance between the various parts of the system. Relational local realism thus fares no better than property local realism. Later on, as we have seen earlier, Smerlak and Rovelli [43] gave strong arguments in favor of interpreting this failure of relational local realism not as a failure of locality but as a failure of realism. According to them, no feature of physical systems, *not even their correlation encoded in quantum entanglement*, has intrinsic reality. Any such quantum feature is *relative* to a certain sequence of experimental procedures worked out by specific observers. In other terms, *even the relations (between quantum subsystems) are relative (to experimental conditions). Just as, in Mādhyamaka, emptiness of own-being is itself empty of own-being, in quantum physics relational being is itself relative.*

This all-pervasive ontological criticism is based on an equally strong criticism of the common sense concept of *time* and *succession*, both in Buddhism and in quantum physics. After all, time, substance and causality are deeply related notions. Substance can be defined as an enduring being, one which does not change in time when (its) attributes change. Conversely, Kant [44] argued that we can perceive time determinations only by reference to (and by contrast with) something permanent that we call *substance*. Similar conclusions hold for causality. On the one hand, causality relies on *succession* of events considered as causes and effects. On the other hand, it has been proposed, in view of the theory of Special Relativity, to base the concept of succession on causal connections [45]. So, we can easily understand that any attack on the ordinary concept of time is bound to have a devastating effect on the concepts of substance and causality.

In Nāgārjuna's work, the criticism of the ordinary concept of time and motion develops in Chapters II, VII, and XIX of *Mūlamadhyamikakārikā*. There, we find roughly three kinds of arguments against the intrinsic reality of time:

It is impossible to detach time and motion from their substrate (if any): The motion, the mover, and the path cannot be separated from one another except verbally and for practical purposes. The mover would not exist without motion, and motion does not exist without a mover. The motion and the mover cannot be separated, even though they cannot be said to be identical either. This kind of argument has been compared with similar ideas of the Greek Megaric philosophers, according to whom the mover cannot be separated from *ongoing* motion.[19]

Motion cannot be defined at a single instant or in a single point. Indeed, one cannot find a point of space where motion *begins*, and one cannot even find a point of space where it merely *occurs*. Nāgārjuna insists on this problem in several parts of his treatise.[20] He also criticizes the very idea of a beginning, by pointing out that a beginning must itself begin[21] and so on ad infinitum. His arguments are remarkably akin to those of the Greek philosopher Zeno of Elea.

Temporal projections going beyond the lived present (to past or future) have no reality. This being granted, the very notion of motion is threatened: "What has been moved is not moving, what has not (yet) been moved is not moving."[22] Here, a partial Western equivalent is St Augustine [46], according to whom time does not exist because the past is no longer, the future is not yet, and the present has no other definition than being a separation between past and future.

In view of this thorough critique of time, there are even more reasons to understand dependent origination as co-arising rather than as a true process of "production" in time.

A similarly articulated criticism of time, succession, and standard causality has been offered as a consequence of quantum physics. The argument was first developed in an often-discussed paper of D. Page and W. Wootters [47] and then amplified in cosmological context by J. Barbour [48]. Their idea is that (a) systematic inventory of simultaneous states can *replace* the study of the succession of states and (b) in quantum mechanics simultaneity is expressed by the *entanglement* of states of systems with the state of a standard called "the clock." The most striking result of these authors is that the equation of evolution of states (e.g., the Schrödinger equation in standard quantum mechanics) can easily be derived by calculating the expectation value of any observable for a given clock state. In other terms, careful consideration of the quantum equivalent of simultaneity *replaces* the concept of evolution in time.

To sum up:

- In quantum physics, time appears as a clumsy statement of a network of relations of simultaneous co-arising of potential values of observables (as expressed by the entanglement of states).

---

[19]Aristotle, *Metaphysics*, 1046b 29.

[20]Nāgārjuna, *Mūlamadhyamakakarikā*, II, 14–15, in J. Garfield, *The Fundamental Wisdom of the Middle Way*, op. cit.

[21]Nāgārjuna, *Mūlamadhyamakakarikā*, VII, 13, 17–20.

[22]Nāgārjuna, *Mūlamadhyamakakarikā*, II, 1.

- In the framework of Rovelli's relational interpretation of quantum mechanics [49], this simultaneous co-arising of potential values holds *relative* to the present project of prediction of an observer.

Here again, time and succession recede to the background, whereas simultaneity, presence, and co-arising tend to replace them.

At this point, we are ready to tackle a recurring argument which seems to undermine the whole parallel we have just established, since it tends to display a major *discrepancy* between Mādhyamaka deconstructive philosophy of *Śūnyatā* and the quantum deconstruction of ontology. This argument is nothing less than quantum indeterminism, interpreted as meaning that certain physical events occur with *no cause at all.* Now, this possibility of occurring with *no* cause is explicitly excluded, together with several other possibilities, by the very first *kārikā* of Nāgārjuna's treatise. Is there any possibility of reconciling these apparently conflicting statements? I believe there is, provided we keep in mind that the ordinary concept of causality is under pressure not only in quantum mechanics but also in Mādhyamaka.[23] We have seen that the central teaching of Madhyamaka is dependent co-arising, itself dependently co-arisen. Ordinary (productive, successive) causality is then criticized in view of the implications of dependent co-arising. Similarly, the central features of quantum physics are (a) entanglement, (b) relativity of entangled states to present projects of prediction of observers, and (c) relativity of the values of observables to the actual workings of experimental devices. A very remarkable fact about quantum physics is that the failure of ordinary causality can be shown to be a consequence of feature (c). Indeed, it was demonstrated by P. Destouches-Février that any predictive theory bearing on phenomena defined *relative* to possibly incompatible experimental contexts is "essentially indeterminist." [50] To sum up, in quantum physics, the invalidity of ordinary productive causality *follows* from the relativity (or context dependence) of phenomena.

7 The Common Root of Quantum Entanglement and Dependent Origination: Deep Cognitive Interdependence!

We can now attempt an explanation of the former analogy between quantum entanglement and Buddhist *Pratītyasamutpāda.* Once again, it must be clear that what we have to explain is not the similarity of two views of the world obtained with two different modes of access (experimental and experiential, technological and contemplative, external and internal). What we must understand is the reason why two radical *deconstructions of metaphysical views* turn out to be so similar; we must

[23]HH the Dalai-Lama, *The Universe in a Single Atom*, op. cit. p. 68.

understand what motivated the strong convergence of these two deconstructions and of these two criticisms.

One simple and general way to formulate the reason why the Buddhist and the Quantum deconstructions of metaphysics are so similar is to notice the similarity of the cognitive situations they have to tackle. In both cases, in the contemplative inquiry of Buddhism and in the experimental inquiry of quantum physics, the knower and the known cannot be entirely separated. In contemplative practice this is most obvious, because the alleged "object" of the study (the workings of one's own mind) is not truly different from its alleged "subject"; so much so that any form of dualism is doomed to failure from the outset. In quantum physics, this is equally clear, in view of the insuperable *contextuality* of properties and entities. Another, more concrete, expression of the same point was offered by David Finkelstein [51]. This author suggested that the similarities of Buddhism and quantum physics might well be due to the hypersensitivity of their respective domains of investigation to cognitive probing. When we observe a thought, it disappears, and when we observe a microstate, it is "reduced," "collapsed," or "projected." In both cases, events cannot be dissociated from the (mental or experimental) circumstances of their occurrence. Since these circumstances are highly variable and are not entirely under conscious or technical control, an irreducible component of change is involved: impermanence in Buddhism and sudden "collapses" in quantum physics.

But with this irreducible component of change, the central human method for identifying properties and entities is at least partly challenged. Indeed, this central method, first described by Kant and universally applied in physics, is the quest for *invariants*; it is the quest for features that do not vary from one situation to another, from one circumstance to another. In view of the irreducible component of change we have just mentioned, the search for invariants must either be renounced to or deflected at a level where changes can be neglected for all practical purposes.

In Buddhism, this strategy of deflecting the search for (approximative) invariants at a coarse level of analysis is well known. Madhyamaka authors accept that we can speak of permanent material bodies, ascribe properties to these bodies, and consider one body as the productive cause of an alteration of the properties of another body, as part of the *conventional truth* useful in everyday life. But when one pays exquisite attention to the flux of appearances, and to the subtle variety of phenomena, it becomes obvious that this kind of truth is indeed only conventional, and that it is based on an approximation.

8 Summary and Conclusion

We have just confirmed that there is indeed a thorough and detailed analogy between quantum contextuality and Buddhist *Śūnyatā* and between quantum entanglement and Buddhist *Pratītyasamutpāda*. But this twofold analogy does *not* show that there is a single essence of reality which can be disclosed by our reason irrespective of whether we rely on an experimental approach or a contemplative approach. The analogy rather shows that, at a sufficiently accurate level of analysis and careful attention, the *negative* conclusions of experimental physicists and contemplative Buddhists are bound to be similar: both realize the approximative status of the reified entities of everyday life, and both must cope with the high amount of instability and lack of self-sufficient existence (*Śūnyatā*) of phenomena. Then, if thoroughly applied, the critical concept of *Śūnyatā* can be seen to underpin the anti-metaphysical stances of both Buddhism and contemporary philosophy of science.

References

1. Einstein A. Oeuvres choisies, 1 Quanta. Seuil: CNRS Editions; 1989.
2. HH the Dalai-Lama. The universe in a single atom. Burlington: Morgan Road Books; 2005. p. 69.
3. Zajonc A, Greenstein. The quantum challenge: modern research on the foundations of quantum. Burlington: Jones & Bartlett Publishers; 1997.
4. Smerlak M, Rovelli C. "Relational EPR". Found Phys. 2007;37:427–445; Bitbol M. "An analysis of the Einstein-Podolsky-Rosen correlations in terms of events". Phys Lett. 1983;96A:66–70.
5. Mohrhoff U. "The world according to quantum mechanics (Or the 18 errors of Henry P. Stapp)". Found Phys. 2002;32(2):217–254; Bitbol M. "Consciousness, situations, and the measurement problem of quantum mechanics". NeuroQuantology. 2008;6:203–13.
6. Zeilinger A. « Foundational principle for quantum mechanics ». Found Phys. 1999;29:631–43; Fuchs CA. « Quantum mechanics (and only a little more) ». In: Khrennikov A, editors. Quantum theory: reconsideration of foundations. Växjo: Växjo University Press; 2002; Grinbaum A. « Elements of information-theoretic derivation of the formalism of quantum theory ». Int J Quantum Inf. 2003;1:289–300.
7. Gröblacher S, Paterek T, Kaltenbaek R, Brukner C, Zukowski M, Aspelmeyer M, Zeilinger A. An experimental test of non-local realism. Nature. 2007;446:871–5.
8. Lévy-Leblond JM, Balibar F. Quantique: rudiments. Paris: Interéditions; 1984.
9. Destouches-Février P. La structure des théories physiques. Paris: P.U.F; 1951.
10. Schrödinger E. "The present situation in quantum mechanics". In: Wheeler JA, Zurek WH, editors. Quantum theory and measurement. Princeton University Press: Princeton; [1935] 1983.
11. Lyre H. "Against measurement? – on the concept of information". In: Blanchard P, Jadczyk, A, editors. Quantum future: from volta and como to present and beyond. Berlin: Springer; 1999; Lyre H "Against measurement? – on the concept of information". In: Blanchard P, Jadczyk, A, editors. Quantum future: from volta and como to present and beyond. Berlin: Springer; 1999; Bitbol M. "Decoherence and the constitution of objectivity". In: Bitbol M, Kerszberg P, Petitot J, editors. Constituting objectivity: transcendental perspectives on modern physics. Berlin: Springer; 2009.
12. Bitbol M. La mécanique quantique comme théorie des probabilités généralisée. In: Klein E, Sacquin Y, editors. Prévision et probabilités dans les sciences. Paris: Editions Frontières; 1998.
13. Einstein A, Podolsky B, Rosen N. "Can quantum-mechanical description of reality be considered complete ?" In: Wheeler JA, Zurek WH, editors. Quantum theory and measurement, Princeton University Press: Princeton; [1935] 1983.
14. Espagnat B. d' (1994), Le réel voilé, Paris : Fayard.
15. Bell JS. Speakable and unspeakable in quantum mechanics. Cambridge: Cambridge University Press; 1987. Bell's theorem establishes the incompatibility of quantum mechanics with certain inequalities (the Bell inequalities) which inevitably result from theories with local hidden variables.

16. Suarez A. Quantum mechanics versus multisimultaneity in experiments with acousto-optic choice devices. Phys Lett. 2000;A269:293–302.
17. Bohm D. Wholeness and implicate order. London: Ark Paperbacks; 1984.
18. Bitbol M. "An analysis of the Einstein-Podolsky-Rosen correlations in terms of events". Phys Lett 1983;96A:66–70; Smerlak M, Rovelli C. "Relational EPR". Found Phys. 2007;37:427–45.
19. Brukner C, Zeilinger A. Information invariance and quantum probabilities. Found Phys. 2009;39:677–89.
20. Scully MO, Drühl K. "Quantum eraser: a proposed photon correlation experiment concerning observation and 'delayed choice' in quantum mechanics". Phys Rev A. 1982;25:2208–2213; Elitzur AC, Dolev S, Zeilinger A. "Time-reversed EPR and the choice of histories in quantum mechanics". In: Proceedings of XXIISolvay conference in physics. Singapore: World Scientific; 2003. p. 452–61.
21. Elitzur AC, Vaidman L. Quantum mechanical interaction-free measurements. Found Phys. 1993;23:987–97.
22. Bruza P, Sofge D, Lawless W, van Rijsbergen CJ, Klusch M, editors. Quantum interaction. Berlin: Springer; 2009a; Busemeyer J, Bruza P. Quantum models of cognition and decision. Cambridge: Cambridge University Press; 2014.
23. Zwirn H. Formalisme quantique et préférences indéterminées en théorie de la décision. In: Bitbol M, editor. Théorie quantique et sciences humaines. Paris: CNRS Editions; 2009.
24. Bruza PD, Kitto K, Nelson D, McEvoy C. Is there something quantum-like about the human mental lexicon? J Math Psychol. 2009b;53:362–77.
25. Stengers I. Cosmopolitiques 4: Mécanique quantique, la fin du rêve. Paris: La Découverte; 1997.
26. Thom R. Prédire n'est pas expliquer. Paris: Flammarion; 1993.
27. Scheler M. Problèmes de sociologie de la connaissance. Paris: Presses Universitaires de France; 1993.
28. Schrödinger E. Nature and the Greeks. Cambridge: Cambridge University Press; 1954.
29. Seamon D, Zajonc A, editors. Goethe's way of science. Albany: SUNY Press; 1998.
30. Kant I. Inaugural dissertation of 1770. Whitefish : Kessinger Publishing; [1770] 2004.
31. Descartes R. Œuvres Philosophiques IV. Paris: Hachette; 1835. p. 30.
32. Descartes R. Les principes de la philosophie II. In: Œuvres Philosophiques III. Paris: Garnier; 1973. p. 195.
33. Schlick M. Philosophical papers II (1925–1936). Dordrecht: Reidel; 1979. p. 26.
34. Schrödinger E. Discussion of probability relations between separated systems. Proc Camb Philos Soc. 1935;31:555–63.
35. Mermin ND. What is quantum mechanics trying to tell us ? Am J Phys. 1998;66:753–67.
36. Mermin ND. The Ithaca interpretation of quantum mechanics. Pramana. 1998;51:549–65.
37. Quine WVO., "Grades of discriminability". J Philos. 1976;73:113–16; Saunders S. "Physics and Leibniz's principles". In: Brading K, Castellani E, editors. Symmetries in physics: new reflections. Cambridge: Cambridge University Press; 2003.
38. Candrakīrti. Madhyamakāvatāra, 6, 17. In: Huntington CN, Wangchen GN, editors. The emptiness of emptiness. Hon- olulu: University of Hawaii Press; 1989. p. 159.
39. Nāgārjuna. Mūlamadhyamakakarikā, I, 7. In: Garfield J, editor. The fundamental wisdom of the middle way. Oxford: Oxford University Press; 1995.
40. Suarez A. « Quantum mechanics versus multisimultaneity in experiments with acousto-optic choice-devices ». Phys Lett. 2000;A269:293–302; Stefanov A, Zbinden H, Gisin N, Suarez A. « Quantum entanglement with acousto-optic modulators: 2-photon beatings and Bell experiments with moving beamsplitters ». Phys Rev. 2003;A67:042115.
41. Stcherbatsky T. The conception of Buddhist Nirvāna. Leningrad: Academy of Science of the USSR; 1927.
42. Cabello A. « Quantum correlations are not local elements of reality », loc. cit. ». Phys Rev. 1999;A59:113–15; Cabello A. « Quantum correlations are not contained in the initial state». Phys Rev. 1999;A 60:877–80; 1999.

43. Smerlak M, Rovelli C. Relational EPR. Found Phys. 2007;37:427–45.
44. Kant I. Critique of pure reason, B278. Indianapolis: Hackett; 1996. p. 291.
45. Mehlberg H. Time, causality, and the quantum theory I. Dordrecht: Reidel; 1980.
46. Augustine S. Confessions. Oxford: Oxford University Press; 1998.
47. Page DN, Wootters WK. « Evolution without evolution: dynamics described by stationary observables ». Phys Rev; 1983:D 27: 2885–92; also, Deutsch D. « Three experimental implications of the Everett interpretation » In: Penrose R, Isham IJ, editors. Quantum concepts in time and space. Oxford: Oxford University Press; 1986.
48. Barbour J. The end of time. London: Phoenix Paperbacks; 2000.
49. Rovelli C. "Relational quantum mechanics". Int J Theor Phys. 1996;35:1637–57; Smerlak M, Rovelli C. "Relational EPR", op. cit.
50. Destouches-Février P. La structure des théories physiques. Paris: Presses Universitaires de France; 1951. p. 260–80; see Bitbol M. "A cure for metaphysical illusions: kant, quantum mechanics and madhyamaka". In: Wallace BA, editors. Buddhism and science. Columbia: Columbia University Press; 2003.
51. Finkelstein DR. Emptiness and relativity. In: Wallace A, editor. Buddhism and science. Columbia: Columbia University Press; 2003.

Chapter 7
Śūnya, Śūnyatā, and Reality in Modern Physics

Herbert J. Bernstein

Abstract Quantum mechanics (QM) is the physics of atoms and their constituents. Under reductionism, QM should provide the solid reality for our world, as Einstein insisted. Instead, quantum properties depend strongly on their observer; they are empty (*śūnya*) until co-dependently created. So physics provides a time-dependent, co-emergent reality (which I designate reali*tty*) reminiscent of *śūnyatā*.

Yet physicists justify QM because "It works," begging for the questions: works for whom? to do what? Responding to these questions and similar ones in many fields, I helped start a small organization, the Institute for Science and Interdisciplinary Studies. Its philosophical program aims to help reconstruct knowledge for progressive purposes. The Institute's analysis provided insight into quantum teleportation and helped me invent a variant that US-NASA currently develops for communication from space.

The author is also a Buddhist Dzogchen practitioner. This essay speculates on the relation of *śūnya* and *śūnyatā* to scientific knowledge. Does the union of emptiness, cognizance, and compassion within dharmakāya [as the trikāya of dharmakāya, saṃbhogakāya, nirmāṇakāya] imply that Eastern concepts from Hinduism and Buddhism can provide a second source of inspiration for reconstructive knowledge?

Keywords Śūnyatā · Quantum reductionism · Quantum properties · Quantum teleportation · Buddhist Dzogchen · Dharmakāya

1 Introduction

Everything is made out of atoms. Atoms obey quantum mechanics, the physics of submicroscopic phenomena. Quantum mechanics, with its emphasis on the observer, on chance, and on what is in principle non-observable, can be intriguing. It is valid

H. J. Bernstein (✉)
Institute for Science/NS, School of Natural Sciences, Hampshire College, Amherst, MA, USA
e-mail: hbernstein@hampshire.edu

S. R. Bhatt (ed.), *Quantum Reality and Theory of Śūnya*,
https://doi.org/10.1007/978-981-13-1957-0_7

for the subatomic nucleus and for all subatomic particles, as well as for atoms. From the time of Planck's first use of quanta in 1900, quantum physics has been the field of "pure science" in which the dictates of logic, reason, causality, and reality itself have broken down in the most spectacular and disconcerting ways. So the orienting question for theoretical and experimental quantum physics remains: Is the micro world really fundamental and solid, or is it really empty and created by observation itself? Hard questions first raised in quantum theory about the limits to knowledge are now encountered in many fields.

My paper covers more than past and current quantum theory. The issues that interest me go deeper. They address the overarching framework of contemporary sciences, their operative epistemologies, and their applications and uses. These issues come up in many twenty-first century fields. As knowledge workers (people who create new ideas, new objects, whole new fields of inquiry and application) when we notice our "thumb on the scale"—the unavoidable observer effect—aren't we required to examine the implications to broaden our sense of responsibility? That question led me many years ago, as a scientist and consultant in Washington, DC, to expand the framework of concerns. My colleagues and I began the study groups at Institute for Policy Studies, Princeton, Santa Cruz, and Cambridge to pose some leading questions: Can we develop a twenty-first century way of knowing that acknowledges the creativity of determining *anything* with scientific precision and conventions? How would we reconstruct the new knowledge out of current practices? What is such a self-reflective reconstructive knowledge and how can knowledge workers in several fields begin to build it up? How does it go beyond the "value-free" myth, responding to complexities of history that are developing even now as we live/speak? How would the effort reconstruct our*selves*—and our societies—in the process? This essay is motivated in part by such questions, especially questions about connections between quantum properties and an emerging moral epistemology.

As a theoretical physicist who practices Tibetan Buddhist meditation, my paper draws on both technical understanding and personal experience. It is not the work of a scholar or theorist of Eastern philosophies but something of a meditation. And it draws on meditative experience, especially on the Tibetan Vajrayana methods of Dzogchen meditation on emptiness. In Dzogchen's highest form, the initiate is shown a glimpse of the so-called natural state of mind, a state which is beyond words, indeed beyond any conceptual understanding. Yet helpful words are uttered and written to describe this profoundly empty (*śūnya*), luminously cognizant state: a state thoroughly suffused with infinite and indiscriminate compassion. All beings, mild and wrathful, friendly or antagonistic, and vicious and good, are held in loving esteem and treated with beneficial wishes for their safety and happiness. Most importantly, for our purposes, this "Buddha mind" is entirely empty, with an emptiness (śūnyatā) that surpasses even the possibility of properties of objects or particles. That very emptiness is said to be both present fresh wakefulness and the ultimate timeless reality.

To many Buddhist theorists, appearance of compound entities such as objects, composed of particles with definite physical properties, is thus illusory: such appearances are caused by the ability of our mind(s) to grasp, even as mind apprehends ultimate reality. Empty Buddha mind is perfectly cognizant and grasps correctly; its nature is the same as the nature of mind in each sentient being. The apparent ordinary reality is there because our minds constantly look for it, grasp at things and properties in the void, forming them while conceiving ourselves as separate entities. When afflicted with grasping—attachment—mind becomes/creates/seems to be ordinary mind, and it sees ordinary reality. All seemingly independent beings and objects, all phenomena and properties, thus arise co-dependently on each other. Co-dependent apparent reality is an illusion, it is ultimate reality, but distorted: reality as grasped by our confused sentient-beings' minds.

This essay does not make an argument or try to prove a point. For the most part, it raises questions rather than answering them. In light of the parallels and despite cross-cultural multiple-century separations, the questions arise, "Is this śūnyatā of the ultimate Reality related to the emptiness of the **quantum** vacuum state?" More simply, "How is the co-dependent origination of properties that emerge from emptiness related to the fundamentally created nature of real properties in quantum physics?" Finally, as sciences leave out the values from their inquiries and suppress human moral judgments, "How is positive emptiness –kindly wishing well to all beings-- related to the emptiness of quantum properties, if at all?" From a meditative perspective, I ask "Does the Natural State of Mind, the presence of compassion and aspirations for the good of all beings implicit in Knowing, provide a model for modern moral or ethical reasoning—for a new more compassionate knowledge?"

1.1 Resources

Two quite personal resources inform the present effort: my book (hereafter ***MT***) *Muddling Through: Pursuing Sciences and Truths for the 21st Century*, co-authored with Mike Fortun, **and** the Institute for Science and Interdisciplinary Studies that we helped to found, together with many collaborators. The Institute is associated with reconstructive knowledge—championed by a leading political philosopher, Marcus Raskin, as a crucially necessary accompaniment of progressive policies.The book describes how today's sciences currently must (and how they actually *do*) work by muddling together several disparate ingredients, hence its title. **MT** reads like an ordinary analysis of sciences as social activity, implicitly full of the rites and rituals that we scientists perform, but it goes further. It characterizes especially the biological and physical sciences by gerunds: articulating, experimenting, judging, and powering/knowing. In *Muddling Through* we coined the term reali*ty*ty.[1] The word is an anagrammatic play on the eerie otherness, the alterity, of science's products.

[1]This was done in a conscious way to parallel French philosophical wordplay on alterité and realit*é*. The French works a little better, but I hope the allusion to that more exact anagram and our elaboration of it do the job, too.

Doing sciences one gets the strange feeling that something—not us, nor our all-too-human product—totally *other* is recognized every time a new scientific topic is successfully investigated. Indeed every time real scientific novelty is created or revealed, even before the investigation is complete. In the universal symbolism of scientific and mathematical literature, the strange italic *t* of our wordplay indicates time. Modern lived reality is not really fixed. It varies with time as scientific research continues to generate new knowledge and the micro-reality changes with *t* whenever measurement occurs or a quantity becomes known by whatever.

MT also reviews lessons learned in the first years of the Institute's operations. The Institute projects fit together as rooms in a veritable "house of experiment" to borrow a term from the time of the scientific revolution. An experiment in changing and reconstructing applied sciences (and of applying sciences reconstructively) for several different cases: in physical and biological sciences, in history and sociology of science, in first world and third world locations, in ecology and economy related fields, and so forth. Our central idea was to try things that changed how questions were generated, how self-consciously the science was developed, and how integrated the end uses were with the underlying theoretical investigations.

We picked projects of all types. The quick list includes military nuclear and toxic waste cleanup, Secoya indigenous aquaculture in Amazonia, monthly Bohm Dialogues,[2] social studies of genomics, and quantum teleportation; this last one leads directly to the physics basis for my discussion.

1.2 Quantum Teleportation (QT)

Quantum teleportation is my scientific subspecialty. It sounds jazzy—a more sedate title might be remote state exchange. QT is simple to summarize. Two participators—conventionally known as Alice and Bob, who are trying to communicate clandestinely and accurately—succeed spectacularly. It's certainly not an effect with *Star Trek* power to "beam me up, Scotty" that feat is quite impossible. But QT is important for quantum communication and quantum computation. It uses a combination of effects to do something really fantastic: a single particle comes to Alice in a rare one-in-a-million state. She never finds out which one it is, but using just two bits of ordinary information, she tells Bob—wherever he happens to be—how to re-create that exact state, on a particle he has in his possession. Alice's choice of what to measure is the crucial ingredient in this feat.

In quantum mechanics whatever an experimenter chooses to measure becomes a real property. Already by 1985, I had wondered about the full moral implication of this power of experimenter choice [1] guessed that the quantum mechanical

[2]Bohm Dialogue was developed in collaboration with Jiddu Krishnamurti, who once—as boy from Andhra Pradesh—was set to be anthroposophists' "world teacher." Physicist-philosopher David Bohm's method reveals the operation of *Thought as a System*.

experimenters' ability to make a property real by *choosing* to measure it would eventually lead to moral reflection.

The surprising way this power of choice worked out in QT is told in the history I recount below. QT presented me with a case of the willies[3] and a chance to ponder the questions of morality that arise from full responsibility for the creation of reali*ty*. When the teleportation article first appeared, the Institute used a combination of social studies of science and literary analysis to see a little more in it than other physicists.

This paper also presents our newest quests. Today, I work on my extension [2] of classic QT. It is called superDense teleportation (SdT); NASA (the National Aeronautics and Space Administration) is funding SdT for communication of quantum information from outer space to Earth. Quite recently our work has gained increased prominence within NASA. The Chinese launch of a quantum-capable satellite in 2016 and accomplishment of a number of *firsts* has spurred a three-way mini space race for quantum communication.

Altogether the journey will lead us to ask even more questions and to learn how physics awakened from its reductionist dreams. These questions are stimulated in part by search for connection between the ground for reality and the emptiness of Buddha mind, the śūnyatā of a Vajrayāna (Dzogchen) path. Does the heightened connection—a perfect melding of knowing and compassion—spell a possible path for future western knowledge for sciences? Is there some direct guidance from Eastern philosophies' concepts of emptiness for today's ongoing scientific research?

We start with some history of twentieth century physical science.

1.3 Reconstructing a History

At first the quantum presented difficult enigmatic problems to the nineteenth-century physical scientists. They struggled to find answers to typical questions posed by a picture of a daily world of objects, composed of tiny particles. Where were those molecules and atoms located? And how fast were they moving? For larger bodies, these two quantities were the necessary initial conditions, knowing both position and velocity allows perfect prediction in classical mechanics,[4] but the indivisible quantum prevents their simultaneous knowledge. Muddling the notions of particles and waves together, invoking probabilities instead of definite answers, and imagining that some classical principles could be (temporarily and briefly) violated or set aside,

[3]The Willies is a colloquial American expression meaning frightened qualms or something more difficult to put in words, a bit like "undefinable moral chills or qualms."

[4]If the particles were spinning on their axes, how much spin did they have, and in what direction was the axis pointing? These were originally particularly interesting and paradoxical questions, but today their answers form the simple example used below to discuss teleportation.

the scientists cobbled together a workable theory. They made remarkable mathematical progress at understanding the atom and all its internal constituents.

The central feature of an indivisible quantum of physical action threw two new elements into the muddle. First, not every conceivable outcome of experimenting is possible. Those that would split a quantum never occur; particle spins along any given axis, for example, can have only a few discrete values. Second, the completely random behavior of a single quantum determines which outcome, from among all possible results, actually occurs. Chance is inseparable from indivisibility, dashing the classical physicist's hope of perfect predictability. In general, quantum theory predicts only the probability of each possible outcome, not which one will occur in a given run of an experiment. Particles have no determined paths, instead they have chance luck. So, together with relativity theory, quantum mechanics upset many of these most fundamental preconceptions of nineteenth-century sciences: determinism, real trajectories, absolute time, and space. To make progress in the face of these difficulties, physicists acknowledged the crucial role of experimenter in observing the micro-phenomena of atomic sciences.

Eventually experiments showed the world, or at least the micro world of physics, really to be indeterminate, spooky, random, and nonlocal. The dream of Einstein and many before him—a dream of a separate independent objective reality—was replaced with something akin to interdependent arising of physical properties: their values only occurred upon looking. Observer and observed were inextricably intertwined. And precisely those sets of things needed to predict determinate outcomes were forbidden by Heisenberg's indeterminacy principle *ever* to be known exactly. Throughout the past century, our sophistication in theory and experiment has grown greatly, but we have never found any exceptions to basic quantum mechanics. By the 1990s, more and more new experimental arrangements were confirming this view, eventually yielding new devices which could actually *operate off of* the quantum realm, rather than simply help interpret it. The baby fields of quantum information, quantum computation, and quantum communication (and my own subspecialty, QT) were born, grew a bit, and began to crawl.

1.4 A Tale of Two Contradictions

As we shall see, the story of mistaken understanding about where indeterminacy comes from—the fundamental emptiness of quantum properties—contains two contradictions. The first contradiction is based on Einstein's classical understanding of reality and of physics as a science which must uphold reasonable requirements of reductionist objective realism. The second actually contradicts Einstein, leaving us where we are today with reali*ty* replacing reality.

Under Niels Bohr's Copenhagen interpretation beautiful physics was made within the limits of the indivisible quantum. The most famous of these limits, the uncertainty Principle,[5] was "derived" in Werner Heisenberg's 1927 paper. Indeterminacy of our knowledge for complementary variables was guaranteed. His final analysis was right, but his derivation seemed to attribute the principle to our own heavy presence in dealing with micro-reality. We are big and ham-handed, but the quantum is small, and so is the atom we are trying to measure. Determining one variable directly disturbed the complementary one.

In 1935, writing the EPR paper with co-authors Podolsky and Rosen, Albert Einstein challenged this "ham-handedness" explanation of indeterminacy by considering *two* particles in a single quantum state. The shared state lets us determine values of one particle's properties by measuring those of its partner, even when widely separated. However strongly we have to interact, the *elements of physical reality* are real; they cannot be affected by such very remote intervention, at least not according to Einstein's definitions of reality. Alas, neither he nor his great debating partner Bohr who intuited and championed interdependence[6] could have foreseen that moving from two to three particles produces a second and much more profound contradiction.

1.5 Einstein's Very Definition of Reality Is Wrong!

Historical accounts begin most often with a description of the EPR paper and for good reason. While most scientific papers have a very brief half-life, the possibilities and problems articulated in this paper have remained relevant to both theorists and experimentalists for more than 80 years. Physicists, philosophers, and combinations of the two returned to it over and over again to reconfirm, redirect, or simply sharpen their questioning.

Even before it appeared[7] in print, the paper made a splash. It became the centerpiece in a lengthy, spirited, and productive disagreement between those two giants of physics at the time, Einstein and Bohr. To understand what was at stake, consider the first two sentences from the body of the EPR paper. They provide a statement of Einstein's working philosophy of science, a clear challenge to Bohr and his colleagues.

> Any serious consideration of a physical theory must take into account the distinction between the objective reality, which is independent of any theory, and the physical concepts

[5]I have intentionally called this *indeterminacy*: Uncertainty seems to imply a knowing subject's limitation. But experiments show these limits are inherent in the very definition of the properties that are complementary. To indicate the objective character of quantum disturbance, without implying a subjective lack of knowledge, the better name is Heisenberg indeterminacy principle.

[6]Bohr is quoted as saying "There is no quantum world. [...physics isn't about] how nature *is*. Physics concerns what we can *say* about nature" [3].

[7]A. Einstein, B. Podolsky, and N Rosen "Can Quantum Mechanical Description of Reality Be Considered Complete?" *Physical Review*, Vol. **47**, (May 15, 1935), p. 777. Even before publication, a New York Times headline read "Einstein Attacks Quantum Theory" 5/4, 1935.

with which the theory operates. These concepts are intended to correspond with the objective reality, and by means of these concepts, we picture this reality to ourselves.[8]

Here Einstein, Podolsky, and Rosen articulate the classical conception of scientific theories which still reigns so powerfully today: reality is objective and independent, distinct from the theoretical representations which "correspond" to it. They go on to say that they don't know what a complete theory is, but at very least one requirement "seems to be a necessary one: every element of the physical reality must have a counterpart in the physical theory."

Later in the paper another deficiency shows up: they state that "a comprehensive definition of reality is... unnecessary for our purposes." But again they suggest a criterion to begin filling this vacancy, a criterion based jointly on an unarticulated notion of the "reasonable" and the "satisfaction" of an inarticulate desire:

We shall be satisfied with the following criterion, which we regard as reasonable. *If, without in any way disturbing a system, we can predict with certainty (*i.e., *with probability equal to unity) the value of a physical quantity, then there exists an element of physical reality corresponding to this physical quantity.*[9]

As we'll see, other physicists could be satisfied with a different conception of "reasonable." And more recent work under the intriguing name "GHZ[10] theorem" shows that what seemed a reasonable definition under some conditions actually fails completely when applied to others. The trick EPR used was to have two particles tightly entangled in one quantum state, and GHZ simply moved to three particles, all entangled together.

When EPR published, Bohr and his Copenhagen associates were thrown into a flurry. Einstein had found a way around the simple-minded idea that our clumsy huge-bodied handling of delicate quanta disrupted reality. Bohr did not anticipate the best retort (using one more particle), but his answer—published in the next volume of the same journal—provided essentially new definitions of "reasonable," new definitions that were *better* not so much because they represented an independent reality better—Bohr was actually moving beyond classical metaphysics[6]—but better because they were *more productive*. Physicists could do more with less, with a restriction on how far they would push ideas from everyday large-scale life in exploring the micro world that can only be reached through extraordinary effort and specialized equipment.So Einstein remained in that hard-to-pin-down but very palpable and forceful state he'd summed up as "unsatisfied." So unsatisfied that he

[8]*Ibid.*

[9]*Ibid*, emphasis in the original.

[10]The work was co-authored by Greenberger, Horne, and Zeilinger, my three long-term collaborating principal investigators for the National Science Foundation's grant to Hampshire College on "quantum Interferometry." The clearest reduction to 3 particles and contradiction is by N David Mermin.

did hard-core physics about it,[11] for Einstein's commitment was no mere battle of wills.[12]

As most physicists practicing today have learned it, the Copenhagen interpretation is basically justified pragmatically, with the slogan "It works. Use it." Not "It's true"—or simple—or even well-constructed. And working meant that the mathematical relations always predicted experimental results correctly if you learned your mathematical manipulations well. But such a slogan begs for further questioning: What works? and for whom? How does it work in the world? Questions that point directly at the acts of experimenting and articulation, perhaps even pointing to the remarkable effectiveness of quantum mechanics in real life: modern electronics, optics, and nuclear physics—all understood by quantum mechanics—appear in forms as diverse as supermarket checkout scanners, power plants, computers, and bombs. For me the switch to "It Works" was eye-opening. As an undergraduate, my quantum teachers showed real honesty about the creative "made-up to serve" character of physical science. As a professor, eager to teach quantum mechanics early and often to my students, it raised questions of morality. What work would my pupils, my colleagues, and I do? To what purposes would that newly minted knowledge be put? For whom would our knowledge work or be deployed? As Bob Dylan's song had it, *you've got to serve somebody*. Who would that turn out to be? Could it be for the common good? If so what questions must we ask? What properties should we choose to make real by deciding to measure them?

1.6 The Spin Doctor

David Bohm made the EPR paper a little easier to think about. Instead of particles moving along a line and working with their positions and momenta, Bohm used spinning particles and focused on measurements of the way the axis of spin is oriented in space. His spin version of EPR applies to any of the particles in ordinary

---

[11]Einstein's EPR paper implies there should be completely deterministic explanation for the quantum phenomena including those that seem probabilistic. In the same year, in the very same journal as the EPR paper, he published a co-authored work showing how two very remote places could be connected by a topological structure: the Einstein-Rosen bridge. Today it's famous for introducing "wormholes" to general relativity. But in a remarkable passage, they suggest it could be a structure of quantum particles, taking the model so seriously as to remark that inside of protons each would need at least two bridges to account for their electrical repulsion, since like charges **repel**. Positive mass bridges—as gravitational structures—can only **attract**. Clearly Einstein was beginning to search for unified theories already in the mid-1930s [4].

[12]The great final irony would come if a nonlocal but geometric theory of general relativity actually entailed quantum mechanics, or both needed quantum mechanics *and* explained it through some self-consistent advanced geometry. The current string theory or the M-theories of quantum field theory may someday lead beyond the current amazing mathematical connections and close back in on quantum mechanics itself—a history-bending, time-twisting outcome if it ever happens. Only time will tell.

matter—electron, neutron, or proton. It also works for the polarization of the photon, Planck's original quantum of light.

In three dimensions, a spin axis can lie in many directions; but there is complementarily an indeterminacy principle for spin. Spin values in two perpendicular directions cannot be determined simultaneously. If one direction in space has a definite value, the perpendicular spin directions are totally indefinite. Even along one direction, values are limited by quantum indivisibility. Ask "what is the spin along a given direction?," and for the usual particles, only two answers ever occur, totally parallel or perfectly opposite to the chosen direction, which makes the idea of quantum luck easy to grasp: if we know the spin value in some direction, any perpendicular direction will be a perfect coin-toss; half the time it is parallel to the field and half the time opposite.

The Bohm spin state idea was much easier to make into an experiment than the original EPR state. The necessary EPR-Bohm state was simply a state with total spin zero, made out of two particles spinning in opposite directions.[13]

With a spin-zero state made from spinning particles, it's also easier to see why the experimental results seem to depend on one another's realit*y*. Whatever is done to each particle separately cannot change the fact that, as a whole system, the two are spinning in opposite directions. So when the problems about being able to measure only one of perpendicular directions is translated into measuring way across the room, at least we have a feeling for what the answer must be over here near us. It will be as if the one near us were suddenly thrown into a state exactly opposite to the outcome at the other particle's experimental station.[14] The correlation is complete. Any of the spin variables you want to know can be determined without direct interaction on the particle in question.

Such a situation of perfect correlation is summarized by the idea of "entanglement" of two particles—the translation given to Schrödinger's term *Verschränkung* for the property exhibited by EPR pairs. (*Verschränkung* actually deserves a more explicit and more neutral translation, one that would intertwine "entanglement," "crosslinked,"—as in crisscrossed arms—and "being tightly mixed up.") Entanglement gives correlations between events at remote locations but doesn't change overall probabilities or pattern of outcomes at one location. We note, therefore, that **because** they are not actions, these correlations cannot be used to send signals faster than light. The particles still have pure coin-tossing luck, but now their individual lucks are completely opposite each other. If one shows heads up, the other is tails.

[13]They are necessarily opposite in all three dimensions. Photons coming from a suitably excited atom would also be polarized in exactly such a singlet state.

[14]I emphasize "as if" in honor of John Bell because what is actually happening with the particles is something that he placed in the domain of the *Unspeakable*. The "Problem then," as he once put it, "is this: how exactly is the world to be divided into speakable apparatus…that we can talk about…and unspeakable quantum system that we cannot talk about?" [5].

1.7 EPR to GHZ and the Alterity of Realitty

By the 1980s experiments had advanced. Spurred in part by Bohm's spin reformulation of EPR, they were testing the strictly statistical implications. And wherever ascribing independent realism to each particle implied measurable results that differed from quantum predictions, the quantum mechanics was right, and strictly local reductionist realism was wrong. But we can imagine Einstein still holding out since his philosophical objections were based totally on determined outcomes, when measurement magnets were either fully aligned or antiparallel. These gave 100% or 0% probability, not the more complex fractional probability relations for remote measurements at strange angles to one another, which were the basis for two-particle experimental disproof of naïve realism.

The real surprises were about to come from considerations involving states of three particles, instead of only two. The first surprise was the fatal flaw in the very definition of Einsteinian elements of reality. When there are three particles, many different elements of reality exist. Enough of them can be found that some actually contradict each other. The second was the possibility of quantum teleportation. Theorized in 1993—by Charles Bennett of IBM's Watson Laboratory and his *five* collaborators—quantum teleportation's first experimental realization came only 4 years later in December 1997, at the Innsbruck laboratory of my collaborator Anton Zeilinger. Quantum teleportation requires quantum states of three particles, as does the disproof of Einstein's definition. The Austrian team next produced laboratory examples of the three-particle states that predict results completely contradictory to Einstein's "reasonable" definition of reality. The contradiction is surprisingly deep. In a sense the *elements of reality* themselves are empty or śūnya.

The disproof uses a special, fully entangled state of three particles. Each particle has two elements of reality corresponding to two perpendicular directions of spin. They are determined (as in EPR) by measuring *only* its entangled partners. Under Einsteinian notions when a certain direction of spin is measured for the first particle and the perpendicular direction for the other two, a definite pattern emerges.[15] That pattern predicts what will happen if we measure the special direction—the copper penny—of all three spins at once. There should be an odd number of parallel results. Either a single head should show or all three. But when we actually do that triple special measurement, completely contradictory results occur. Half of the time there are two heads; the other half shows *no* heads, all tails (when all three particles are spinning opposite the special direction). The definition of *element of reality* leads to contradiction with experiment. It is just plain wrong.

[15]Random luck still operates as in most quantum effects, so the first particle lines up with field (Heads) half of the time. But when it does, the two others have identical luck in the perpendicular field. When the first comes up tails, the other two have opposite luck. See appendix for full details. Calling the first direction a copper penny and the second silver shilling makes the contradiction easy to see. Appendix to Muddling Through & on request from the author, hbernstein@hampshire.edu

Einstein would no longer be pleased and definitely not satisfied. There is no preexisting, independent local reality for microscopic variables, just as the Copenhagen interpretation had argued all along. These developments in quantum theory remind us of the double reading of reali*ty*. Quantum reali*ty* is inextricable from our acts of measuring and knowing; and at the same time it surprises us with its absolute alterity, behaving in utterly unexpected ways, following paths picked by overall coordinated luck instead of being pre-determined. But what's truly remarkable about this nondeterministic, radical alterity is that it can be used in perfectly reliable ways—say, to "teleport" the properties of a particle from one place to another.[16]

1.8 Quantum Leaps

In twenty-first-century America and throughout the world, familiarity with teleportation is a given. Kids, grown-ups, and nonagenarians alike know it through movies, comics, TV shows, short stories, and novels. The original coining of the word "teleportation" apparently comes from spiritualist literature as what we now call *psychokinesis*. But somewhere in the history of science fiction, probably in its golden age, teleportation acquired the modern connotation: moving things by dissolution in one spot and re-materialization elsewhere. *Star Trek* never uses the word. There (or *then*, in the twenty-fourth century), teleportation is so commonplace that the particular technology used to achieve it is called, simply, the "transporter."

If scenes from movies and TV are supposed to be in real time, then distance apparently presents no difficulty, as long as you are within range of the transporter. The object being transported begins to appear almost immediately at the new location, going from planet surface to orbiting starship instantaneously, no matter how much time light would take to travel between them. Those scenes contradict special relativity, which dictates that no signal, and no object, can ever go faster than the speed of light.

Quantum teleportation (**QT**) is slightly different. No object is transported from here to there (or anywhere). Only information is transmitted, and not even that travels faster than light.[17] And the information cannot be "beamed" or directed to a particular location. The classic 1993 QT paper has as its central feature the entanglement of quantum particles. In their most general form, these entanglements "assist in the 'teleportation' of an intact quantum state from one place to another, by

[16]Charlie chooses a direction for his spin-selecting magnet and sends a single particle to Alice. Charlie's state is reproduced by her partner Bob with perfect fidelity. Alice only needs to tell Bob which of four things to do. He can get the correct chosen state using a very short message, only two bits long. That's thousands of possible settings of Charlie's magnet condensed into only four possibilities, How? The trick is that Bob and Alice have previously shared an EPR pair between them. [The appendix to **MT** has a simple explanation of QT.]

[17]In all the correlations and "passions at a distance" of quantum weirdness, this prohibition on the speed of information transfer is always respected.

a sender who knows neither the state to be teleported nor the location of the intended receiver" [6]. So QT can be seen as one effect in the larger field of operations known as quantum information processing. This prompts another of my questions to stay close to: How is quantum teleportation itself entangled with another technique, quantum cryptography, a method of sending coded messages that are impervious to eavesdropping and impossible to break?

When we at the Institute first heard of teleportation, you can imagine our delight and excitement: finally a place where entanglement, the choices involved in experimenting, and the constructions of realit*ty* combine to achieve something practical/possible. And **what** a something: The long-held dream of teleportation. But the entanglement of quantum teleportation with quantum cryptography—that's an invitation to the willies.

1.9 The Willies

Quantum teleportation is physics at its exciting, challenging, mind-bending, and realit*ty*-creating best. But this amazing world of realit*ty*-invention also gives some of us the willies. Getting the willies is *sort of* like having an ethical or moral dilemma: you worry about what the right thing to do is. But the willies are more haunting because they are less clear, more muddled, and *spookier*— than some famous historical examples of science, ethics, and morality.

There's no question that any history of physics in the twentieth century will have the Manhattan Project and the work of physicists on nuclear weapons at its temporal and moral midpoint. That Project often has personal resonance. My career (like that of Michio Kaku and several other currently active physicists and science writers) was strongly influenced by my parents' wartime and Manhattan Project experiences.[18] And in the middle of the Manhattan Project, historians will always find J. Robert Oppenheimer. And in the middle of Oppenheimer's life, they will always find two interesting utterances of his. These familiar reflections on physics and what it had become responsible for are remarkably concerned with morality in two cultural contexts: One of them he attributed to the **Bhagavad Gītā**, I am become death, the destroyer of worlds. The other is less poetic: "We physicists have known sin." In some sense Oppie was prescient.

In the postwar period new federal institutions were created in the USA to channel immense new levels of support to the physical and other sciences: the National Science Foundation (NSF), the Office of Naval Research (ONR), the Office of Scientific Research and Development (OSRD), and a host of others. The national laboratories at Los Alamos, Brookhaven, Livermore, Oak Ridge, and other federal

[18]The story of my father and his absolute joy at narrowly having declined Edward Teller's call to join the project in Oak Ridge appears in *New Ways of Knowing*, Marcus Raskin and Herbert Bernstein (Totowa, NJ: Rowman& Littlefield, 1987), Chap. 2.

sites took in ever larger amounts of money, personnel, territory, and equipment. “In the fifteen years following the war,” as historian Paul Forman has documented, “the central fact of scientific life in physics was unprecedented growth based upon military funding [7].”

Questions arise from this rather crude juxtaposition of two key features of the postwar landscape in physics. What are the connections between moral and social responsibility, funding, the creative impulses of physicists, and the military? How can or should a physicist think about these linkages between the work that he or she does at the blackboard and what happens on the other side of that blackboard? These questions seem particularly troubling in the areas of inquiry which focus on the quantum level where, as we’ve seen, the physicist may be “responsible” for choice of experiments—or even more strikingly (and somewhat megalomaniacally)—is said to participate in the “creation” of the quantum world.

What once might have been a relatively straightforward ethical choice on the part of the scientist confronting the national security state—“I’m not going to work on nuclear weapons,” or “I won’t accept Defense Department money”—is now a more tangled affair. This is hard enough when you know the work would be used for ends which you judge to be illegitimate, unethical, or immoral. But what happens when you *don’t* know what your physics is going to become or what things it will lead to? What happens when the grant money you’ve accepted is “clean” but the meaning of that word has been muddied beyond easy recognition, because the worlds of the military, the national security establishment, and “pure” physics are now so thoroughly entangled after decades of work, billions of dollars, and constant traffic of people, ideas, and equipment between these worlds? What happens when you’re not only worried about the future use or application of your science but about what possibilities are being built into it now at the supposedly abstract, theoretical level?

What happens is you get the willies. You sense that something is wrong, but you’re not quite sure what. You hear a call to responsibility, but you’re not at all certain what responsibility in this situation involves. You think there’s something strange and spooky happening on this dimly lit road, but your companions tell you you’re imagining things.

And there’s no way to reconcile these contradictions.

1.10 Worlds Entangling

At the Institute we got our first, rather mild, case of the willies relatively early in the process, reading the scientific literature on quantum teleportation closely. The language of coded information, “unknown” quantum states, tightly controlled protocols for information exchange, “unspeakability”—a language which is part and parcel of the recent history of quantum theory—suddenly suggested the qualities of a spy narrative to us. But that’s absurd, right? Everybody knows that spy stories don’t get published in the leading physics journals.

Yet in earlier papers on quantum cryptography, I had seen the same characters "Alice" and "Bob" who were teleporting particle states involved in the theories of creating perfectly secure codes and message transmission that unequivocally ruled out any form of eavesdropping. And there it was clear that those names had come from the literature on "classical" cryptography.

At least one physicist I know decided not to work on quantum cryptography because of its entanglement with institutions like the National Security Agency. It's an ethical decision which we respect, but at the same time, the boundaries here are simply not that easy to draw. As a scientist, the minute you get interested, if you are any good at all in your field, that interest itself provides a valuable commodity: listeners and questioners who make a researcher think. Whether the researcher is you, your student, or some innocent third scientist, you are already contributing to the field. Your interest begins to articulate an effort, to power an approach, and to stimulate new experimenting and new thinking; so continuing to work on quantum teleportation means contributing directly and indirectly to work in the areas of quantum cryptography and quantum computing.

It's a lesson in disciplinary entanglement that was confirmed in a couple of ensuing experiences of the willies, both connected to conferences. In the first case, a friendly casual conversation turned my mathematical paper answering the question "can everything be measured?" [8] into a proposal for secure communication on the battleground of the future. I was shocked when my acquaintance needed to attach a hardcopy to his proposal. It seemed the defense agency—or its reviewers—had no access to the leading physics journal. This was more than a little eye-opening and willies-inducing. I thought we were exploring, at a fundamental theoretical level, how the experiments and ideas of the quantum physicist help create reali*ty*. I didn't think someone would right away go out and try to make a reali*ty* entirely different from the one I would have liked.

The second incident was somewhat deeper. For a few years I had been attending small conferences and workshops on quantum information theory, comprised mostly of the initial cadre of theoreticians and a few pioneering experimentalists. Arriving a little late at a conference in DC, I peered in the doorway of the room. Someone had opened the rear wall to double the size of the seminar. Who were these extra people? I found out at lunch that most were mathematicians and physicists from the National Security Agency just up the road and some were from the Department of Defense; all were now paying close, personal attention to an area of research they had been monitoring from some small distance. The difference was that a Bell Labs researcher had proven that a quantum computer could solve certain mathematics problems with relative ease.

Ah, and what problems! they were the so-called one-way problems of factoring a product of two huge prime numbers. Given the product it is devilishly hard to find the factors, but the other way—given the primes—you just have to multiply. One-way problems are the basis for the encryption of all truly secure messages now sent. Quantum computers promise the technology capable of breaking *any* such public-key code. After that DC meeting, the inflow of interest and money from civilian and military sources was immense and serious. Our field became a veritable industry of its own, or at least a vital sub-discipline of physics. What a great time that was! And in many ways that era continues to this day: gratifying and, yet with a little chill, a bit willies-inducing.

2 The Missing Characters and Doing (SuperDensely) More by Trying Less

What do you do when *not* knowing where participation in science production will take your work? There are few examples of how to behave. Norbert Weiner, coiner and developer of cybernetics, famously wondered, post-World War II, what societal changes his ideas might produce. He even considered suppressing his ideas and whether "…I ought to exercise a right of personal secrecy, parallel to the right of governmental secrecy assumed in high quarters."[19] But when he recognized that those ideas were not uniquely his own—they belonged to his field and his times—he decided to stay in the flow of groundbreaking mathematics. And to model for other scientists how to point out social and moral dangers, his decision to "ride the bucking bronco" of science-technology consequences is similar to what we have done so far. And our work has had some powerful help in the form of science studies.

Take the case of teleportation, to which The Institute applied literary analysis. The story is encapsulated by three diagrams from a semi-popular talk. Leaving all technical details aside, we can focus on the cartoon characters. In Fig. 7.1, the QT diagram as usually construed features the two communicators Alice and Bob—here editorialized and drawn[20] as spies (*cf.* pg.131: THE WILLIES).

The wiggly line at the left is the particle whose state Alice will transmit to Bob. It comes in as "unknown" to both of them. Alice needs to send only a short, two-bit, message because she has previously shared a pair of EPR particles, denoted by the dotted lines coming from a source **S̲**. But a quantum state cannot be unknown to everyone in the world. It is created by the observer who chooses it; without such choice, there cannot be any communication at all. So we redrew QT with additional characters (Fig. 7.2).

We have introduced a character Charles, the chooser. He picks any one of a million different spin states, setting his selection magnet to any direction in space. One-in-a-million and yet the two bits that Alice sends to Bob enable him to reconstruct the chosen state every time. That's remarkably dense communication! But it still is secret: in fact, Bob gets completely random, 1-in-4 messages from Alice, each appearing 25% of the time, no matter what Charlie has chosen.

[19]Trevor Pinch, [at note 11, p. 186].in *The Social Production of Scientific Knowledge* (Dordrecht: D. Reidel, 1977), E. Mendelsohn et al., *eds.* Other famous scientists had similar moral qualms: see, e.g., Lanouette/Szilard's bio, *Genius in the Shadows*.

[20]Skillfully drawn by our former student Usha Lingappa, now a CalTech PhD student in exobiology.

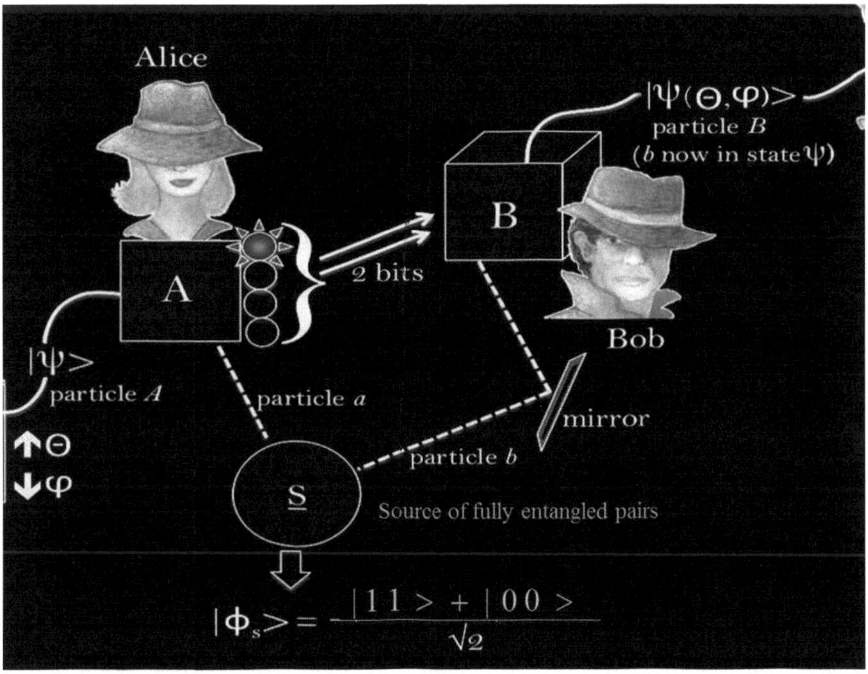

Fig. 7.1 Cartoon depiction of teleportation, emphasizing espionage narrative; for Greek letter and math symbols, see appendix, available on request

Not just one but two additional characters are needed, because someone (or some device) also has to detect, deploy, or do **some**thing—Diana the detector/deployer. So we put Diana in the drawing, signifying the end user, and made our overall point. Bringing in the high tech spies means physicists are linked to the state, depicted here as British Royals. Our diagrams reveal a political/moral dimension.

These characters also led to a new piece of physics, the SdT. Putting Charlie in the picture meant we could play with his positioning, in particular to put him *between* the source S̲ and Alice.

Moving to Fig. 7.3, we let Charles get his hands on Alice's EPR particle *a*—the dotted line from S̲—before she does. That gave me an idea. It turns out that Charles can modify the particle's state and have that affect what she sends to Bob. Charlie has to be careful and use only the wave aspects of the state, which is just half of the quantum information. But this lets Alice use fewer bits of information to signal Bob what to do. He's able to make *b* have the state that's exactly what Charles originally chose (as shown by the thinking in the bubble above his head).

Doing out all the math, I found that SdT has advantages over QT; an advantage that increases with every dimension of the particle's state. Particles with more spin have high dimension and so do fancy varieties of photons, the particles of light. As

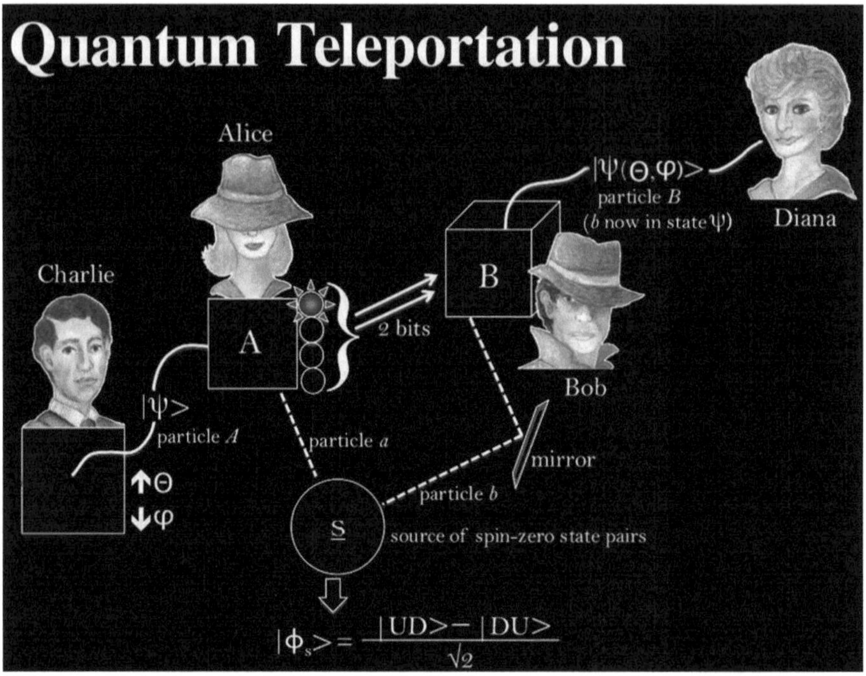

Fig. 7.2 A more complete diagram has four characters. It reflects a cryptography fact. Spies often serve the state, here two British Royals

the dimension gets higher, the improvement in number of bits grows. Eventually, Alice sends twice as much quantum information so the effect is called SuperDense.[21]

Now NASA is always looking for factors of ten in improvements, but they will take doubling capacity for a start, especially for bits that carry information, since bits from outer space are so hard to come by. At first I was dubious. Why should NASA want to send quantum information anyway? But now that China has launched a satellite with proven quantum communication capability, I understand. There's a three-way mini space race with China and Europe. And NASA aims once again to be "not the first, but the best" as they were in the post-Sputnik era.

[21]I originally thought of it as semi-teleportation since only half the state parameters were sent; Charlie Bennett dubbed it SdT after a previous effect, superDense coding, which carried twice as much classical information per photon as classical binary coding.

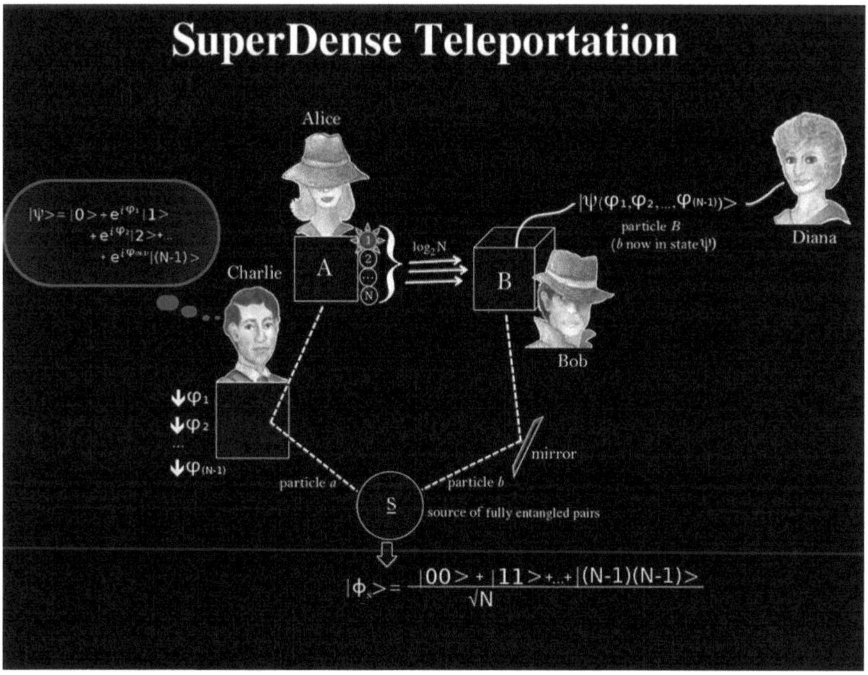

Fig. 7.3 SuperDense teleportation (SdT) uses Charlie's extra capability when he can get his "hands" on *a*, the entangled particle headed to Alice. As number of dimensions N goes up, SdT sends more quantum information than QT, ultimately approaching twice as much per bit

3 Back to the Future

All of which brings me back to the heightened need for social responsibility of science. I've shown the mixture of literary, social, and historical analysis in the Institute's work on quantum teleportation, but these investigations and framing inquiries into science investigation are rarely enough. Can Hindu and Buddhist philosophy help? Can the emptiness of quantum properties and need for clearer values connect fruitfully to certain aspects of śūnya and śūnyatā?

3.1 Śūnya, Śūnyatā, and Moral Epistemology

3.1.1 Buddhism and the Quantum

Many physicists and science writers have popularized connections between Buddhist philosophy and quantum mechanics, almost from the beginning of that science. Both Schroedinger and Heisenberg were given to such analogies in the early years, at least in nontechnical discourse.[22] Wolfgang Pauli sought to acknowledge the subjective wholeness he took from notions of Eastern philosophy. He was impressed with Jungian archetypes and synchrony and wanted to balance the extreme materialism of objective, reductionist science—probably feeling its limits if only by foresight or intuition.[23]

Eugene Wigner famously wondered if the human consciousness of an observer—and only such a consciousness [9]—definitively turned the random luck of quantum particles into the definite deterministic solid and predictable outcomes we experience in our macroscopic lives. In the 1970s, many of us were delighted to see Zukav's *Dancing Wu Li Masters*[24] and Capra's *Tao of Physics*.[25] These books had defects, however, that relegated them to interesting but perhaps *too* popularized glosses. They did not ultimately shape physics or other sciences. Capra was wedded to a form of elementary particle theory (so-called nuclear democracy) already losing favor by the time it appeared; today it's been replaced by a rather hierarchical standard model. And Zukav's book seemed to physics-trained readers much more authoritative and correct on Eastern philosophy than in his physics where mistakes, small and big, reduced its credibility. Those who studied or practiced the Eastern concepts he'd used as analogies, however, thought that physics was quite excellent, whereas his grasp of Hinduism, Buddhism, Zen, or Tao is "not so much."

None of these earlier commentators attempted to draw upon the Eastern ethical wisdom to inflect Western sciences or their applications. They were intrigued by apparent analogy and sympathetic cross-cultural resonance across the centuries rather than changing their enterprise's morality. Today the political-philosophical search of reconstructive knowledge for a moral epistemology may open to the key concepts of śūnya and śūnyatā for such a change. Does this key pair open another opportunity to further heighten the social responsibility of science? This possibility is especially interesting to practicing meditators like me and like a growing number of other scientists.[26]

[22]Early quantum physicists were perhaps drawn to Buddhism more by the intellectual popularity of Schopenhauer's versions than by their own direct experience or practice.

[23]Pauli's correspondence with Carl Jung was so mystical that a committee of overseers feared it might tarnish his scientific reputation: VF Weisskopf on K. von Meyenn's archival work (private communication, circa 1987). See also "The Innermost Kernel" by Suzanne Gieser. Weisskopf's forward to the Pauli Letters points out that he had contributed to *every* initial advance in quantum theory.

[24]New York: Morrow, 1979.

[25]London: Wildwood House Ltd., 1975.

[26]These include an American quantum field theorist who presides over a Kagyu Sangha in Santa Fe, NM, a Zen meditator who serves as the principle investigator of the most relevant experiments for SuperDense quantum protocols, and a solid state/condensed matter theorist at the University in our local consortium, who practices Vipassana. In addition of course to East Asian, Indian, and other Himalayan national/native meditators like our former colleague, a student co-author in several of our papers, now in industry, to many—perhaps most—of those in our undergraduate classes who meditate, and many more, both of Asian and of western extraction.

3.2 Śūnya, *Theory of Everything,* Śūnyatā, *and Dzogchen/ Tibetan Buddhism*

To the well-trained physicist, śūnya (empty, void) immediately calls to mind the universal quantum vacuum, the lowest energy (or ground) state. It is written $|\mathbf{0}\rangle$, an Arabic numeral written between vertical bar and angle bracket, further connecting etymologically to zero. But the uncertainty principle means the energy of the state cannot be precisely zero. There are countless pairs of mediating particles virtually present at all times. The vacuum is quite lively; its state $|\mathbf{0}\rangle$is śūnya—truly void yet marvelously productive. And it is the ground for every possible state of all real particles in their fullest potentiality. Moreover, $|\mathbf{0}\rangle$is the indispensable basis for the long-sought "Theory of Everything." In great measure it is actually a theory of zero, specifically the theory of what the vacuum state $|\mathbf{0}\rangle$contains. ToE is the theory of what is created out of the vacuum and what returns to void, in the continual expression of quantum phenomena, from and within the vacuum. Current quantum field approaches to a Theory of Everything are exactly about this, including current high-dimensional string theories.

Traditional Eastern epistemologies quite understandably turn away from the physical aspect of objects and actions. But, as Pauli may have intuited, their concepts have much to offer modern science. Inward explorations could supply the balance to an outward focus, especially now that physics has recognized a form of emptiness. And seeing ego as illusion may provide a non-egotistic antidote to some worst aspects of twenty-first-century sciences. In Buddhism, as I understand it, the awakened Buddha mind is triply embodied as dharmakāya, saṃbhogakāya, and nirmāṇakāya. These three aspects of the ultimate reality occur as one: the emptiness of suchness entails both radiant cognizance **and** a manifold capacity for compassion. The union of wisdom with compassion is perfect and seamless. Might this not be a model of knowledge *with* values?

Dzogchen (great perfection, or total completion, in Tibetan) is the form of Buddhism that I practice. The Dzogchen tradition holds that it is possible for the prepared student to see the nature of mind and experience rigpa, the recognition of original fresh wakefulness, knowledge of the ground of being. The qualified master can show this ego-less and thought-free, yet cognizant, alert awareness to such a student, which the student can from then on find for him/herself. My experience of Buddha mind is one that is not so much experiential as "recognitional" and is still perhaps elementary; it comes from receiving the "pointing out" instructions of a Dzogchen master.[27] The view is that of emptiness of all apparent phenomena, the meditation is the non-meditation of resting in the natural state of nature of mind. Rigpa happens in small glimpses at first—hopefully repeated many times—and the fruition is to be dwelling in rigpa at all times, or at least much of the time.

---

[27]In my case, first recognition and being in rigpa was followed by over 24 h of heightened compassion, confirmed by a high lama.

But time is not exactly the correct measure. Neither is experience the right term for what happens in rigpa. The full recognition happens (if my own case is any correct measure) totally outside of space and time. Indeed ultimate reality is hard to describe. It is beyond thought or words; neither of these occur in rigpa. It surpasses ego, not just particular ego traits like to whom we were born, what ethnic or national group one is part of, or where we learned what language(s); it surpasses age gender, status, etc. True, one partakes of universal humanity, beyond all the particulars. But in some ways of talking about this "substrate consciousness," rigpa is beyond it; it is beyond both duality and non-duality. Even the idea of ego is impossible. There is bright awareness but no "me"—not a particular personality nor a general human being who is "at one with all humanity," no *one* subject-thing (however far beyond identity). There are no object-things either.

First recognition is, of course, not the end of spiritual development. Taking the fruition as the path, one nevertheless has many stages to continue through in order to reach the ultimate full awareness. And several (many[?]) practices, techniques, text studies, and ceremonies exist to help one's progress. For me there is an additional dimension to Dzogchen Buddhism. It gives new meaning to incorporating values into my sciences and those of the projects we undertake. While previous undertakings have had political-philosophical shadings in their reasons, their outcomes, – their raison d'etre, one can now think about incorporating the compassionate aims of bodhisattva aspiration to benefit beings. Of course this is not a simple task. It needs more study and preparation fully to know even what the idea means.

Emptiness, cognizance and compassion are properties of the natural state of the nature of mind; their inclusion in that recognition are conditions afforded, or – more emphatically, are all necessarily entailed in the realization of emptiness in Dzogchen practice. Does this present a possible and natural line of investigation for modern science to pursue? Is it related to the social studies of science? Can it be done as part of an action program for the common good or universal Human Rights to society and its most pressing needs?

Can the East provide Western knowledge systems with the needed combination of values and fact-seeking? Is Buddhism's wisdom of an emptiness entirely suffused with compassion a key to the future of such reconstructive knowledge work? Could a small science institute like ours play a role in such a transition?[28] How do we go forward? Should the previous politico-philosophical motivation be joined with or supplanted by spiritual-philosophical direction? If so how?

[28]Some suggestions and precedents readily occur: benefitting all beings points to an ecological focus. Arthur Waskow's Shalom center models how to draw on (Jewish) religious values and practices to implement political action. Could there be similarly, motivated science research? How would one square the aim to show one's enemies and friends equal compassion with the reconstructive knowledge impetus toward alliances, judgments, and politics?

References

1. Raskin MG, Bernstein HJ. *Cf.* New ways of knowing. Totowa: Rowman & Littlefield; 1987. p. 61–4. I even imagined that some striking application of experimenter choice would make it clear that we are quite responsible for our lived reality
2. Bernstein HJ. SuperDense quantum teleportation by remote state preparation. Quantum Inf Process. 2006;5(6). Doi: https://doi.org/10.1007/s11128-006-0030-5, Springer Verlag. www.springerlink.com/content/r0773t4138510q21/
3. Petersen A. The philosophy of Niels Bohr. Bull At Sci. 1963;19:13 [emphasis in the original].
4. Einstein A, Rosen N. The particle problem in the general theory of relativity. Phys Rev. 1935;48:73–6.
5. Bell JS. Speakable and unspeakable in quantum mechanics. Cambridge: Cambridge University Press; 1987. p. 171.
6. Bennett CH, Brassard G, Crepeau C, Jozsa R, Peres A, Wootters WK. Teleporting an unknown quantum state via dual classical and Einstein-Podolsky-Rosen channels. Phys Rev Lett. 1993;70 (13):1895–9.
7. Forman P. Behind quantum electronics: national security as basis for physical research in the United States, 1940–1960. Hist Stud Phys Sci. 1987;18(1):149–229.
8. Reck M, Zeilinger A, Bernstein H, Bertani P. Experimental realization of any discrete unitary operator. Phys Rev Lett. 1994;73:589.
9. Good IJ. Remarks on the mind-body question. In: The scientist speculates. London: Heinemann; 1961.

Chapter 8
Quantum Reality and the Concepts of Infinity, Infinitesimal, and Zero in Mathematical and Vedic Sciences

Radhey Shyam Kaushal

Abstract Absolute reality in Nature has two aspects of its existence—one is only realizable, and the other is describable with an element of realization. While the objective sciences follow the second route to understand the absolute reality, the subjective sciences, on the other hand, go mostly by the former. It will be argued in this paper that the quantum reality, manifesting through the subjects of mechanics of microscopic systems and quantum field theory, is not the final step in approaching the absolute reality. Like other cases in the history, it only offers a rung in the ladder and that too strictly in the domain of analytical description vs. accurate measurement. The understanding of quantum reality, in fact, brings in the concepts of infinity (*ananta*), infinitesimal, and zero (*śūnya*). Further, these concepts while are necessary in precise mathematical terms in objective sciences, in philosophical terms in Vedic (subjective) sciences, however, these concepts are found to have much deeper meanings. Some mathematical tools for this purpose are pinpointed here which can act as a guide for analytical studies of these concepts in Vedic literature.

Keywords Infinity · Infinitesimal · Zero · Mathematical science · Vedic science · Macroscopic systems · Quantum field theory

1 Introduction

As far as the search for "absolute reality" or "ultimate truth" in Nature is concerned, the Vedic literature has been advocating it through the words *neti*, *neti* (not this, not this) since time immemorial. In spite of all this, efforts have been there to realize this truth in the self and also in Nature in one way or the other depending upon the taste of an individual and the level of development of his/her faculties of understanding (FOUs). It is the level of development and training of one's FOUs that apparently create divisions and hierarchy even in this same absolute entity, and accordingly one

R. S. Kaushal (✉)
Department of Physics & Astrophysics, University of Delhi, New Delhi, India

S. R. Bhatt (ed.), *Quantum Reality and Theory of Śūnya*,
https://doi.org/10.1007/978-981-13-1957-0_8

realizes it through manifest (objective reality) or unmanifest (subjective reality) form of Nature or through both. While common man is generally ignorant about this ultimate truth and Yogi always lives with this truth, the other two category of persons (see glossary of words in [11]), namely, philosopher-scientist (having spiritual approach to subjective aspect) and scientist-philosopher (having scientific approach to objective aspect), have their own ways to have a feel for this reality in Nature [5, 6]. For one it is only realizable, and for the other, it is quantifiable as well as describable with an element of realization.

Whether it is subjective or objective aspect of absolute reality, each has huge spectrum of rungs in the ladder (see [5], Chapter 7), depending upon the level of development of FOUs of the practitioner in the first case and on the resolving power (measuring capacity in terms of accuracy) of the measuring instrument in the second case. There are various methods and systems suggested in Vedic literature to achieve the highest (and possibly the best) of subjective reality, whereas the objective science, supposed to provide the objective reality and develop mostly in the West, is still struggling for achieving progressively better accuracy of measuring instruments. Every now and then, there is a progress on this front, and accordingly the worldview of scientist changes every decade.

If one goes with the three-world concept of Karl Popper and Eccles (first, the world which exists; second, the world which we perceive; and third, the world which can be modeled by scientists), then it is the third world whose understanding is changing with a view to achieving the second, i.e., the perceived world. The next stage would be to go further from the second perceived world to the first—the world that exists (absolute reality) by way of realization only. Quantum reality noticed mainly from the study of microscopic systems like molecules, atoms, nuclei, etc. is only one of the plateaus in the search of objective aspect of reality which is further strengthened from the studies of phenomena in quantum field theory. In the process several interesting concepts like vacuum, infinity, emptiness or void, degenerate vacua, etc. arise which, in fact, are found to have much deeper and generalized meanings in Vedic literature.

In the next two sections, we discuss the quantum reality as can be extracted from the study of the dynamics of microscopic systems and from its field theoretic generalization. In Sect. 4, we discuss the concepts of infinity and vacuum in physical sciences. All-pervading cosmic consciousness field as advocated in Vedic literature is briefly analyzed in the light of tools of quantum field theory in Sect. 5. With a view to giving a broader perspective to the study, in Sect. 6, we suggest a kind of mathematical guide to study the hidden concepts of *śūnya* and *ananta* in Vedic literature. Appendix A gives details to this effect. Finally, concluding remarks are made in Sect. 7.

2 Quantum Reality as Revealed from the Studies in Quantum Mechanics

Prior to this plateau of quantum reality[1], there has been yet another plateau of "classical reality" through classical mechanics in the history and that prevailed for several centuries and explained the reality as perceived by our five senses of knowledge. In this regard, for example, the outcomes of the principle of virtual work or of the D'Alembert principle, Hamilton principle of least action, etc. cannot be ignored. An account of consciousness in the study of classical systems [8, 12, 13] further adds to value the inherent classical reality. In fact quantum reality manifests through both nonrelativistic and relativistic quantum mechanics.

2.1 Nonrelativistic Quantum Mechanics

After noticing the limitations of classical mechanics (which basically emerged from the work of Newton about 400 years ago) in describing the dynamics of microscopic systems like molecules, atoms, nuclei, etc., a new mechanics, called "quantum mechanics" (QM), was developed about 100 years ago for this purpose. In this case, in the light of some plausible postulates in physical and mathematical terms, a beautiful mathematical setting[2] of a partial differential equation (called "Schrodinger wave equation") was carried out. This equation, in turn, explained the nonrelativistic systems with reference to both their bound and scattering states. The wonderful agreement found between the theoretical predictions and experimental results forced one to believe in the quantum reality that was inherent in the basic postulates made in advance toward the foundation of QM. An important ingredient in this wonderful mathematical setting was the quantum wave function satisfying the partial differential equation and interpreted as the "probability amplitude" on the one hand and as a vector in the infinite dimensional Hilbert space on the other. It is the scattering/collision mechanism that has supplied a great deal of information about the subatomic world including the structure and substructures that exist in microsystems.

2.2 Relativistic Quantum Mechanics

It was during the same period when the Schrodinger equation was discovered and its limitations with reference to the description of fast-moving (with relativistic speeds) and spinning particles were noticed. Using the tenets of special theory of relativity of Einstein, Paul Dirac and several others came forward with other beautiful mathematical settings (known as "Dirac equation" for spin half-multiple particles and "Klein-Gordon equation" for integer-spin particles). These equations, in corroboration with experimental results, in fact, exposed another aspect of quantum reality realized only by a few in those days.

2.3 *Some Inferences Drawn About Quantum Reality from Present Developments*

Whether it is nonrelativistic or relativistic QM, in both cases, the window of quantum reality is characterized by some universal constants like Planck constant ($h = 6.6 \times 10^{-31}$Js), electron charge ($e = 1.6 \times 10^{-19}$coul.) and mass ($m_e = 9.1 \times 10^{-31}$ Kg), and velocity of light ($c = 3 \times 10^8$ m/s). In fact, the aspects of reality which can be extracted from these studies are as follows:

(i) *Discretization or quantization of continuum*: Note that action, energy, or for that matter any other dynamical variable that takes continuous values during the evolution of the system in classical mechanics is now allowed to take discrete values, e.g., changes or jumps in energy now take place in the units of some minimal packets, called "quantum" and at the scales of Planck constant, h.

(ii) *Dual nature of matter or the wave-particle duality*: The same material object (electron) can behave as a wave (double-split experiment) and also as a particle (photoelectric effect) depending upon the nature of the phenomenon. In this connection, the concept of matter wave was advanced by Louis de Broglie ($\lambda_{\mathrm{dB}} = h/p$)..

(iii) As a feature of motion, there is an entanglement between space and time which is translated into dynamical coupling [12, 13] at microscales through the so-called uncertainty principle due to Heisenberg. This principle, in fact, sets the limit on the simultaneous measurement of a particular pair of physical variables (called "canonical" pair) in terms of the Planck's constant h. It may be mentioned that these aspects again define a boundary line for the extraction of space- and time-mediated quantum reality. In fact there exists an inherent connection between the uncertainty relations and the wave-particle duality (Ehrenfest theorem).

(iv) *Probability entering through the quantum wave function (Ψ)*: The most plausible interpretation sought for the quantum wave function Ψ is due to Max Born, i.e., its modulus square defines the probability of finding the particle in a given region. In some sense, while this concept supports the wave-particle duality, it, however, forces one to believe that "God never plays a dice" in accordance with Einstein's belief. In brief, this offers a testimony of confusion. Clearly, the proposed concept of probability in the description of Nature is a reflection on the limitations of observer/apparātus with reference to accurate description.

(v) *Zero-point energy*: Contrary to classical mechanics, the quantum systems do have some energy even in the ground (unique and most stable) state. This concept again is an outcome of uncertainty relations and supported by several experiments.

(vi) *The presence of magnetism in microscopic systems*: Among several others, the main outcome of Dirac theory has been to establish the presence of magnetism in the spinning microscopic systems (like electron and proton) which otherwise was not possible from the Schrodinger quantum mechanics.

(vii) An important outcome of Dirac theory was the prediction of negative energy states, which, in turn, led to the concept of sea of occupied states, and the same is defined as the normal state of vacuum[3] consisting of infinite density of negative-energy electrons (or Fermi particles in general).

The reality as revealed from quantum mechanics is already of subtle category and is not easily realizable in one go by an average physicist. In the next section, we give some glimpses of quantum reality of still subtler category as realized through the subject of quantum field theory.

3 Quantum Reality as Revealed from the Studies in Quantum Field Theory

The four fundamental interactions in physics, namely, strong, electromagnetic, weak, and gravitation, have been studied rather elegantly through their "field" character. These space- and time-mediated interactions are characterized by their range and strength. While strong and weak are of finite range and active mostly in the nuclear domain, electromagnetic and gravitational are of infinite range and are spread all over the Universe. The strength however is a function of energy involved in the studies. It has been found that at energies of the order of 10^{15}GeV, first three interactions attain the same strength, and accordingly a grand unified theory has been developed. Attempts have been there to accommodate the finest possible gravitational interaction in this general scheme of study just by bringing a modification in the geometry or in the space-time structure of Nature up to the distances of the order of 10^{-33} cm. Whether it is the question of studying the behavior of coupling strength with energy (for strong, electromagnetic, or weak interactions) or the spatiotemporal distance toward the modification of geometry in the case of gravitation, in either situation, two entities (physical particles in the former and geometrical points in the latter) are considered. It is the "force" character of these interactions that explains the phenomena and suffices at the classical level. This provides only a limited aspect of reality. At the quantum level, however, the "field" character of these interactions is capable of providing better glimpses of absolute reality. In fact the spirit of both these aspects of study can be extended in an analogous manner to human systems [10] or to cosmic character of all-pervading consciousness field [1, 3, 10–12] as described in Vedic literature. We shall return to some of these details in the forthcoming sections.

Basically the physical field is considered as a continuum—an all-pervading medium which consists of above mentioned infinite member of entities (quantum or particles). It is described by the so-called field variable $\varphi(\vec{x}, t)$—a measure of displacement of particle at position $\vec{x}$ and time t. Not only this, the method of "second quantization" (the first being the quantization of dynamical variables in standard quantum mechanics) turns out to be an important mathematical tool to study the continuum of the field once again as an infinite system of discrete (granular) entities. For this purpose, different quantization schemes have been used [4]. The most convenient one is the canonical quantization in which the quantized entity carries a particle-like character described locally by a harmonic oscillator. In this case one writes the field variable as the combination of two terms—corresponding to particle-antiparticle (or particle-hole) description of the system in view of the Dirac concept of negative energy. This in turn introduces the creation-annihilation operators in the description.

This methodology, while successfully employed to study a variety of physical fields, however works on the basis of some assumption[5]. For example, *locality* assumption requires an idealization of space-time measurements in arbitrarily small regions. Further, as required by *Lorentz invariance* and a weak form of *causality*, these measurements separated by space-like interval cannot influence one another. In other words, (a) local observables exist, and (b) local observables, corresponding to space-like separated regions, can be measured simultaneously (commute). It may be mentioned that these assumptions, restoring the micro-causality and relativistic invariance, in conjunctions with other invariances like time reversal, parity, and charge conjugation (which indirectly conform to the existence of antiparticles), are found to work up to distances $(\hbar c/\sqrt{s}) \sim 10^{-15}$cm (for $\sqrt{s} \sim 60\,\mathrm{GeV}$) or up to nuclear sizes (see Chapter 4 in [16] (for a discussion on as to how the discretization of continuum in physical theories in classical sense is possible, we refer to Sect. 2.3 of [9]).

As far as the applications of these concepts to cosmic consciousness field (Φ_{cc}) of Vedic science are concerned, we shall return to such discussions later.

4 The Concepts of Infinity and Vacuum (*Śūnyatā*) in Physical Sciences

In historical perspective, the concept of discrete-continuum conversion was also used by Newton and Leibnitz when developing the subjects of integral and differential calculus about 400 years ago (cf. case (a) in Appendix A). Besides infinity and zero (*śūnya*), the concepts of limit and continuity were introduced through the so-called "infinitely small" (or infinitesimal) quantities in these developments. Still better understanding of these concepts has been possible only through the study of functions of complex variables or through the so-called $\in$-δ language, after the advent of the computer-based numerical techniques. For analytic functions a connection between pole-type singularities (infinities) and zeros of the functions is well known (cf. case (c) in Appendix A).

Whether it is the concept of infinity or of vacuum, in physics both are defined in a contextual sense [15].This is mainly because physics deals with space-time structure of Nature and that too in a measureable domain. No doubt the idealization is also a part of these studies in theoretical terms at the mental level (e.g., for a physicist the gases are ideal, surfaces are smooth, conductors are super, bodies are perfectly black, etc., or in the same vein the concepts of "infinity" and "zero" are ideal mental constructs for a mathematician), but in practice this concept has come down to a quantifiable or measurable level. Therefore, the accuracy in the measurement goes hand in hand with the ever-developing technologies on both experimental and computational fronts. In view of such a methodology of studies in physics, the phrases like "infinitely big" or "infinity small" have specific meanings at different scales. Again mathematician also talk of terms like countable or uncountable infinities or of zeros from left or from right in the sense of limit, but we refrain ourselves from going into such details here. Note that the word "infinitely" not only sounds as instrument-based but also depends on the level of development of the FOUs of the observer, e.g., for a common man, scientist, philosopher, or a yogi, the same word carries different meanings. While the first two categories of person move around space-time-related infinities, the last two categories, however, think of infinities beyond space time. In fact, it is the yogi category of persons who can visualize the absoluteness of infinity or of *śūnyatā*.

The realization of these concepts of infinity and zero in physics is slightly different. In physics one deals with the symmetry of space with reference to the energy/matter distributed in it. The particle number thus present constitutes the infinity, while the absence of matter/energy in space constitutes the vacuum (*śūnyatā*).

In practice, the word vacuum, emptiness, or void is defined differently in different contexts and at different levels of understanding. For example, at school level, the absence of air in a spatial region is defined as vacuum. In QFT, the absence of a particular physical field in a region defines the vacuum with respect to that particular field, while other fields could very well be there. In the same vein, one can as well define the (absolute) vacuum with respect to the absence of all the four physical fields and thus leaving only the space (container) behind. Alternatively, in physics, a medium, called "ether" (perhaps subtler than physical fields), in which the physical fields propagate, has also been postulated, but its experimental verification is still questionable. In other words, "ether" and "space" are not synonym to each other in physics. In QFT, such a vacuum has offered several interesting concepts mainly to give a "self-consistent" or "perfection" flavor to the developed theory. For example, the existence of "Higgs boson" (corresponding to a mechanism leading to the mass generation of a particle) and of the "self-energy" of a particle due to vacuum polarization are a few such studies. Experimental support to the theoretical ideas so developed, in some sense, suggests the Nature's liking for these concepts.

In Vedic science, one of the *mahābhūtas* is "ether" (out of the five, namely, earth, water, air, fire, and ether), and the same is identified with "space" (or *Ākās'a)* in which other *mahābhūtas* are contained. Now the next question is as to where all these five *mahābhūtas* of Vedic science or the "space" of physics are situated or else as to how the "space" is created? Naturally, there is still something more fundamental and subtler field than the ether in which not only the *pañca mahābhūtas* of Vedic science are situated, but it also creates the space, time, and geometry at the cosmic level as a result of certain mechanism occurring possibly in this field. Such an infinite, eternal, and immortal field is postulated here as the cosmic consciousness field, Φ_{cc}. Note that at microcosm level, it is the human consciousness (as a quantized version of Φ_{cc}) which, in conjunction with one's mind, intellect, and ego, creates space and time at a local level.

5 Cosmic Consciousness Field of Vedic Science and the Concept of *Śūnyatā*

5.1 Some Vedic Concepts About the Creator and the Creation at Primordial Stage

A lot has been said in Vedic literature about the Creator, the Grand Creation (*Sṛṣṭi*), and their role and interplay in the origin of both animate and inanimate objects in the Universe, at primordial stage. Here, however, following the work of Swami Dayanand Saraswati (see *Satyārtha Prakāśa*, Samullasa 8), we cite only a few verses from Vedas[6] and Srimad Bhagvad-Gīta[7] (SMBG) which clearly give some glimpses of absolute reality on the one hand and the existence of a physics-like field, called cosmic consciousness field, Φ_{cc} and a QFT-like process of "second quantization" on the other.

One of the verses in ṚṚgveda (Mandal 10, Sūtra 129, Mantra 7) states that "O (Anga) Man! know only him and not any other who is creator, sustainer and destroyer and also responsible for enlightening this multifaceted, all –pervading Creation (*Sṛṣṭi* derived from the word '*Sṛjana*') in which the entire Universe (or Universes) also constantly undergo creation, sustenance and destruction" and another one [11] that "Prior to creation, this entire Universe in its micro but ether-like form was completely dark, neglected, not even knowable during darkness and was locally embedded in the Creation." Yet there is another verse in Yajurveda (Ch.3)[8] which states that "O man, one who is present in all in totality, imperishable, Master of all Beings and who is beyond the concept of animate and inanimate, is the Creator of past, present and future of this time bound Universe."

To this effect, several are verses in SMBG which also give a similar message. For example, the Verse 9.4 clearly states that "All this is permeated by me in my un-manifest aspect (as the ice by water) and all Beings abide in the idea within me. Therefore, really speaking, I am not in them," and the Verse 9.6 emphasizes that "Just as the extensive and all-pervading air (which is born of ether) always remains in ether, likewise know that all Beings (sprung as they are from my thought) abide in Me."

While the detailed implications of the hidden meanings of these verses are discussed elsewhere [13] from the physics point of view, here we just draw a one-line conclusion that there exists an all-pervading, infinite, eternal, finest possible cosmic consciousness field (*Sṛṣṭi*, denoted by Φ_{cc}) in which not only the five cosmic elements earth, water, air, fire, and ether (toward the creation of inanimate) but also the cosmic mind, cosmic intelligence, and cosmic desire (toward the creation of animate) are embedded. While ether contributes to the creation of space, time is created with the origin of a particular Universe. In summary, note that the present physical theories through force and field versions of interactions, in fact, are exploring only this limited aspect of reality associated with space-time-mediated creation of Nature in concurrence with possible experiments in the laboratory and that too within the limitations of the formulated laws. A vast aspect of study of Φ_{cc} however remains untouched by modern scientist mainly because they are afraid of treading into the domain of consciousness.

In SMBG there are some more verses which clearly support the above theory of Creation (see, e.g., SMBG: 2.6, 9.7-9.10 and several other verses) and offer the spirit of the method of second quantization of QFT. For example, Verse 13.16 states that "though indivisible (like ether), It stands as if divided among Beings. That knowable substance is sustainer of Beings (as *Viṣṇu*), destroyer (as *Rūdra*) and creator (as *Brahmā*)." In another verse (SMBG: 10.39) the Lord states that "Arjuna, I am even that which is the seed of all Beings. For there is no creature, animate or inanimate, which exists without me." In other words, it is the same element of consciousness which is the seed (or cause of existence) of both Beings and non-Beings in the Universe. (For similar details, also, see SMBG: 13.27 and 15.17).

5.2 *Lessons from QFT and Cosmic Consciousness Field: Possibility of a Workable Model*

When applying the tools of QFT to study cosmic consciousness field, it is worthwhile to recall the assumptions made in developing the QFT itself, namely, locality, Lorentz invariance, and the weak form of causality in the assumed arbitrarily small region. As a matter of fact, all these assumptions move around the finiteness and constancy assumption of velocity of light in vacuum which in turn defines the space- and time-like regions. The canonical quantization which allows the identification of the quantum of the field with a particle, however, is equally appealing in this respect. For a moment if one relaxes all these assumptions in QFT, then analogous application of these tools of QFT to the cosmic consciousness field of Vedic science can provide several interesting clues about life, its origin in Beings, and the world order in general. Toward such studies, it is worthwhile to list here some meaningful lessons drawn from QFT which may be useful in analyzing the Vedic concepts of Sect. 5.1 regarding the Creator, Creation, and the World order, viz.:

(i) There exists a finest and subtlest possible, fundamental all-pervading cosmic consciousness field with the properties discussed above and denoted by Φ_{cc}.
(ii) In Φ_{cc}, there exist and embedded epi-fields not only of five cosmic elements but also of microcosm of human Being [11, 13] in primordial (seed) form.
(iii) These epi-fields in $\Phi_{cc}$are there in their infinitely dressed (with consciousness) versions [11] and merged with Φ_{cc}. Also, their partially dressed and nascent versions are present in Φ_{cc} but at different scales of fundamentality.

In this connection the particle-hole picture of QFT provides a direct link between the quantum (soul) and the ocean of cosmic consciousness, Φ_{cc} . In fact both of them (particle and hole) are always in entanglement during the life span of a Being, and this entanglement is possibly responsible for the *prārabdha karma* of a human Being. For further details we refer to our earlier [11, 13] works.

6 Understanding Zero and Infinity in Mathematics and Vedic Literature

There are plenty of situations in Vedic literature where the verses hint directly or indirectly to modern scientific and mathematical concepts including those of zero, infinity, and infinitesimal. To this effect, while some glimpses are already given in Sect. 5.1, we further refer to earlier works [7, 14]. The only thing is that in such attempts one has to read between the lines and analyze the meaning(s) of the verse appropriately. Beside the language, the reader also has to be a subject expert, and his degree of expertise in fact matters a lot in this endeavor. Note that such wisdom obtained from Vedic literature can be of two types: the direct and the derived one. In the former case, the scientific or mathematical wisdom reflects directly from the literary beauty of the verse even to a common reader. In the second case, however, there is enough scope for philosophers and yogi category of persons to realize or extract the hidden truth in the verse. Other persons use analogies in the extraction of appropriate meaning(s) of the verse. In this case also the quality of knowledge extracted is affected by the nature (order) of analogy [6, 7] used. We briefly list some disconnected places in mathematics in Appendix A from where the gathered wisdom on zero and infinity can possibly help in understanding the concepts of *śūnya* and *ananta* in Vedic literature better and in an alternative way.

7 Summary and Concluding Discussion

It is argued that the absolute reality in Nature can be approached through two seemingly different styles of study, namely, subjective and objective, which, in turn, correspond to studies of unmanifest and manifest aspects of Nature. The quantum reality noticed in recent times for microscopic systems basically belongs to the objective aspect of study. The ever-increasing accuracy in computational and measurement techniques, in fact, is responsible for unfolding such great vistas in Nature, particularly in the realm of microscopic systems. With a view to exploring the quantum reality as it exists today, a brief survey of subjects of quantum mechanics and quantum field theory is carried out with reference to their philosophical foundations. It is suggested that the concepts of zero-point energy in quantum mechanics and that of vacuum in QFT can possibly have some links with the concept of *śūnyatā* in Vedic literature.

These are the concepts of *śūnyatā* and *ananta* which have been talked about elaborately in Vedic literature with reference to the Creator, Creation, and the world order. A brief survey of such situations in Vedic sciences is presented in Sect. 5. In analogy with the existing concepts in QFT, attempts are made here to identify, study, and analyze the cosmic consciousness field, Φ_{cc}, of Vedic science as inherent in several Vedic verses and in SMBG. In this context not only the concept of vacuum is investigated, but the discretization of continuum of Φ_{cc} through a process like "second quantization" of QFT is found an interesting answer to the question of division of Φ_{cc} and origin of soul in an individual. It is suggested that the concept of embedding of epi-fields in Φ_{cc} can give new dimensions to these studies.

As a matter of fact, there is yet another powerful language through which Nature expresses Herself better, and that is the language of mathematics. Both zero and infinity have been the part and partial of studies in this subject, particularly in conjunction with physics. To this effect, a brief account of some situations from mathematics literature is given in Appendix A with a view that such a ready availability of tools might help in exploring the concepts of *śūnyatā* and *ananta* in Vedic literature better, of course, in an analogous manner.

Acknowledgments The post-retirement association with Ramjas College and the Department of Physics and Astrophysics, University of Delhi, is gratefully acknowledged. Thanks are also due to Professor S. R. Bhatt for encouragement and many inspiring discussions.

1. In fact, the quantum reality, in philosophical terms, is described in various ways in different contexts. There exists a huge literature on this subject. We cite here only a few. See, for example, [1, 3, 18].
2. See, for example, [17].
3. See, for example, ([2, 16]: [4]).
4. See, for example, ([2, 16]: [4]).
5. See, for example, ([2, 16]: [4]).
6. Here we cite the spirit and only meaning of a few verses quoted in *Satyartha Prakash* by Swami Dayanand Saraswati.
7. See any edition of Srimad Bhagvad-Gītā, Gītā Press Gorakhpur (here abbreviated as SMBG).
8. See Footnote 6.

Appendix A: *Śūnya* and *Ananta* in Mathematics vis-à-vis Analogy-Based Vedic Wisdom

In continuation of Sect. 6, here we pinpoint some cases from mathematical analysis relating to zero and infinity that may be useful in analyzing the analogy-based Vedic wisdom on *śūnyatā* and *ananta*.

(a) *Discrete-continuous conversion*: Recall the definition of differential coefficient (d*f*/d*x*) for the function *f* in $y = f(x)$, viz.,

$$\frac{df}{dx} \equiv \lim_{\Delta x \to 0} \left(\frac{\Delta y}{\Delta x}\right), \qquad (8.\text{A1})$$

where Δx and Δy are small but finite changes in the independent variable x and dependent variable y, respectively. In the limit when the change $\Delta x \to 0$, the ratio $(\Delta y/\Delta x)$ becomes a continuous function (d*f*/d*x*) of x. At the computation stage since the computer does not understand the philosophical idealized concept of limit or zero, one introduces the "$\in - \delta$ language" at a subtler level. Sometimes the identity

$$\frac{\mathrm{d}f}{\mathrm{d}x} - \lim_{\Delta x \to 0} \left(\frac{\Delta y}{\Delta x}\right) = 0 \qquad (8.\text{A2})$$

is also used in setting the accuracy in computation.

(b) *Algebraic equation, identity, and symbolic relation* versus *śūnyatā:* When one talks of objects or the operations on the set of objects, then there are various ways of arriving at zero (or *śūnyatā*). As a naïve example, consider the following algebraic relation among (say) two variables x and y:

$$ax + by = 0. \qquad (8.\text{A3})$$

Some plausible meanings of this relation are as follows:

(viii) Adjustment of fraction a of the item/operation x and of fraction b of the item/operation y is such that the net effect of both operations neutralizes or vanishes. (A case when (8.A3) is a symbolic relation).

(ix) For fixed nonzero parameters a and b in (8.A3), variables x and y vary in such a way that the net effect of all arithmetical operations involved on the left-hand side in (8.A3) is zero. Not only this, x and y in this case are expressible in terms of each other, i.e., they are dependent variables (A case when (8.A3) is an algebraic equation).

(x) If x and y are not linearly dependent and left-hand side of (8.A3) has to be zero, then only way out is that $a = b = 0$ (A case when (8.A3) is an identity).

These interpretations of (8.A3) suggest that the concept of *śūnyatā* in Vedic literature can be understood in a variety of ways. Therefore, the context hinted by the given verse is important.

(c) When studying the integration of a particular type (logarithmic derivative) of complex analytic function within a closed contour in the theory of functions of complex variable, recall that there exists a definite relation between the total number of zeros and total number of pole-type singularities (infinities) of this function (Cauchy's theorem). Such an observation may be relevant to find a link between *śūnyatā* and *ananta* in Vedic literature.

(d) *Case of Fourier transform*: Out of a large number of integral transforms defined in applied mathematics, it is only the Fourier transform that is physically realizable. It is this transform which has explored the one after the other substructure(s) deep inside the matter at microlevel in terms of structure functions or form factors. Symbolically, one writes this transform in the form

$$g(p) = \int f(x)\, e^{ipx}\, \mathrm{d}x, \tag{8.A4}$$

where the variables x and p constitute a canonical pair in the sense of Hamiltonian dynamics. This transform connects two functions $f(x)$ and $g(p)$ which are the functions of their respective arguments. An interesting property of this transform is that the behavior of the function $f(x)$ for small values of x correspondingly describe the behavior of $g(p)$ for large values of p or the vice versa. Keeping such a property in mind, one can use the Fourier transform in psychophysics in correlating one's behavior in the material and spiritual worlds through the meditation variable μ (see glossary of words in [11]). (Recall the behavioral fact that small μ corresponds to worldly involvement, whereas large μ corresponds to spiritual involvement.)

(e) *Case of Dirac delta-function*: It is worthwhile to add another interesting case of Dirac delta function $\delta(x)$ here following the work of [14]. This function (or rather "functional"), defined as

$$\delta(x) = \begin{cases} 0, & \text{for } x \neq 0, \\ \infty, & \text{for } x = 0, \end{cases} \tag{8.A5}$$

with

$$\int_{-\infty}^{\infty} \delta(x)\, \mathrm{d}x = 1, \tag{8.A6}$$

was first introduced in quantum mechanics by Paul Dirac in 1928. Later, this functional has given birth to a new branch in mathematics called "distribution theory." Prof. Pandey calls this peculiar function as "Vedic functional" and has offered a number of interpretations of Eqs. (8.A5) and (8.A6) and other properties of

$\delta(x)$, which, in turn, support the immortality of soul and the theory of rebirth in Vedānta philosophy.

(f) *Null set and universal set in set theory*: One can find wonderful ways to interpret *śūnyatā* and *ananta* in Vedic literature by using the concepts of null set and universal set along with various operations on them in pure mathematics.
(g) *Zero and infinity under simple arithmetical operations*: In numerical analysis sometimes one comes across certain situations when it becomes difficult to extract the exact values/meanings of certain computed quantities. These quantities, often quoted in mathematics literature as "indeterminate forms," are $\frac{0}{0}$, $0.\infty$, and $\frac{\infty}{\infty}$. While there are ways and means to handle these situations in mathematics, in philosophical terms, the meaning(s) of these quantities has been attached with the roles of *prakṛti* and *puruṣa* in the maintenance of world order.

What we have presented above are some representative cases from mathematics in which the roles played by zero and infinity can analogously be extended to the philosophical meanings of some verses in Vedic literature but in a judicious manner.

References

1. Goswami A. Quantum mechanics. Dubuque: Wm. C. Brown Puablisher; 1992.
2. Itzykson C, Zuber JB. Quantum field theory. International Student Edition; 1986.
3. Jammer M. The philosophy of quantum mechanics: the interpretation of quantum mechanics in historical perspective. New York: Wiley; 1974.
4. Kaku M. Quantum field theory: a modern introduction. New York: Oxford University Press; 1993.
5. Kaushal RS. Structural analogies in understanding nature. New Delhi: AnaMāyā Publishers; 2003.
6. Kaushal RS. Abstraction and structural analogies in mathematical sciences. In: Grattan-Guinness I, Yadav BS. History of the mathematical sciences. New Delhi: Hindustan Book Agency; 2004. p. 33–52.
7. Kaushal RS. Structural analogies, abstraction and mathematical concepts in vedic sciences. Indian Philos Q. 2006;33:125–46.
8. Kaushal RS. Classical and quantum mechanics of complex Hamiltonian systems. Pramana J Phys. 2009;73:287–97.
9. Kaushal RS. Discretization of characteristics: a concept prevailing in mathematics and behavioral sciences. Indian Philos Q. 2010;37:149–73.
10. Kaushal RS. The science of philosophy: theory of fundamental processes in human behaviour and experiences. New Delhi: D.K. Print World; 2011.
11. Kaushal RS. Psychophysics and human interactions: art and science of integrating body, mind and soul. New Delhi: New Age Books (Motilal Banarasidass); 2015.
12. Kaushal RS. The psychophysics of Chevreul hand-held pendulum. J Conscious Stud (UK). 2016a;23(9–10):134–52.
13. Kaushal RS. All-pervading cosmic consciousness field of vedic science and the quantum field theory of modern physics, to appear in proceedings of international conference on WAVES-2016, held at Bharatiya Vidyā Bhavan, New Delhi during 15–18 December 2016, (Abstract No. 53). 2016b.

14. Pandey GS. The vedic concepts of infinity and infinitesimal system. In: Grattan Guinness I, Yadav BS, editors. History of mathematical sciences. New Delhi: Hindustan Book Agency; 2004. p. 65–82.
15. Podolny R. Something called nothing: physical vacuum what is it? Moscow: Mir Publishers; 1986.
16. Ryder LH. Quantum field theory. Cambridge: Cambridge University Press; 1987.
17. Schiff LI. Quantum mechanics. 3rd ed. New York: McGraw-Hill Book Co.; 1968.
18. Stapp HP. Mind, matter and quantum mechanics. 3rd ed. Berlin/Heidelberg: Springer; 2009.

Chapter 9
Concepts of Reality and Śūnya from the Perspective of a Physicist

Kashyap Vasavada

Abstract This paper discusses the philosophical concepts of Reality, Śūnya, and Śūnyatā which are integral parts of many Hindu and Buddhist scriptures from the perspective of a physicist. Amazingly, these concepts find strong parallels in areas of modern physics such as quantum physics and cosmology. These concepts in modern physics will be explained, and it will be shown that they have strong resemblance to these concepts mentioned in Indian scriptures. Especially intriguing is the finding of the lack of objective reality, subjective role of observer, and nonrealistic interpretation of quantum phenomena. Some speculations, on how this close agreement between such completely diverse areas could come about, will be also presented.

Keywords *Brahman* · *Śūnayatā* · Cosmology · Quantum physics · Veda · Consciousness · Ādi Shankaracharya

1 Introduction

Concepts of ultimate reality in Buddhism, namely, Śūnyatā, and in Hinduism, namely, Brahman, are discussed from the perspective of modern physics. We find that there is an astonishingly close parāllelism between the two completely diverse fields. Some speculations are presented suggesting how this could happen. We also discuss universal consciousness as suggested by the two religions.

Deep philosophical concepts of ultimate reality called Brahman, Śūnya, or Śūnyatā are integral parts of many Hindu and Buddhist scriptures. Brahman is the name given to the formless, shapeless, abstract, omnipresent, invisible, eternal, transcendental, and immanent form of ultimate reality or God in Hinduism. The literal meanings of Śūnya and Śūnyatā are zero and emptiness, respectively. Buddhism, which arose in India, agrees to a large extent in essential details with the

K. Vasavada (✉)
Indiana University-Purdue University, Indianapolis, IN, USA
e-mail: vasavada@iupui.edu

S. R. Bhatt (ed.), *Quantum Reality and Theory of Śūnya*,
https://doi.org/10.1007/978-981-13-1957-0_9

original Hindu religion of Vedas and Upaniṣads.* Mathematically, the Buddhist concept of Śūnyatā and Vedic concept of Brahman would correspond to zero and infinity, respectively. But as we will see, the two concepts have a lot in common. In fact both Buddhist monks and Hindu yogis try to realize Śūnyatā and Brahman, respectively, by emptying the mind during meditations. Amazingly, these concepts find strong parāllels in areas of modern physics such as quantum physics and cosmology.** Founding fathers of quantum physics such as Bohr, Schrodinger, and Heisenberg were deeply impressed with Eastern religious philosophy.***

Pre-twentieth-century physics is called classical physics. Its basis is the material world we experience with the help of the eyes and other sense organs. Then, at the beginning of the twentieth century, physicists found several phenomena, especially in the study of atoms, which forced them to change their concepts of the structure and reality underlying the universe. This resulted in what is known as quantum physics. It took years and years of painstaking experimental, theoretical, and mathematical work to understand the new phenomena. When compared with experiments, quantum physics works to an astonishing accuracy of about one part per billion or better! But the meaning of the equations cannot be understood in terms of our everyday life experiences which are classical! Anyway, this resulted in complete change of physicists' world view of underlying reality. The debate as to the real meaning of all these equations has been going on for more than 90 years without any consensus. But some ideas are becoming clear beyond any reasonable doubt. The difficulty physicists encounter in describing quantum physics in everyday language is very similar to the difficulty of expressing various ideas about Brahman or Śūnyatā from the Eastern religious scriptures in everyday language. Hindu sages (called Ṛṣis) say that the only way of understanding Brahman is by going the route of "Neti, Neti (not this, not this)." This says that when you discard everything in the world that you see, the remaining concept is Brahman! Buddhist concept of Śūnyatā is similar. In the following we will discuss the ideas of quantum physics and compare with the philosophical ideas of ancient Hindu and Buddhist scriptures. In this article, when "quantum physics" is mentioned, we will include both the nonrelativistic quantum mechanics as proposed by the pioneers mentioned above and the relativistic quantum field theory which developed later.

2 Cosmology, Quantum Physics, Śūnya, Śūnyatā, and Brahman

It is well known that the idea of Śūnya as zero in the number system originated in India. This development of the decimal place system containing zero was of fundamental importance in the advancement of science in the Western world. These ideas had obviously some connotation with the philosophical ideas of Śūnya and Śūnyatā.

Next, we discuss the concept of Śūnya and Śūnyatā in detail. Initially we will take the literal meaning as zero, emptiness, void, or vacuum, although philosophers have attached deeper meanings to these such as the totality of reality. The Buddhist philosopher Nāgārjuna[1] had deeply contemplated on the meaning of Śūnyatā or emptiness. In his opinion everything in the universe is empty in the ultimate analysis. To Nāgārjuna the entire universe came up from Śūnyatā!

We will see that this meaning finds parāllels in modern physics. Modern cosmologists believe that our known universe started as an extremely small patch of vacuum some 13.8 billion years back. It did not have any material particles in it. We can call it Śūnyatā. But it is not right to say that it means that something came up from nothing! The quantum vacuum is not a true emptiness or nothingness or the absence of everything! It has a very complex structure involving fluctuating quantum fields in it. The present material world had origin in quantum fluctuations of this vacuum. There were no particles initially, and because of absence of light, it must have been dark.

Now let us compare this modern cosmology picture with the astonishing description of origin of universe from Vāyupurāṇa,[2] a Hindu scripture:

> In the beginning, there was nothing in the universe. The Brahman (the divine essence) alone was everywhere. The Brahman had neither color nor scent; it could not be felt or touched. It had no origin, no beginning or no end. The Brahman was constant and it was the origin of everything that was destined to be in the universe and the universe was shrouded in darkness.

This is very impressive! They realized that it must have been dark because visible light was not created yet! Nāsadīya Sūkta,[3] another Hindu cosmological verse, also says that it was dark before the universe was created!

An excerpt from ṚṚgveda,[4] one of the four original Hindu scriptures, called Vedas, says:

> The universe is brought about by the collapse of fullness in the transcendental field in which reside all the laws of nature responsible for the creation of the entire manifest universe. How is the transcendental level functioning? It is functioning from its unbounded nature to point to itself. He who does not know that initial pure consciousness state, ultimate reality, what can the laws of nature accomplish for him? He who knows it, remains established in evenness, unity, wholeness of life.

Since Brahman was by itself, it is clear that it interacted with itself, i.e., self-referral, and eventually manifested in every particle of the universe. In an interesting parāllel, according to modern cosmology, a particle or quantum field called inflaton may be the origin of everything in the universe. Strictly speaking the word "manifestation" rather than "creation" is used in Vedic cosmology with a subtle meaning. They

---

[1]For a comprehensive article describing Nāgārjuna's philosophy and references to his works, see, for example, http://www.iep.utm.edu/nagarjun/

[2]Hinduonline.co/Scriptures/Purāṇas/VayuPurāṇa.html

[3]https://en.wikipedia.org/wiki/Nāsadīya_Sūkta, ṚṚgveda (10:129).

[4]The ṚṚgveda: ṚṚgveda I.164.39]. Several articles from Maharishi University of Management, Fairfield, Iowa, give similar translation of these Vedic verses.

allude that Brahman did not create the world, but it manifested itself in a world that was somehow implied *in* Brahman.

It should be emphasized that physicists built up the model for the universe arising from quantum vacuum after centuries of wrong concepts and models. It is an astonishing testimony to the ancient Ṛṣis that the scriptures had this idea without having recourse to the experimental or mathematical methods used in modern times!

It is now clear that the material world we see around is not really made out of rigid, solid brick-like substance if you go to the submicroscopic scale. Physicists found that as you go deeper and deeper, there is vacuum and vacuum all the way down. Materials are made out of atoms which are in turn made out of nuclei and electrons with a lot of empty spaces in between. As an analogy, if an atom is of the size of a football stadium, the nucleus would be a football at the center, and most of the electrons will be spectators in the stadium with an empty space in between. This emptiness continues as we go deeper and deeper until we get to the fundamental particles with whimsical names like quarks, gluons, etc. According to quantum physics, none of these have rigid, solid structure.

3 Quantum Reality, Absence of Objective Reality

From the beginning, physicists found that the so-called particles have a dual nature, viz., waves and particles. Whether you see experimentally a wave or a particle depends on your method of observing. So the question "Does matter consist of waves or particles?" cannot be answered without context of the experiments and the mathematical machinery of quantum mechanics. The particles are in some sense both here and there at the same time and are described by a wave function, a superposition of mathematical functions with seemingly contradictory properties. Absolute Square of this function gives probability (not certainty) of observing various properties of the "particles." For example, one cannot say certainly that the particle is at a particular place. It's location at any place is given only in a probabilistic way. Such a description is very similar to the description of Brahman, e.g., in the scripture Īśopaniṣad[5]:

> It moves and it moves not; it is far and it is near; it is within all this and it is also outside all this.

Nāsadīya Sūkta[3] talks about neither existence nor non-existence at the beginning of universe which would be a superposition of contradictory concepts! Nāgārjuna[1] also talks about simultaneously existence or non-existence, both or none!

Schrodinger, one of the founding fathers of quantum physics, had already noticed a peculiarity of combined wave function of two or more particles in certain situations. He realized that it cannot be factorized as a product of wave functions of each

[5]Ishopanishad: sanskrit.org/WordPress/wp-content/uploads/2015/10/IshaEnglish, text 5.

of the particles. He called this occurrence "entanglement." One consequence is that, for the entangled particles, measurement of an entangled property of one particle is enough to predict the properties of the other particles without actually measuring their properties. Einstein, Podolsky, and Rosen[6] then argued that measurement of positions or momenta for the entangled particles would lead to contradictions with quantum theory if one insists on the particles having actual real properties before measurement, the reason being that these variables cannot be simultaneously determined because of uncertainty principle. To them this meant that quantum theory was incomplete since they believed that particles had real properties before measurement. This is known as EPR Parādox[6]. Then, in a very interesting development, John Bell[7] analyzed cases in which particles would be produced in quantum entangled spin states (spins are correlated) and sent in different directions, including consideration of hidden variables which are unknown. Subsequently their spins in different (random) directions would be measured by two different observers. Based on classical logic, he derived a mathematical inequality. A large amount of experimental data subsequently proved that Bell's inequality is very badly violated, and the results agree completely with quantum physics predictions. Moreover local hidden variables were also ruled out. Over the years the experiment has been repeated many times with similar results. Experiments have been also done with particles which were never in actual contact but have been in contact with other particles which were in contact (called entanglement swapping). These results prove that a particle does not have any definite real value of spin in a particular direction (or position or momentum) before it is measured! Once it is measured, it is found to be definitely correlated with the spins, positions, and momenta of the other entangled particles! The particles have some kind of suspended, unreal existence before measurement! This conclusion resolves the EPR Parādox also. The fact of a lack of real properties before measurement matches very well with the Buddhist concept of Śūnyatā as the absence of "Svabhāva" (inherent or intrinsic nature). There is also a current theory that possibly the whole universe may be a web of entanglement, probably due to Schrodinger.[8] It would make our universe a holistic rather than a collection of totally independent entities. This would be consistent with the central message of Vedas and Upaniṣads also. It should be noted that realistic interpretation (with nonlocal hidden variables) of Bell's results such as Bohm's interpretation has been essentially ruled out. Bohm's concept of nonlocality contradicts the main axiom of Einstein's special theory of relativity that no signal can be propagated faster than the velocity of light. Thus local nonrealistic interpretation is the only one which has majority consensus.

This is the stark world of quantum reality. It reminds one of the ideas of Māyā (loosely translated as illusion or delusion) covering the whole universe, as Ādi

[6]https://en.wikipedia.org/wiki/EPR_Parādox

[7]https://en.wikipedia.org/wiki/Bell's_theorem

[8]S. Rammohan has pointed out that Yajurveda (one of the Vedas) (6.3.7) is in agreement with this theory. See New Delhi Conf. Proceedings.

Shankaracharya said "**Brahman Satyam, Jagat Mithyā**": Brahman is the only truth; the world is not real.[9] Māyā, jagat, is the mistaken impression that appearances are real, like mistaking a rope in the dark for a snake. Quantum physics teaches us that it is a mistake to accept the world as we perceive it to be real. Instead, the appearances arise from something that is beyond our ability to intuit, much like the nature of Brahman. Now, just like us, Shankaracharya must have seen solid bodies, rigid walls, trees, etc. I believe he realized that all of this disintegrates and goes away. Then it cannot be fundamental reality. Thus, because of the covering of Māyā, one does not see the underlying reality of Brahman. It would be similar to the fact that we do not see wave-particle duality with our sense organs. Just as modern physics says that everything, living and nonliving, is made out of the same elementary particles, Hindu scriptures say that Brahman is present in everything. We can conclude that the ancient Indian culture came to this fundamental insight by its means just as we have arrived at the same insights in modern times by our means.

An important conclusion of quantum physics is that there is no observer-independent reality. Reality, whatever it is, is strictly subjective. In the West, this position is called "idealism" and was long rejected by realists, but now, the evidence of quantum physics forces a type of idealism into science. As mentioned above, whether you see particle nature or wave nature depends on your subjective choice of experimental setup. The basic nature is modeled as probabilistic, and each observer will get a different result. Only after a large number of observations, the results are found to be consistent in a statistical sense. Such a subjective implication is found in various Upaniṣads also. It is said that true Samādhi is achieved when the observer and the observed merge with each other. There is a theoretical concept in high energy physics that the usual four forces in nature, which look very different at low energy, would be unified at very high energy. The symmetry is broken at low energies and may be restored at very high energy. In this respect Samādhi may be like a mental particle accelerator, and it goes up the energy scale and fuses the observer and the observed into a unified state. From this unified state, the reality of Brahman becomes accessible. Otherwise, the observer and the observed appear to be different to our minds. In a "meta-mind," the observer and the observed are unified.

In quantum physics, there is a long-standing unresolved debate whether, ultimately, the observer has to be conscious or an inanimate machine can also be an observer! The main reason for this particular controversy is the Western (Abrahamic religious) concept that only humans have souls and consciousness. We will return to the discussion of consciousness later.

[9]https://en.wikipedia.org/wiki/Vivekachudamani

4 Everyday Logic, Modern Physics, and Scriptures

Both the ideas of reality in quantum physics and Hindu and Buddhist scriptures defy not only our subjective intuitions of the world but also the everyday logic with which we are familiar. As an example of violation of conventional logic, let me mention one situation. Consider a simple logical inference such as: if in a roomful of 200 people, 50 have brown eyes, then 150 do not have brown eyes. Bell's inequalities are based on such logical statements when applied to quantum objects. They are violated by experimental results on atomic systems. The conclusions from the quantum theory agree completely with the experiments. This means that the assumption in classical logic that the quantum objects have real permanent properties before measurement just like the eyes have permanent color before measurement is not valid!

Such a situation would correspond to Upaniṣads' idea that Brahman cannot be understood by logic. As mentioned previously, it can be identified only by the words "Neti, Neti (not this, not this)." There is a story about two yogis. One meditated all day. The other one read scriptures all day. At the end of every day, the second one always said "I do not understand. I do not understand." Then one day, to the great surprise of the first one, the second one said loudly, "I understand. I understand." The first yogi asked him with astonishment as to what suddenly happened. "Now you understand everything suddenly!" The second yogi replied "I now understand that this cannot be understood!"

As a related idea, one can consider Gödel's incompleteness theorems.[10] They say that "no consistent system of axioms whose theorems can be listed by an effective procedure (i.e., an algorithm) is capable of proving all truths about the arithmetic of the natural numbers" and "the system cannot demonstrate its own consistency." One runs into similar problems in considerations of quantum theory and ideas of Brahman and Śūnyatā in that it may be impossible to understand them by using our everyday logic.

To me, it does not appear surprising that when one tries to put the mathematical ideas of modern physics in human languages, they look similar to the philosophical ideas of ancient Ṛṣis and Buddhist monks obtained after deep meditations. In fact it would be shocking if they did not agree. Physicists and the Ṛṣis (or Buddhist monks) would not agree if the world was described by classical physics. I think the nature of reality is such that both parties were led inevitably to adopt these ideas because it reflects the deeper truth of the nature of reality. As for Ṛṣis (or Buddhist monks), it is not clear when the analog of the classical to quantum transition took place or indeed if such a transition even took place. Some parts of Vedas are full of worship of natural elements like wind, water, fire, etc. and also picture Gods that look like human beings in the form of avatars. In some parts of Vedas and many Upaniṣads, we see clearly concept of abstract, omnipresent, invisible, eternal, transcendent, and immanent Brahman who has qualities unfamiliar in our everyday life. Just as many

[10]https://en.wikipedia.org/wiki/Godel's_incompleteness_theorems

things in everyday world are described by classical physics, concept of deities in the scriptures would correspond to classical concepts. There is nothing wrong with that. The concept of Brahman would correspond to the nonintuitive abstractions of quantum concept. Nonetheless, unlike the case with physics, the deeper abstractions of the nature of reality seem to have been present since the earliest origins of Hindu and Buddhist thought.

Both modern physics theorists and Ṛṣis (or Buddhist monks) who presented their ideas in Hindu (or Buddhist) scriptures were using their thought processes in the brain. Obviously, the human brain evolved as the human body evolved in nature. So one possibility is that some cognitive information about nature may be stored in the brain. A question for physics is that why mathematics works so well when our intuition based on everyday life fails. After all, mathematics is also a creation of our human minds. In fact, as I mentioned before, Bell's inequalities bring out in a superb way that conventional logic fails in quantum theory. Thus somehow mathematics describes systems which are outside our everyday experience. Similarly, why Ṛṣis and Buddhist monks realized something in their meditations which went beyond their everyday intuitions? They were living in the classical world like everybody else in cottages, carrying out the usual human activities. In both cases the analysis looks irrational from the conventional logical point of view. Perhaps deep within our brain, there is some component which goes farther than experiences in everyday world. It is somehow sensing the so-called ultimate reality which physicists have arrived at by using sensory experiments and mathematics and Ṛṣis and Buddhist monks arrived at by deep meditations induced by the methods of yoga and Samādhi. The means are different, but the results are so overwhelmingly similar that something very fundamental must be going on.

Sometimes one hears the argument that internal world is different from external world. I was never convinced by that argument, and what is discussed above paints a different picture. World is world. How can internal laws be different from external laws? My internal world is actually somebody else's external world! If we assume that the internal world cannot be independent of the external world, we have to conclude that this similarity in philosophical statements of modern physics and ancient Indian mysticism is not a coincidence. It must be the "ultimate reality" of nature. Most of the Western scientists and some Indian scientists believe this to be merely coincidence without any significance. In fact some Western scientists have ridiculed such similarities as accidental and misleading. I have been myself wavering for a number of years about whether this is accidental and forced association or genuine. Now I am convinced that it is a real association. I suspect there is an all-pervading holistic nonlocal layer which we may call Brahman or Śūnyatā. Part of our consciousness may draw on this. Alternatively, some have argued that our very consciousness is Brahman, the living light that brings not illumination but being. The connection between the infinite being of consciousness (Brahman) and our individual lives as seemingly autonomous beings may be what is called Atman. Nonlocal entanglements of atoms may be also related to this layer. Local interactions which one sees in physics and also in everyday life may be superimposed on this. These are much more prevalent at our large scale made of innumerable countless quantum events. The main point of this article is that, somehow, the sensory and non-sensory aspects of the universe are in fundamental agreement. A complete answer to these puzzles will come when we understand consciousness and its relation to the nature at large.

5 Unsolved Problem of Consciousness

Physics (science in general) understands matter and energy very well. The great stumbling block is understanding of consciousness. Despite years and years of efforts, science has not made much progress in understanding it. Here perhaps the ancient wisdom seems to be right. There is a saying in Vedas that consciousness is singular. Hindu scriptures suggest Brahman as a universal cosmic super consciousness: **Prajñanam Brahman** – Consciousness is Brahman.[11]The other well-known Vedic sayings express similar concept about individual and Brahman: **Ayam Ātma Brahman** (This Self (Atman) is Brahman[12]), **Tat TvamAsi** (That Thou art[13]), **Aham Brahmāsmi** (I am Brahman[14]), and **Sarvam Khaluidam Brahman Tajjalān Iti śāntam upāsita** ("All this (collectively) is Brahman, indeed: what evolves from That, what dissolves in That, what breathes or functions in That, should be closely and calmly studied. . . ."[15]). In a number of verses in the Bhagavad Gītā, it is mentioned that God (Brahman) is present in every animate and inanimate object.[16] So the source of consciousness appears to be external yet is experienced as *the* internal reality. There is a universal cosmic consciousness, and we are reflections of it as different faces of a single crystal would reflect multiple images of a single object. This may explain why consciousness per se is wholly uniform in its nature in spite of the fact that all of our bodies look different and the contents (Vrittis) in consciousness vary from being to being. Our everyday sense of reality is drawn from our everyday experience which is undoubtedly approximated by classical physics. But fundamental reality has to be quantum.

There are some models (especially by Hameroff and Penrose [1]) which consider processes in neurons in the brain to be quantum mechanical. An interesting point mentioned by these authors [1] is that consciousness may come in discrete number of flashes. According to them, the best known temporal correlate for consciousness is gamma synchrony EEG, 30–90 Hz (cycles/sec). They also mention that Buddhist monks in meditation know about such flashes coming as 40–80 times a second! In addition, there is some experimental support for resonant frequencies in neurons.

[11]AitareyaUpanishad 3.3 of the Rig Veda.
[12]Mandukya Upanishad 1.2 of the Atharva Veda.
[13]ChandogyaUpanishad 6.8.7 of the Sama Veda.
[14]BrhadaranyakaUpanishad 1.4.10 of the Yajur Veda.
[15]*Chandogya Upanishad* 3.14.1 of the Sama Veda.
[16]Verse 10.39 and several other verses of the Bhagavad Gītā.

More recently Matthew Fisher (unpublished) has proposed that the nuclear spins of phosphorus atoms could serve as rudimentary (entangled) "qubits" in the brain – which would essentially enable the brain to function like a quantum computer. A number of other authors have tried to get a quantum mechanical model of consciousness without looking at the parts of the brain (e.g., SubhashKak and collaborators [4, 5] who consider veiled reality and nonlocality and Henry Stapp[17]). Hagelin [6, 7] has introduced a concept of a unified field of consciousness. Such a field of consciousness is also discussed in the paper of RadheShyam Kaushal.[18] In addition, recently, Tononi and colleagues [8, 9] have developed an integrated information theory (IIT) which is a scientific theory of what consciousness is, how it can be measured, how it is correlated with brain states, and why it fades when we fall into dreamless sleep and returns when we dream. It seems that, although some progress is being made, almost 100 years of neuroscience research, while producing voluminous knowledge of brain structure and function, has taken us little beyond the correlations discovered between the brain and consciousness in the 1920s. The entire subject is full of controversies, and it will be some time before various questions are resolved. It is likely that many questions pursued to link the brain and consciousness will eventually resolve by being seen to be wrong questions as new knowledge accumulates giving us insights we cannot now imagine.

6 Summary and Speculations

In this article we have suggested that there is a strong parāllelism between concepts of reality, Brahman, and Śūnyatā in the Eastern religious scriptures and modern physics. These scriptures describe Brahman as *the* universal cosmic consciousness. It is manifested in some form in every object of the universe. This concept is distinct from the concept of God in many other religions which assume that God is outside the universe and is creator of everything in the universe. Modern physics suggests that everything in the universe is composed of some fundamental particles and the universe arose from some fundamental quantum field present at the beginning of the universe. It seems very likely that the two concepts (Brahman-Sunyata and concept of reality in modern physics) are closely related.

If the concept of Brahman or Śūnyatā as universal cosmic consciousness present in everything in the universe is true, then an unavoidable consequence is that there is some degree of consciousness in every fundamental particle in the universe. This view is called "panpsychism" and goes back in the West as far as Leibniz' theory of monads circa 1700. From the point of view of physics, I would like to have a program to understand consciousness starting with atoms or fundamental particles, i.e., a bottom-up approach rather than a top-down approach. Starting with

[17] https://en.wikipedia.org/wiki/Henry_Stapp

[18] RadheShyam Kaushal, New Delhi Conf. Proceedings.

fundamental particles, inanimate objects like rocks, etc. and then cells, plants, bacteria, animals, and finally human beings would have progressively increasing consciousness. The ultimate Brahman would have infinite amount of consciousness. Consciousness in various forms of living and nonliving objects in the universe may arise in various amounts from this source. "Amount" may not be the right parāmeter, perhaps. It may be the complexity of the forms that consciousness can take. Nonetheless, the idea of a hierarchy of consciousness in all objects of the universe is gaining increasing scientific support, to the chagrin of realists and materialists. At our present level of understanding of consciousness from the Western scientific perspective, we are on somewhat speculative ground. But it is worth investigating. If the idea is fruitful, it will take science to an unbelievable level. It is tempting to draw a parāllel from biology. Biologists know that genes can be turned on or off (gene expression). One can speculate that units of consciousness can be turned on or off in different systems. This could be a reason why we do not seem to see or cannot appreciate the presence of consciousness in primitive living systems or nonliving entities.

We have previously discussed that Ṛṣis and Buddhist monks found correct philosophical ideas about the universe by the yogic methods of meditation, i.e., non-sensory, purely mental means. On the other hand, physicists came to these conclusions after performing experiments, making mathematical models, and verifying by further experimentation, i.e., basically by sensory means. So it seems that quantum physics may be a bridge between sensory and non-sensory parts of the universe. Future research will tell if these speculations are correct. In the meantime, we can marvel at the astonishing similarities between the two completely diverse areas of human endeavor. Finally, I wish to expess my thanks to Prof. Donald Degracia for careful reading of the manuscript and a number of suggestions.

Footnotes

† This article was presented (as an invited talk) at a conference in New Delhi, India (Dec. 9–10, 2016) on "Quantum Reality and Theory of Śūnya." This conference is referred to as the New Delhi Conf. in the references.

*Hindu religion, in the form prescribed in the original scriptures ("Vedas and Upaniṣads") should be preferably called "Sanātana Dharma" meaning universal and eternal way of life or a set of responsibilities and obligations for the followers. The name Hindu came up as a distortion of the word "Sindhu" which is the name of a river (Indus) in the northwest Indian subcontinent. Persians called the people who lived on the banks of the river "Sindhu" as "Hindus"! Since the word Hindu has been universally adopted, unfortunately we have no choice but to continue using it!

**We will describe physics-related ideas in a nontechnical way as far as possible. More technical details can be found in my guest blog:

http://motls.blogspot.com/2014/04/hinduism-for-physicists.html

***Quotes about Vedānta and Buddhist Philosophy from three pioneer physicists:

Niels Bohr

I go to the Upanishad to ask questions.

For a parāllel to the lesson of atomic theory...[we must turn] to those kinds of epistemological problems with which already thinkers like the Buddha and Lao Tzu have been confronted, when trying to harmonize our position as spectators and actors in the great drama of existence.

Erwin Schrodinger

The unity and continuity of Vedānta are reflected in the unity and continuity of wave mechanics. This is entirely consistent with the Vedānta concept of All in One.

The plurality that we perceive is only an appearance; it is not real. Vedāntic philosophy... has sought to clarify it by a number of analogies, one of the most attractive being the many-faceted crystal which, while showing hundreds of little pictures of what is in reality a single existent object, does not really multiply that object...

The multiplicity is only apparent. This is the doctrine of the Upaniṣads. The mystical experience of the union with God regularly leads to this view, unless strong prejudices stand in the West.

Werner Heisenberg

On conversations with Rabindranath Tagore, as quoted in *Uncommon Wisdom: Conversations With Remarkable People* (1988) by Fritjof Capra, who states that after these "He (Heisenberg) began to see that the recognition of relativity, interconnectedness, and impermanence as fundamental aspects of physical reality, which had been so difficult for himself and his fellow physicists, was the very basis of the Indian spiritual traditions."

"After the conversations about Indian philosophy, some of the ideas of Quantum Physics that had seemed so crazy suddenly made much more sense".

More recently, Fritjof Capra [2], Gary Zhukav [3], Subhash Kak [4, 5], John Hagelin [6, 7], Amit Goswami,[19] and George Sudarshan[20] have written articles and books and given talks about parāllels between Hindu and Buddhist scriptures and modern physics.

Bibliography

There are many nontechnical popular books on cosmology and quantum physics (quantum theory). The two with interesting historical accounts are:

(i) *Cosmos*, Carl Sagan, Random House Inc. New York, 1980
(ii) *The Quantum Moment: How Planck, Bohr, Einstein, and Heisenberg Taught Us to Love Uncertainty*, Robert P. Crease and Alfred Scharff Goldhaber, W.W. Norton & Company New York, 2014

---

[19]Amit Goswami; references on books and articles can be found on the website, www.amitgoswami.org

[20]See, for example, paper by E.C. George Sudarshan in "On Mind and Consciousness: Selected Papers from the Micon 2002 Conference," Indian Institute of Advanced Study, Shimla, Amazon Books, 2003.

References

1. Hameroff S, Penrose R. Consciousness in the universe: a review of the 'Orch OR' theory. Phys Life Rev. 2014;11(1):39–78.
2. Capra F. The tao of physics. Boston: Shambhala publications, Inc.; 1975.
3. Zhukav G. The dancing Wu Li masters, an overview of the new physics. New York: William Morrow and Company; 1979.
4. Kak S. The universe, quantum physics, and consciousness. J Cosmol. 2009;3:500–10.
5. Kak S, Chopra D, Kafatos M. Perceived reality, quantum mechanics, and consciousness. J Cosmol. 2014;18:231–45.
6. Hagelin J. Is consciousness the unified field? a field theorist's perspective. Modern Sci Vedic Sci. 1987;1:29–87.
7. Hagelin J. Restructuring physics from its foundation in light of Maharishi's Vedic Science. Modern Sci Vedic Sci. 1989;3:3–72.
8. Tononi G. PHI: a voyage from the brain to the soul. New York: Pantheon Books; 2012.
9. Koch C, Massimini M, Boly M, Tononi G. Neural correlates of consciousness: progress and problems. Nat Rev|Neurosci. 2016;17:307–21.

Chapter 10
Intrinsic Property, Quantum Vacuum, and Śūnyatā

Sisir Roy

Abstract In modern physics, the properties like charge, spin, etc. of elementary entities like electron, proton, photon, etc. are considered to be "intrinsic properties" of the entity. Intrinsic properties are those properties that a thing possesses, irrespective of whether or not there are other contingent things. In Buddhist philosophy especially in Mādhyamik philosophy, no such concept of "intrinsic property" or svabhāva exists. The problem of origin of the universe baffled the scientists and philosophers for many centuries. Within the framework of general theory of relativity as discovered by Einstein, the origin and structure of the universe were discussed in a comprehensive manner. According to the recent formulation of cosmology (i.e., the origin and structure of the universe), the universe originates from the fluctuations of the quantum vacuum. Vacuum in modern physics is not exactly nothing but rather a "something called nothing," meaning that it is replete with activity governed by the principle of quantum theory. From philosophical perspective, what is significant is the division of creative conceptions into those which assume that the universe arose from "nothingness" in the strong ontological meaning of the word vs. those which lead to the conclusion that it was originated from a certain "poorer" physical reality, usually called "quantum vacuum" or space-time endowed with fluctuation. This vacuum or ontologically speaking a substratum exists which is devoid of any matter but full of activities or full of potentialities. Special theory of relativity is based on two axioms, one of which is the speed of light taken as constant and maximum.

Keywords Intrinsic property · Quantum vacuum · Śūnayatā · General theory of relativity · Special theory of relativity

---

S. Roy (✉)
National Institute of Advanced Studies, Bengaluru, India

S. R. Bhatt (ed.), *Quantum Reality and Theory of Śūnya*,
https://doi.org/10.1007/978-981-13-1957-0_10

1 Introduction

Over the last few decades, following the revolutionary scientific contributions to the theory of Einstein's relativity and quantum physics, many scholars have been inspired to discover striking parallels between worldviews of Vedānta and modern physics. Few attempts have been made toward the comparative study of various branches of science, studying parallels between modern physics, especially, between quantum theory, cosmology, and Buddhist concepts.

In the present paper, our aim is not to discover the findings of modern physics in Buddhist framework or in Vedānta or to show how modern physics can also be claimed as *spiritual*. On the contrary, an attempt has been made to critically analyze the claims of convergence between the important concepts of quantum theory and concepts of Buddhist philosophy. The limited knowledge of the present author on various schools of thought of Buddhist philosophy restricts the present discussions within the domain of Mādhyamika and Yogācarā school of Buddhism.

The question of the origin and constitution of the universe lies at the center of most ancient natural philosophies: how did the world come into existence? Was it created from a *sort of nothingness*, or does it originate from a basic substance or principle that still underlies its existence? In the early twentieth century, Albert Einstein discussed about the origin and structure of the universe in a comprehensive manner within his framework of general theory of relativity. According to the recent formulation of cosmology (i.e., the origin and structure of the universe), the universe originates from the fluctuations of the quantum vacuum. Vacuum in modern physics is not exactly nothing but rather a "something called nothing," meaning that it is replete with activities governed by the principle of quantum theory. From philosophical perspective, what is significant is the division of creative conceptions into those which assume that the universe arose from "nothingness" in the strong ontological meaning of the word vs. those which lead to the conclusion that it was originated from a certain "poorer" physical reality, usually called "quantum vacuum" or space-time endowed with fluctuation. This vacuum or ontologically speaking a substratum exists which is devoid of any matter but full of activities or full of potentialities. It has striking similarity with the concept of Tathāgatagarbha in Yogācarā Buddhism. Scientists are yet to understand how the behaviors and attributes of all the entities in this physical universe are being originated from these potentialities inherent to quantum vacuum: for example, how the elementary entities like electron, protons, quarks, etc. get their specific properties like mass, charge, spin, interaction character, etc. and make atoms, molecules, etc.? In modern physics, the properties like charge, spin, etc. of elementary entities like electron, proton, photon, etc. are considered to be "intrinsic properties" of the entity. This leads to new debate whether "these intrinsic properties" are really intrinsic since they are created out of the fluctuation of vacuum. In Buddhist philosophy especially in "Mādhyamik philosophy," no such concept of "intrinsic property" or svabhāva exists at the ultimate level. Here, I shall elaborate this issue of lack of existence of intrinsic attributes for elementary entities in quantum vacuum and compare with Nāgārjuna's philosophy of emptiness or *Śūnyatā*. It is worth mentioning that in the process of understanding the characteristics of quantum vacuum, one needs to understand the quantum theory along with special theory of relativity. Special theory of relativity is based on two axioms, one of which is the speed of light taken as constant and maximum. This constancy of speed of light does not depend on other external conditions. This is in contrast to the doctrine of "pratītyasamutpāda" or "dependent origination" in Buddhist philosophy. The existence of multiverse can explain the contingency of speed of light, i.e., in our present universe, the speed of light is constant and has the fixed numerical value which might be different in other universe. The basic issue is whether Buddhism permits the existence of multiverse or not.

This substratum or "quantum vacuum" is not devoid of anything but full of some attributes which give rise to the physical characteristics of various fields like gravitational, electromagnetic, etc. present in the physical universe [1]. The challenge is to understand how these physical characteristics of various fields emerge from the seeds or attributes embedded in the substratum. Conceptually, the term "quantum vacuum" in modern cosmology refers to the same reality as depicted by the term "Tathāgatagarbha" in Yogācarā Buddhism [2] where an ubiquitous physical substratum and its fluctuations give rise to the physical universe. This will be discussed in Section II.

In modern physics, matter is supposed to be composed of molecules, atoms, electron, proton, etc. If we further down to the scale, i.e., to smaller scale of length and time, all the elementary particles like electron, proton, and neutron are supposed to be composed of particle called "quarks." The behavior of elementary entities like electron, proton, neutron, photon, etc. is governed by an equation known as Schrodinger equation, whereas the object of everyday experience like a stone, basketball, etc. is governed by an equation known as Newtonian equation. It is to be noted that the properties of these elementary entities like mass, charge, spin of electron, proton, etc. are considered to be intrinsic in nature. Intrinsic properties are those properties that a thing has, irrespective of whether or not there are other contingent things. Moreover, the spin can take discrete values. This has been verified in the early twentieth-century laboratory experiment. It raises serious philosophical debate especially if one considers Mādhyamika school of Buddhism. In Buddhist philosophy especially in "Mādhyamika philosophy," no such concept of "intrinsic property" or svabhāva exists in contrast to Vedāntic view, where the "operational self" (ahamkāra) is underpinned by the "ontological self" (*jīvātman*), which denotes the intrinsic identity. I will discuss this aspect in Sect. III.

It is worth mentioning that various fields exist in this universe, for example, gravitational, electromagnetic, weak, and strong field. It is the dream of scientists (rather dream of Einstein) to find a theory which unifies all types of fields, and hence theory of everything may be constructed. To construct theory of quantum field, one needs to consider simultaneously the quantum theory and relativity theory. In special theory of relativity, Einstein started with two axioms: the speed of light is constant and maximum and the space is homogeneous and isotropic. The first axiom is based on the experimental observations, and so far no experimental observations contradict this assumption. So the contingency of speed of light, i.e., the constancy of speed of light, does not depend on any other conditions in the present universe. It raises serious philosophical debate within Mādhyamika school of Buddhist philosophy. The contingency of speed of light can be envisaged if and when one considers existence of multiverse theory. This will be discussed in Sect. IV. Concept of nonlocality plays an important role in modern-day quantum theory. The concept of interconnectedness in Buddhist philosophy and the nonlocal connection or correlations are critically discussed in Sect. V. Finally the concept of quantum reality and Śūnyatā are discussed in view of creating possible dialogue in Sect. VI.

2 Tathāgatagarbha, Quantum Vacuum, and Ontological Issues

The concept of *ālayavijñāna* plays a central role in the development of Buddhist dialectical thought. *Nāgarjuna* is considered to be the first Buddhist thinker to introduce a dialectical system as a means of developing progressive philosophical views. However, he never mentioned the term *ālayavijñāna*. Asanga, the propounder of the *Vijñānavāda* doctrine, wrote the first systemized theory of causation by the ideation storehouse or *ālayvijñāna*. It is not clear from the text whether the storehouse consciousness should be considered as a universal source of individual consciousness or as a pluralistic principle per se performing at the bottom of each individual human mind. The text of Asanga considers *ālaya* as the seat of pure subjectivity, out of which objectivity develops. However, the text does not imply any sort of ultimateness in the storehouse consciousness as such. Vasubandhu, brother of Asanga, adhered to the doctrine as the basis of subjective consciousness and the source of all phenomenal multiplicity. In Lankāvatāra sūtra, ālaya is compared with infinite ocean (Verdu) on whose surface waves roll on permanently aroused by the winds of ignorance, but the depth remains unmoved, that is, the *ālaya* body itself "subsists uninterruptedly, quite free from fault of impermanence.... and thoroughly pure in its essential nature." Later development toward more ontological formulation, the sūtra identifies the concept of *ālaya* with the concept of Tathāgatagarbha, the [5] matrix or womb. The latter one is the most cherished Kegonian expressions which designates the ultimate reality of Buddha nature itself or *tathatā* or suchness. In 1974, Verdu [3] in his book on *Dialectical Aspects in Buddhist Thought* raises an important issue which is much relevant in the context of modern science. He wrote "The Lanka, however, leaves a vital aspect of its doctrine shrouded in deep mystery." This is the question concerning the character and origin of the agency that stirs the waves of change upon the ocean of the *garbha*. This agency is metaphorically designated as the winds of ignorance. What principle are these winds related to? What is the ontological status of this principle as primordial stimulator of the growth within the womb (*garbha*) of the *ālaya*? Is it extrinsic to the *garbha*, or is it internal to the *garbha*, in which case it is the very source of error and of its subsequent evils?

Now if it is considered to be external to *garbha*, then the concept of *garbha* cannot be treated as ultimate one [4].

There exists striking similarity to the idea of creation of the universe by the fluctuations of quantum vacuum in modern cosmology and the concept of Tathāgatagarbha [5]. In modern cosmology, the quantum principle is considered as the fundamental principle existing at all levels ubiquitously. So the issue is whether this quantum principle is extrinsic to the quantum vacuum? The answer is no. The very concept of quantum vacuum is intimately related to this uncertainty principle. Indeed it is intrinsic to the quantum vacuum.

Let us now look into the solution of the enigma raised in the context of Tathāgatagarbha by Yogācarā Buddhists. A highly speculative scripture *Mahāyānaśraddhotpāda (Awakening of Faith in Mahāyāna)* usually attributed to Aśvaghoṣa offered an interesting solution to this enigma. Briefly speaking, Tathāgatagarbha is considered as the universal flux or storehouse of all reality including both its *samsāra* and *nirvāṇa* aspects, whereas *ālayavijñāna* is considered as the storehouse of all knowledge. Phenomenologically if the enlightenment is the noetic aspect of the *ālaya*, the Tathāgatagarbha is its ontic aspect. Edward Thomas [(6)] stated "It becomes superfluous to ask whether this mind or store-consciousness (ālaya) is universal or individual. It is conceived as the one reality beyond all differentiation....This is Tathāgatagarbha ..... in which all reality and difference is embraced."

According to modern cosmology, the universe arises in the process of "fluctuation of quantum vacuum," not out of "nothing" but out of a certain physical reality. This vacuum or ontologically speaking about the existence of substratum exists devoid of any matter but full of activities or full of potentialities. These potentialities give rise to various attributes of the physical universe. For example, there exist four fundamental forces in physical universe, namely, the electromagnetic force, gravitational force, strong nuclear force, and weak force. These fundamental forces are associated with corresponding fields – like electromagnetic, gravitational fields, etc. Each field has its own distinct characteristics. These characteristics are supposed to have some kind of imprints or seeds in "quantum vacuum" (via symmetry breaking). So these characteristics or attributes arise out of the fluctuation of the "quantum vacuum" with potentialities. Here, the scientists are yet to find the answer to the question of how the attributes of the physical universe arise out of "fluctuation of quantum vacuum with some seeds or potentialities" or a space-time endowed with fluctuation without having specific "attributes" of the various fields.

It is worth mentioning that though several concepts of vacuum like "true vacuum" and "false vacuum" are proposed in modern cosmology, none of them is "nothingness" in the strict ontological sense of the word. In the context of these theses on the creation of the universe out of "nothingness," "tunneling out of nothingness" or "fluctuation of nothingness" is not purely scientific theses. Now even if one does not consider the concept of energy, mass, and stress tensor as characteristics of nothingness but that of space-time (which we call dynamic geometry), then one should recognize the existence of a law of nature at the "very beginning" according to which "nothingness creates the world." Clearly, such a law would appear to go beyond any known physical laws, and presumably it is subject to causal laws that invoke some world of logic and mathematics, because we believe (almost religiously!) that an explanation of the origin of the universe cannot do without assumption of structure of rationality. This entails that, in understanding the question of the "beginning of universe," the involvement of philosophy is inevitable in the following sense. Cosmology as an empirical science is predicated based on the structure and evolution of only a small observable part of the universe (like correctness of inflation), but not the universe as a whole – for example, we are not able to check the correctness of a rule for initial conditions from a scientific point of view. It is to be noted that extrapolations of local physics onto the whole observable universe cannot do without cosmological principles, about which what we know is basically of a philosophical character. It immediately raises the pertinent question: "Is the emptiness of quantum vacuum similar to the emptiness of Buddhist philosophy?"

3 Intrinsic Property in Quantum Theory and Dependent Origination

Nāgārjuna is considered to be the first Buddhist thinker and propounder of dialectical system as a means of developing progressive philosophical views. Nāgārjuna's philosophy consists principally of two aspects:

1. Exposition of a concept of reality [Śūnyatā] *according to which fundamental reality does not consist of independent, substantial components, i.e., absence of "svabhāva" in contrary to other Indian traditions.*
2. *The second aspect is an answer to the inner contradictions of four extreme concepts of reality (catuskoti, meaning "four corners").*

It insists that there are four possibilities regarding any statement: it might be true (and true only), false (and false only), both true and false, or neither true nor false. Nāgārjuna introduced the concept of Pratītyasamutpāda which has two parts in Sanskrit – Pratītya and Samutpāda – where Pratītya is a continuative meaning "having dependent" and Samutpāda is an action noun meaning "arising."

Pratītyasamutpāda refers not just to the things which arise from causes but also to nonproducts since their existence is relative. According to this doctrine, all phenomena are dependent arising. Since everything is interdependent, nothing can be self-defining and exist inherently. The *notion of interdependence* leads us directly to the idea of emptiness in Buddhism, which does not mean nothingness but the absence of inherent existence. Intrinsic are those qualitative properties that a thing has irrespective of whether or not there are other contingent things; all other qualitative properties are extrinsic or relational.

According to western perspective, the world consists of independent individual things that are embedded in space-time. These things are individuals because:

- They have a spatiotemporal location.
- They are a subject of the predication of properties each.
- There are some qualitative properties by means of which each of these things is distinguished from all others (at least the spatiotemporal location is such a property).

In modern physics, the properties, for example, charge, spin, etc., of elementary entities like electron, proton, photon, etc. are considered to be "intrinsic properties" of the entity. Elementary particles like electron, proton, photon, etc. are described by:

- Electron has definite mass, charge, and spin.
- It belongs to a category called fermions.
- Fermions have spin of 1/2, 3/2, 5/2, etc.
- Photon belongs to a category called boson; bosons have spin of 0, 1, 2, etc.
- Boson obeys Bose statistics and hence known as boson.
- Rest mass of an electron: $9.10938215(45) \times 10^{-31}$ Kg.
- Charge of an electron: $4.80320451(10) \times 10^{-10}$ esu
- Spin of an electron: 1/2.

Scientists discovered many such elementary particles, out of which the particles called "leptons" (like electron, muon) and "quarks" are considered to be the building blocks of all other particles and matter. It means matter and all other elementary particles like neutron, proton, and mesons are composed of leptons and quarks. These "quarks" possess also intrinsic properties like spin, charge, etc. However, if we carefully analyze the situation, these "intrinsic properties" no longer can be identified as "intrinsic property." This can be visualized in the following way:

In a phenomena called pair creation in modern physics, it has been shown from observations that a pair of electron (negative charge, fixed mass, and definite spin) and positron (opposite charge that of electron but the same mass as that of electron and same spin) is created out of the fluctuation of vacuum or gamma ray (quanta of light) which may also create such a pair. Here, when it is created from gamma ray (which does not have charge, rest mass but has spin one), how the attribute like spin of electron and positron (spin half) is created from a spin one gamma or photon quanta (which has spin one only) or the charge of electron and positron arises from chargeless gamma? In physics we consider the law of conservation of charge or angular momentum. Ontologically it gives rise to the problem of understanding the intrinsic property of elementary entities at the level of quantum vacuum. Further analysis is needed to clarify the present situation.

4 Special Theory of Relativity, Constancy of Speed of Light, and Contingency

Since the perception of quantum theory, the debates started regarding the ontological issues present in quantum theory. According to Copenhagen interpretation, one cannot say anything about the reality unless one observes it. Essentially no observer-independent reality exists in this framework, and hence it is epistemic in nature. During the process of development of quantum theory, the field like electromagnetic field has been shown to be quantized. Once we have quantized field, we do not have concept of particle as such in the framework, but the quanta of this field are nothing but the photon associated with electromagnetic field. The enigma of wave and particle was resolved within the framework of quantum field theory. This is one of the most successful theories in terms of experimental observations in the domain of modern physics. Here, one needs to construct a framework of quantum theory along with special theory of relativity. The fields have infinite degrees of freedom and have more intuitive ontological understanding. Einstein introduced two axioms in formulating the special theory of relativity: one the speed of light is constant and maximum. This is based on observational findings. Many laboratory experiments as well as experiments performed in the space have been performed to measure the speed of light. No results violate this axiom so far.

The speed of light has a definite numerical value (3×10^{10} cm/s) which can be measured in the experiment as a consequence. It gives rise to new philosophical debate: why this particular value is prevalent in our present universe? Can we think in terms of contingency (contingency means dependent on other conditions)? One possible solution has been proposed by Peter Mittlestaedt [7] regarding the contingency of speed of light. He suggested that if we take the principle of quantum theory as the most fundamental principle in nature, then there is a finite probability of having this particular value in our universe. Its value may be different in other universe. So his solution demands the existence of multiple universe. Intuitively speaking, one can think of an ocean where bubbles are being created and dissolved. These bubbles are nothing but many universes. This ocean is nothing but the quantum vacuum where fluctuations give rise to many universes or bubbles.

Now, the issue is whether any school of Buddhism propounds the manifestation of many physical universes from Śūnyatā.

5 Concept of Nonlocality and Idea of Interconnectedness in Mādhyamika Philosophy

Buddhists see the world, not as a collection of isolated objects but as a network of phenomena that are fundamentally interconnected and interdependent. The Buddhist worldview, therefore, is holistic because it sees the world as an integrated whole rather than as a dissociated collection of parts. It recognizes the fundamental interdependence of all phenomena in the world, and *everything is dependent.*

Now, let us look into phenomena called quantum entanglement and nonlocality in quantum theory. Quantum entanglement, also known as quantum nonlocal connection, is a property of quantum mechanical systems. This connection contains two or more objects which are linked in such a way that it is not possible to describe the quantum state of a constituent of that system without fully mentioning its counterparts, even if those individual parts are spatially separated. Erwin Schrodinger [8] coined the term "entanglement" for this phenomenon and used it in his three-part article published in Proceedings of Cambridge Philosophical Society. He proposed that entanglement arises due to the interactions of two particles (i.e., two quantum objects) through the evolution of wave equation (popularly known as Schrodinger equation) and called this phenomenon as the characteristic trait of quantum theory. Schrodinger, also, realized that this character, rather peculiar nonclassical correlations to quantum systems, could be used to steer a distant particle into one of a set of systems having a certain probability.

But, Einstein disliked this approach at first and called quantum entanglement "spooky action at a distance" in the famous EPR debate. In 1982, Aspect et al. [9]. first showed that nonlocal interactions do exist and occur even in the laboratory.

Two quantum objects, when entangled, initially, both will have an undetermined state. When one object's state is determined, the other's state is instantaneously known to be the opposite, no matter how far apart they are. Quantum theory permits that the states of quantum systems are entangled. Instead of speaking of entangled states, one can also talk directly in terms of entangled systems. However, since entanglement is state-dependent, it seems more appropriate to use the notion of entangled states. Examples for this phenomenon are position and momentum and also spin angular momentum in any direction. Instead, there are only correlations between the state-dependent properties of the quantum systems in question, encoded in a joint probability distribution determined by the joint state. Quantum theory does not include any property of a quantum system taken separately which is a supervenience basis for these correlated probability distributions. Furthermore, these correlations – and thus entanglement – are independent of spatiotemporal distance. Because of this property, i.e., entanglement, quantum physics seems to exhibit some sort of a holism.

Next, let us summarize the concept of nonlocality in the following manner:

> *The states of two or more systems are non-sePar$\bar{a}$ble if and only if it is the joint state of the whole that completely determines the state-dependent properties of each system and the correlations among these systems (to the extent that these are determined at all).*

According to Ithaca interpretation of Quantum Theory (QT)

The only proper subjects of physics are correlations among different parts of the physical world. Correlations are fundamental, irreducible, and objective. They constitute the full content of physical reality. There is no absolute state of being; there are only correlations between subsystems.

So the correlation or interdependence is fundamental in understanding quantum reality. It seems to have striking similarity with the concept of interdependence in Mādhyamika's view. But it does not preclude the existence of intrinsic property of the quantum entity. Moreover, the correlations for quantum entities in quantum entanglement have to be constructed by certain measurement process. Of course, once the correlations occur, it will be there whatever the distance the entities are separated. Again, any entity in quantum world is not connected with any other entity, i.e., electron may be correlated with another but not with other particle like neutron. However, further investigation is needed to understand this situation of correlations in quantum world and the interdependence in Mādhyamika philosophy.

6 Quantum Reality and Śūnyatā: Possible Dialogue

In physics, the scientists generally follow the "top-down" approach. Since the dawn of civilization, curious minds are engaged in looking into stars and galaxies in the sky. Kepler discovered the law of planetary systems, i.e., how the planets are moving around the sun. Newton formulated the laws of gravitation, i.e., why the apples fall toward the earth. Their efforts were mainly concentrated in understanding the behavior of large objects like planets and everyday objects like falling of apples. The developments of research in the beginning of the twentieth century changed the paradigm. Scientists tried to understand the behavior of smaller and smaller objects like "atom" as the smallest unit of the matter and went on further down to find the fundamental particles like "electron," "proton," etc. These are the entities which constitute the microscopic world. So the journey is from macroscopic (for larger objects) to microscopic world. The behavior of the entities of the microscopic world is governed by different law of nature called Schrodinger equation, whereas the law governs the behavior of macroscopic objects by Newtonian equation. It does not mean that the discovery of quantum paradigm refutes Newtonian paradigm. We need to understand different levels of reality – at one level, Newtonian paradigm is applicable, whereas at another level of reality, quantum law is applicable. One may argue that as the quantum law which governs the behavior of microscopic entities, the behavior of the whole universe should be described by the law of microscopic world. This is far from understanding of the scientists even in the twenty-first century.

The developments of modern physics in the twentieth century raise epistemological and ontological issues which may be relevant in Buddhist framework. They can be summarized in the following manner.

- Issue of intrinsic property of the objects in quantum theory and Mādhyamika philosophy
- Issue of interconnectedness and concept of nonlocality and quantum entanglement in the microscopic world

- Concept of quantum vacuum – as ultimate reality and Tathāgatagarbha in Yogācarā Buddhism

Great Buddhist scholar Nāgārjuna in his famous text *Mūlamādhyamaikakārikā* attributes two truths to Buddha "the Dharma taught by the buddhas is precisely based on the two truths: a truth of mundane conventions and a truth of the ultimate."

The doctrine of two truths lies in the core of Buddhist philosophy. One is conventional truth which involves our everyday experience and understanding the way the phenomenal world appears and functions. According to Buddhism, this conventional truth is deceptive, and this deception is called inherent existence and is identified as the root cause of sufferings. Mādhyamika school of Buddhism asserts that no independent phenomena exist. Phenomena are neither self-created nor self-enduring but arise in dependence on other conditions without a nature or essence of their own. For example, if the atoms producing an object say pitcher are discrete entities as Buddhist maintains, then simply the aggregates of the atoms will not produce pitcher.

In modern science one needs concept of cohesion to produce such an object. Jaina scholars emphasized that the atoms are being capable of being connected to become as concrete object. This is strikingly similar to the discovery of Pauli's exclusive principle which solves the mystery of periodic table of various elements in nature. Hydrogen and oxygen combine to produce water. Hydrogen and oxygen are independent entities – the combination occurs due to their propensity to each other based on the abovementioned exclusion principle. In modern science, scientists are looking for fundamental entities or fields by which the whole universe can be described. Again the fluctuation of a substratum called quantum vacuum has been studied to understand the origin of this physical or phenomenal universe. The various attributes of the microscopic entities like spin, charge, and mass are considered as intrinsic properties at certain level, but they are considered as merely potentialities at the level of quantum vacuum. This is one of the challenging issues in modern cosmology how these attributes arise out of vacuum fluctuation where "so-called intrinsic property" does not exist. So we do have the concept of two truths in modern science too similar to Mādhyamika school of thought. Recently, physicists constructed "quantum ontology" by relaxing some of the axioms of classical ontology from very general perspective. Within this new perspective, further analysis is needed in finding the convergence and divergence in modern physics and Śūnyatā or emptiness in Buddhism.

References

1. Duquette J. Towards a philosophical reconstruction of the dialogue between Modern Physics and Advaita Vedānta: an inquiry into the concepts of *ākāś a*, vacuum and reality:, Ph.D. Dissertations, University of Montreal; 2010.

2. Roy S. Towards reconstruction of the dialogue between Modern Physics and Buddhist Philosophy: an inquiry into the concepts of Quantum Vacuum and Ālayavijñāna. Bihar: Nava Nalanda University; 2015.
3. Verdu A. Dialectical aspects in Buddhist thought, International Studies, East Asian Series Research Publication, Number Eight, The University of Kansas; 1974.
4. Brown BE. The Buddha nature: a study of the Tathagarbha and Alayavijnana. Delhi: Motilal Banarasi Das Publishers Pvt. Ltd; 2010.
5. Roy S. An inquiry into the concept of Quantum Vacuum in Modern Physics and Akasha in Advaita Vedānta; Prajnanam- self awareness. Kaivalyadham; 2015. p. 36–56.
6. Thomas E. The history of Buddhist thought. London: George Allen & Unwin Ltd; 1963. p. 234.
7. Mittelestaedt P. Rational reconstruction of modern physics. Dordrecht: Springer; 2011.
8. Schrodinger E. Discussion of probability relations between seParāted systems. Proc Camb Philos Soc. 1935;Vol. **31, p** 555–63; ibid.Vol. **32** (1936) p 446–451.
9. Aspect A, Grangier P, Roger G. Experimental realizations of Einstein-pdolsky-Rosen-Bohm Gedanken experiment: a new violation of Bell's inequalities. Phys Rev Lett. 1982;49(2):91–4.

Chapter 11
Beyond Vedānta: Speculations of a Quantum Realist

Rajendra Prasad Bajpai

Abstract Hindus consider Vedānta as the ultimate knowledge of reality whose emphasis is on human behavior. Sciences also determine reality that is evidence based but is not restricted to human behavior. Vedānta and Sciences should, therefore, have overlapping region and convergence of their worldviews. We got a glimpse of overlapping region by identifying *Brahman* and *Jagat*, respectively, with quantum and classical versions of reality. The convergence in their worldviews will, probably, require Vedānta to come out of its mold and to interact with scientific thinking and biological data. There appear some contact points between Sciences and Vedānta that can catalyze their interaction, trigger the evolution of Vedānta, and enlarge the overlapping region. These points belong to relativity theory, quantum field theory and unexplained holistic features of living systems.

Keywords Hindu wisdom · Prārabdha · Rebirth · Maya · Ātma · Parāvidhya · Aparāvidhya · Ākāśa · Vedānta

1 Introduction

Vedānta[1] is the gist of worldview in Hindu wisdom tradition about the aspirations and tribulations of human life. Hindus claim it to be the ultimate truth and the end of knowledge. The word Vedānta means end of knowledge, and it is interesting that the Hindus speculated about the end of knowledge eons before the west started talking about the end of History and Sciences. Modern Sciences have evolved from

[1]The primary sources of the exposition of Vedānta are Brahma Sutra, Gītā, and Upaniṣad. Gītā is least cumbersome, lucid, and the best exposition. The wisdom traditions in other great civilizations, e.g., Greek, Chinese, etc., have also developed worldviews of the triode. There is a substantial common overlapping region among different worldviews, which implies a consensus about the validity of knowledge contained in the overlapping region among wisdom traditions.

R. P. Bajpai (✉)
Vimal Research Society for Agro-Biotech & Cosmic Powers, Rajkot, Gujrat, India

S. R. Bhatt (ed.), *Quantum Reality and Theory of Śūnya*,
https://doi.org/10.1007/978-981-13-1957-0_11

successes in the searches of reality, and they claim to be inching toward the limit of knowledge. The successes have endowed them with a worldview that also covers aspirations and tribulations of human life. The worldviews of Vedānta and Sciences show significant differences, which call for investigations to reconcile the differences and to integrate the two worldviews. These investigations may take us beyond Sciences and Vedānta. Sciences have the tradition of accepting impalpable doctrine if it is experimentally verified by many peoples. Accepting a properly debated doctrine was also not against the ethos of Vedānta. In fact, the knowledge contained in Vedānta was revealed through or got crystallized in erudite debates called *Satsanga*. The paper considers some important differences in the two worldviews and discusses which ones can be eliminated or reconciled and which ones need further investigations. We begin by considering the difference about nonhuman life forms, whose references in Vedānta appear incidental and arising from afterthoughts. Sciences have found that all life forms are carbon based and have many commonalities, e.g., genes and biochemical reactions. Sciences consider all life forms on equal footing with human beings. The obvious step is to treat references to non-carbon life forms in Vedānta as mere metaphorical. Equal footing is not problematic because Vedānta envisages 8.4 million different life forms for a human being, and Sciences estimate nearly 8 million species. The problematic part is rebirth. Sciences do not consider the purported evidence of rebirth as credible. Sciences view rebirth as a ploy to resurrect the observed failure of classical causality. Future events are unpredictable and do not adhere to strict causality. The deviations from the causality need to be accounted. Vedānta does it holistically by introducing the twin concepts of *Prārabdha* and rebirth. *Prārabdha* is behind unpredictability and keeps track of accumulation and adjustment of deviations from the norm. The inability of the *Prārabdha* to balance the deviations to zero in any imaginable accounting system necessitated the concept of rebirth. How individual accounts are kept and by whom are not properly spelled out. Sciences deal with unpredictability by considering life form as a composite structure of large number of molecules of quantum nature. Unpredictability arises from the inability to determine the quantum state of life form and inherent uncertainty of quantum processes. The two approaches could be complimentary, and their integration, if achieved, may proclaim that *Prārabdha* expresses quantum nature and quantum state is the ledger folio containing individual account.

Vedānta deliberates upon the nature, relationship, interaction, and dynamics of three basic elements called *tattvas*. They are *Jīva* (living being), *Jagat* (cosmos), and *Jagadīśa* (God). They constitute its basic triode. Sciences, perhaps, because of the influence of Greek Philosophy, use a slightly different triode with basic elements thing, being, and God, for similar deliberations. Sciences and Vedānta attribute different properties even to common elements of their triodes. As a result, their worldviews differ. We give a brief description of how Sciences perceive the three basic elements of Vedāntic triode. *Jīva* is perceived as being, which can perceive, experience, feel, imagine, think, analyze, model, and narrate happenings. The degree of ability in each of these functionalities varies in different *Jīvas*. Human beings constitute the most significant subset of *Jīva*, and its members have most of the above noted functionalities. Vedānta contains the knowledge acquired by this subset using these functionalities. This knowledge is, obviously, from human perspective but is claimed to have universal validity. There is no record of how this knowledge was acquired and got accumulated. The process of acquiring knowledge, perhaps, began at some point of time when the members of human beings started sharing their narrations. The sharing created a superstructure in which new attributes were discovered, e.g., commonalties in the shared narratives about some objects might have led to the existence of the object with specific properties. It is pointed out that it was a scientific method of acquiring knowledge in which any result obtained by many different observers or perceivers implies the reality of the result. The members of the human subset might have recognized after a while that some objects are localized and tangible, while other objects are non-localized and intangible. They became preceptor, observer, detector, and measurer of the properties of tangible objects. They called the entire arena of tangible objects as *Jagat*. The concept of *Jagat* was a testimony of their ability to integrate many pieces of information into an overall picture. It is this ability that probably prompted them to propound the concept of *Jagadīśa* as the creator and controller of every activity of *Jīva and Jagat.* They must have studied the properties, behavior, constitution, aggregation, and inter- and intrarelationship of the three elements of the triode. They called the outcome of their studies as *Vidyā* or *knowledge*. They divided the knowledge into two classes, mundane or *Aparā Vidyā* and supramundane or *Parā Vidyā*. The former contains properties and activities of entities contained in *Jagat* that can be described without the direct involvement of *Jīva* and *Jagadīśa*, while the latter contains properties and activities, whose descriptions require the involvement of *Jīva* and/or *Jagadīśa*. The collective pronoun *Tat* or that is used for entities studied in *Aparā Vidyā*. *Aparā Vidyā* provides objective third-person descriptions, and Sciences could belong to it. *Parā Vidyā* contains first-person descriptions of *Jīva* as *Aham* and second-person description of *Jadish* as *Tvam*. Vedānta envisaged convergence of first-, second-, and third-person descriptions to a scenario by attributing all activities to the grace of *Jagadīśa*. *Vedānta* proclaimed that the realization of this convergence is the ultimate objective of *Jīva*. *Parā Vidyā* provides strategies for achieving the ultimate objective and its consequential effects. These strategies depict commonality in attempts and perceptions and have scientific flavor, to which Sciences are attracted. Sciences try to identify grace neutral contents of *Parā Vidyā* and develop new narratives for them. The grace neutral contents can then be transferred to *Aparā Vidyā*. Successful alternative narratives enthrall Sciences to aspire for developing alternative narrative of the entire *Parā Vidyā*.

An alternative narrative of widespread consequences suggested by Sciences identifies Ākāśa and *Kāla* of *Parā Vidyā* with space and time, respectively. Space and time make up the arena for all activities of objects and entities. Vedānta envisaged *Ākāśa* and *Kāla* to be independent and beyond the reach of *Aparā Vidyā*. *Ākāśa* was thought to be homogeneous, isotropic, eternal, and three dimensional. Three ordinal coordinates specify the location of a tangible object in *Ākāśa*. The objects or their properties do not affect it. *Kāla* was thought to be unidirectional homogeneous flow that assigns a changing coordinate to an object, and its value indicates whether the description belongs to the past, present, or future. This coordinate and its flow can be synchronized for objects at different locations. The synchronization ensures unique assignment of this coordinate in extended objects, assembly of objects, *Jīva* or *Jagat*. The change in one or more properties of an object with this coordinate constitutes a dynamical activity of the object. The objects and their properties do change with *Ākāśa* and *Kāla*, but *Ākāśa* and *Kāla* remain unaffected. Vedānta investigates and makes statements about the patterns in the activities of objects; so do the Sciences. The most important claim of Vedānta is that activities of objects manifest the grace of *Jagadīśa*, and they arise from the interplay of free will of *Jīva* and *Līlā* or playful act *of Jagadīśa*. *Srīmadbhāgavata*, considered to be the best exposition of Vedānta, contains examples of activities showing two intriguing features: one, a few *Jīvas* had the capability to view past, present, and future of objects, and two it was possible to experience *Kāla*-embedded versions of activities of some objects. These features have serious implications for Sciences. The viewing of past activity implies that information about the activity is stored in some retrievable form; the viewing of present activity implies instantaneous communication of the information about the activity from the location of its occurrence to the viewer, and the viewing of future event implies either a deterministic world or some form of retrocausality. The experiencing of *Kāla-embedded* reality implies the possibility of influencing the way *Jīvas* remember past events; this phenomenon is currently attracting some attention and is christened Mandela effect [1]. Phenomena with either feature are ruled out in Newtonian Sciences but not in quantum theory. Vedānta, therefore, suggests that Newtonian Sciences do not give the correct description of some phenomena, and quantum theory should be formulated in a manner that accommodates phenomena with the above intriguing features.

Sciences also investigated patterns in the activities. They directed their investigations to grace neutral activities and speculated about possible grace neutral origin of activities. They found many grace neutral activities and discovered the nature, relationships, and interactions of elements implicated in grace neutral activities. Vedānta did not ignore the findings of Sciences. It took cognizance of the discovered patterns in activities, called them *Māyā*, and shifted the explorations to the *Aparā Vidyā* for further study. Vedānta remained perplexed with *Māyā* and called it *Thagini* (deceptive) meaning slippery and illusory. Vedānta refrained from any systematic exploration[2] of *Māyā* and failed to discover any grain of eternal truth in its shroud. Sciences explored the patterns attributed to *Māyā* systematically and discovered some grains of truth. Sciences desire these grains to be incorporated in Vedānta as eternal truths. The incorporation is likely to trigger many collaborative endeavors of sharing and validating each other's grains of truth. The conservative protagonists of Vedānta resist it; they relegate the grains of truth discovered by

[2]Hindu wisdom tradition has other concepts of Māyā, and to distinguish them, this concept may be called Nārāyaṇi Māyā, in which Māyā is a dressing of Jagadīśa. A much widely prevalent concept calls Māyā as Śhakti, the power and vitality of Jagadīśa. There is also the concept of Ardha-Nārīśvara, in which Māyā and Jagadīśa are on equal footing being subsumed into each other and become one.

Sciences as findings suitable for *Aparā Vidyā*. They want to keep *Parā Vidyā* as a branch of knowledge containing matters related to *Jagadīśa*, the grace of *Jagadīśa*, and *Chetnā* (life and consciousness). They claim that the gap between *Parā* and *Aparā Vidyā* is unfathomable. Sciences not only dispute the claim but also nurture the ambition to bridge this gap. The successes of Cognitive, Biological, and Physical Sciences have kindled the hope of achieving the ambition and conquering the formidable frontier of ignorance pertaining to life and consciousness. The conquest, when complete, will authorize Sciences to examine and edit worldviews. Some scientists, in anticipation, have already started identifying foundational facts or *Sat* and basic premises of the worldviews, a *Cit* fact to be verified and premises to be validated.[3]

Newton concretized the concepts of space and time in a mathematical framework for formulating object-centric Science that describes the motion of tangible objects. The Newtonian Science has closure and completeness; all objects are localized in space with measurable attributes, and the attributes of an object measured at one time determine its history and future trajectory. The influence of space and time on the motion of objects is in the form of conservation laws of energy, momentum, and angular momentum. These conservation laws are consequences of the homogeneity and isotropy of space and time. They need not to be considered as manifestations of the grace of *Jagadīśa*. The closure of Newtonian Science ensures the separation of *Aparā* and *Parā Vidyā*. Newton discovered gravitational theory. The gravitational theory showed that the movements of celestial objects are grace neutral, which shifted celestial objects and their movements to the domain of *Aparā Vidyā*. Newton also made a revolutionary proposal that the validity of a premise, concept, or theory should be checked by repeatable observations. He claimed that observations are the ultimate arbiter of validity; a single contradictory observation is sufficient to demolish even a well-entrenched theory. Newton's proposal should be extended to concepts of *Parā Vidyā* by treating mental experiences as observations of a live human detector. Popper supplemented Newton's proposal by suggesting that the criterion of falsifiability be applied for checking scientific validity of a premise, concept, or theory. The two criteria – verifiability and falsifiability – are responsible for the rapid progress of Sciences. Two scientific theories with large overlap with Vedānta are relativity and quantum field theories. Both theories have not been countered by any experimental observation so far. In addition, a vast amount of biological data is being accumulated, from which a valid biological theory must emerge. Scientists are looking everywhere for clues of a biological theory. They are looking toward Vedānta as well for it uses an elaborate metaphor for life and consciousness. The handshake of Sciences and Vedānta may benefit both of them; Sciences may get desired clues of formulating biological theory, and Vedānta may get its metaphors upgraded and validated. The paper prepares the ground for the handshake by

[3]We are tempted to suggest a reinterpretation of Saccidānanda as an edict proclaiming that the source of bliss (Ānanda) lies in identifying valid basic premise of foundational facts.

enumerating probable contact points of Vedānta with the two scientific theories and biological data.

2 Contact Points Between Vedānta and the Theory of Relativity

The theory of relativity transformed static space and time of Newton into dynamic space-time. Space and time were independent, universal, and rigid, while space-time has structure that varies with the presence of objects and their kinetics (in the language of Vedānta, it has acquired *prapañca*). Space-time must be shifted to *Aparā Vidyā*. The shifting is bewildering, and phenomena emanating from it, perhaps, belonged to sacred hidden knowledge or *gudha rahasya*, revealed only to a few *Jīvas* but not specified in the scriptures. The unearthing of this knowledge is a creditable act of Sciences. The theory of relativity also obliterated the difference between matter and energy and made them interconvertible. They affect space-time in the same manner. Matter is tangible and localized and falls in the domain of *Aparā Vidyā*. Energy is intangible and extended; it should have footprints in *Parā Vidyā*. The interconvertibility suggests that *Parā Vidyā* and *Aparā Vidyā* may not have unfathomable gap; they may have overlapping region amenable to scientific investigations. Cosmogenesis is one such overlapping region. Many scenarios of cosmogenesis consistent with the theory of relativity have been envisioned in Sciences. They all predict the same age (or date of birth) of space-time and assign the eternal attribute cosmological constant to space-time. The value of cosmological constant differs in different scenarios. Cosmological constant determines many other properties of space-time, e.g., average energy density, curvature, rate of expansion, etc. These predictions of relativity contradict speculations about cosmogenesis made in many wisdom traditions other than Vedānta. Vedānta remained unscathed because it did not prescribe any specific scheme of cosmogenesis. Some *Mīmāmsā* thinkers did ponder over cosmogenesis and concluded that the world is beginning less, and the assumption of its creator is both problematic and unnecessary [2]. Vedānta is flexible enough to accommodate scientific cosmogenesis in its fold. The accommodation is reflected in calculations of the age of the universe based on stray descriptions in Vedic and other literature. The calculated age of universe agrees remarkably well with the value estimated in models of scientific cosmogenesis [3]. The agreement with the age does not imply that Vedānta was aware of scientific cosmogenesis because many other entities such as neutron stars, black hole, microwave background radiation, dark matter[4], and multiverses [4] associated with scientific

[4]Dark matter is the mysterious substance introduced for accounting nearly 79% of matter in the universe. Its existence is inferred from astrophysical observations of processes such as galaxy formation and dynamics, which appear to be governed by gravitational forces exerted by dark matter. It does not interact with light and is, therefore, called dark matter. The standard model of Particle Physics does not have any particle representing dark matter.

cosmogenesis are not mentioned in Vedic and other literature. Only a vague mention of the concept of multiverse seems to occur in the description of *Virāta Svarūpa*, but other concepts probably remain hidden in *gudha rahasya*. Loss of eternity of space-time is problematic for an entity having eternal reality. The perception of such entity in a non-eternal space-time framework will contain some blemish. Even *Jagadīśa*, perceived as the ultimate eternal reality, cannot be free from some blemish. Perhaps, *Māyā* is the blemish in *Jagadīśa*. *Jagadīśa* is roughly depicted as a multilevel hierarchical structure with comprehensible and partially comprehensible lower levels and slightly comprehensible top level. There are two possible worldviews; one ascribes all lower levels to *Māyā* and top level to *Jagadīśa*, while the other ascribes the entire hierarchical structure to *Māyā*, the observable blemish of *Jagadīśa*. It is a new perspective to the perennial debate on duality and non-duality. Another consequence of the loss of eternity is the need for recognizing validity interval of a reality. There can be many realities: each one eternal in a limited period and all of them on the same pedestal. Sciences are adept in identifying and handling realities of different time scales.

Vedānta, perhaps, missed out some part of knowledge by ignoring time-bound realities and dumping them as *Māyā*. In our humble opinion, eternal reality should be investigated along with other time-bound realities and not in isolation. It will integrate *Māyā* and *Jagadīśa* in a single meme, which when expressed is *Māyā* and *Jagadīśa* otherwise. Sciences are veering toward a hierarchical structure for the living system. The bottom level of the hierarchical structure is well understood in terms of biomolecules and genes. Its next level is partially understood in terms of composite motifs of genes and biomolecules. It has a few higher levels that are poorly understood. They are holistic in character and need organismic properties for their description. Sciences are fast comprehending them. The progress of Sciences in comprehending these levels is like a journey from the state of knowledge characterized by it is not, it is not or *Neti*, *Neti* to the state characterized by it is, it is or *Yeti Yeti*.

3 Contact Points Between Vedānta and Quantum Field Theory

The theory of relativity shifted considerable amount of cultural baggage of grace neutral phenomena from *Parā Vidyā* to *Aparā Vidyā* but failed to bridge the gap between the two because its framework was essentially Newtonian, i.e., local, continuous, and deterministic. In contrast, the framework of quantum field theory is non-local, discrete, and probabilistic. Bohr, Einstein, Heisenberg, and Schrödinger, the pioneers of quantum revolution and discoverers of quantum framework, were bewildered by the possible scenarios in the new framework. They got attracted to Vedānta, which also has a vast repertoire of equally bewildering scenarios [5]. They thought that similar scenarios, perhaps, indicated common grounds and therefore wondered about the possibility of bridging the gap between *Parā Vidyā* and *Aparā Vidyā*. It did not happen, and the attraction of the pioneers did not evolve into a relationship. The following were the main reasons constricting the blossoming of their relationship: ethical considerations ruled out repetitions of critical experiments on human systems, quantum theory resorted to cumbersome numerical computations for obtaining results in complex systems, and procedures for binding discrete quantum results into a smooth continuous picture did not exist. The situation is fast changing. Neuronal imaging techniques are poised to determine neuronal correlate of sensations, images, feelings, and thoughts [6] in a noninvasive manner, and data mining techniques can extract reliable information of complex systems from discrete numerical data. Many counterintuitive issues of the body, mind, and intellect appear to be amenable to Quantum Sciences and will, probably, be shifted from *Parā* to *Aparā Vidyā*. Quantum theory posits three counterintuitive features in a particle: its presence can be detected at many places, it can interact instantly with far away entities, and its attributes acquire values that depend upon the observer. These features are also associated with *Jagadīśa*. Does this association imply quantum nature of *Jagadīśa*? Neither the answer nor the legitimacy of the question is clear. Perhaps, the need of the concept of *Jagadīśa* arose for binding disjoint elements in the classical description of observed behavior of *Jīva* and *Jagat* of quantum nature. The concept contained seeds of quantum revolution, which Vedānta failed to recognize. The quantum nature of *Jīva* and *Jagadīśa* is an interesting possibility that may help in identifying some more grace neutral phenomena. The key for identification lies in the fact that all features or attributes of a quantum entity are quantitative properties, calculable in a well-defined and heavily tested procedure. The main ingredients of this procedure are statistical outcome, wave function, superposition, entanglement, and measurement. The potentiality of each ingredient in identifying grace neutral phenomenon is briefly discussed below.

Statistical character of a property of quantum entity implies that its value obtained in repeated measurements fluctuates, but fluctuations do not indicate lack of permanency. They neither negate the existence of the property nor of the quantum entity. They contain information about the state of the quantum entity, which is beyond Vedānta. Fluctuating values and extractable information are counterintuitive aspects; they inflate the domain of *Aparā Vidyā* to an extent, so that it overlaps with the domain of *Parā Vidyā*. A pertinent example is the spontaneous emission of mainly visible range photons by almost all living entities including human. The whole body of live human being incessantly emits photons which can be counted in many small intervals of say, 50 ms from any portion of the body. The number of emitted photons counted in different intervals fluctuates, and the fluctuations determine eight properties of the photon signal. The eight properties have been measured at many anatomical sites of the same and different human beings. Three properties have the same values in all signals. They are universal and perhaps contain clues of the origin of life. The remaining five properties have different values in different signals. The variation of these five properties over the body of a human being is linked to the physical, mental, and psychological health of the human being [7]. The existence of these properties supports quantum nature of life, while the linkage with mental and psychological health makes them relevant to *Parā Vidyā* [8].

Wave function is a mathematical function used for describing the coordination in the change or disturbance in some property occurring in a large region of space for a long time. Coordinated disturbance is a characteristic feature of a ubiquitous phenomenon of wave motion. A few Parāmeters usually express a wave function though it comprehensively describes the disturbance in the property at all space-time points. Compactness of description is a hallmark of coordinated functioning, and another name for it is coherence. Schrödinger introduced wave function as an intermediary construct needed for calculating properties of nonrelativistic particle (material object) in quantum mechanics. The wave function and quantum mechanics associated with it have been successfully extended for describing many relativistic and nonrelativistic particles of different kinds. The extension has transformed wave function into state vector in Hilbert space. The status of state vector has become primary, while that of material object has become subsidiary. The state vector everywhere is well defined, but the material object it describes does not have a definite position in space. Even the so-called point material object is not localized but spreads over the entire space and exists here, there, and everywhere with differing probabilities. Material object transcends the limitation of space in quantum description and so should a *Jīva* because of material content. The information contained in the state vector of an object is complete and includes past, present, and future activities of the object. The state vector gets updated after every skirmish of the object with the world. The relationship between an object and its state vector resembles strongly to that between *Jīva* and its soul (*Ātmā*) envisaged in Vedānta. The complete record of the interaction of *Jīva* with the world is contained in its *Ātmā*. The record is kept *as* impressions or *Vāsanā*, which is like the values of Parāmeters specifying state vector. The activities of *Jīva*, like that of material object, are observed in space-time, while information about the activities is stored in *Ātmā*. *Jīva* and Ātmā have distinct identities but are inseparable. All attempts to locate *Ātmā*, an entity of Hilbert space, in space-time of *Jīva* will be futile. State vector may turn out to be an independent rediscovery capable of concretizing the concept of *Ātmā*. The possibility calls for a relook and updating of important concepts propounded in Vedānta. As an illustration, we present a speculative updating of the interpretation of a famous edict of Vedānta *Brahma Satyam Jaganmithyā*. Its conventional interpretation is "*Brahman* is the eternal reality while all activities in *Jagat* or world are temporary or transient realities." Our updated interpretation is "*Brahman* stands for quantum description while *Jagat* stands for classical description; quantum description is correct, classical description is an approximation."

Vedic scriptures mention some phenomena that appear esoteric to standard quantum theory. Some of them have anecdotal evidence. They usually require some amount of retrocausality, i.e., effect of a cause moves backward in time as well. So that what happens "here and now" may have active causes "there and then" in the future. Vedānta envisages that some *Jīvas* recognize future active causes of present events. The examples of clairvoyance and precognition are anecdotal evidence of retrocausality. Standard quantum theory is built on the premise of past causes for present events. The incorporation of retrocausality in quantum theories is a challenge. One particle quantum mechanics cannot meet this challenge because state vector is unique, and its all representations are unitarily equivalent. Quantum field theory permits unitarily in equivalent representations of a state vector [9]. It may be possible to incorporate retrocausality in some representation. Bohmian dynamics [10] is a legitimate quantum theory that allows retrocausality even in its one particle quantum mechanics version. In this formulation, a matter particle needs two wave functions for its correct description: the usual Schrodinger wave function and a pilot wave function. Both wave functions are the solutions of the same equation but with different boundary conditions. The usual wave function is obtained from initial boundary condition, i.e., from the knowledge of the state at an earlier time. So that effects of a cause propagate forward in time; the wave function contains past information of the universe and is said to depict explicate order. The pilot wave function is obtained from final boundary condition, i.e., from the knowledge of the state at later time. The influences of a cause propagate backward in time; pilot wave depicts future information of the universe and said to depict implicate order. A material particle is, thus, connected to the entire universe, contains information of the universe, and is enfolded in the universe. Similar statements frequently occur in Vedāntic deliberations.

Another characteristic feature of waves is superposition. It is the property because of which any number of waves can coexist in a region of space-time and are not affected by each other; however, the value of the property or disturbance caused by the waves at any space-time point is the algebraic superposition of property or disturbance caused individually by all waves at the space-time point. The state vectors also obey superposition like in wave motion and give rise to observable consequences that are counterintuitive and difficult to comprehend in the classical framework. These consequences constitute definitive evidence of the validity of the description of material particle in the quantum framework. In the quantum framework, a particle moving along a path is depicted by one state vector and along different path by another state vector. The same particle moving from one location to another location without specifying the path is depicted by yet another state vector, which is the superposition of state vectors corresponding to the moving of the particle along all possible paths. It means that the classical trajectories of a particle are not only probabilistic but they superimpose as well. Superposition of trajectories produces interference effects, which are considerably enhanced in situations where periodic distribution of high probability regions occurs, e.g., in crystalline solids. Both light and matter waves produce spectacular results on encountering such regions. Attempts are being made for creating periodic structure of light for observing interference effects with quantum waves of matter particles [11]. If successful, it will bring out a novel way of transferring information to specific places, which may clarify some more concepts mentioned in Vedānta.

Entanglement is a purely quantum feature of composite system whose parts maintain their identity. The entangled parts exhibit correlation in properties and appear to act in a coordinated and cooperative manner without any explicit mechanism of information/instruction transfer among them. Consider the two parts of entangled composite system. Both parts contain equivalent information about the composite system, and the measurement of a property of one part determines the property of the entangled part as well. The parts can be matter or photon field or both. We prefer to model a human being (or any living entity) as an entangled quantum system of material body and photon field. The model explains many aspects of human beings deliberated in Vedānta. The equivalent information about a human being is distributed differently in its material body and photon field, localized in material body and spread out over large region in space in photon field. The information in photon field is accessible from a distance and justifies remote viewing. The model opens up the possibility of measuring holistic qualitative features as health, happiness, meditativeness, viability, etc. The model also suggests new ways of manipulating these holistic features for corrective purposes; they can be manipulated by altering material contents, photon field, or combination of the two. The manipulation of photon field is especially interesting for it can be affected from a distance and may provide recipes for remote intervention.

Measuring the property of an entity is the least understood aspect of quantum theory. The quantum state of the entity becomes known only after the measurement, and the measuring process causes an uncontrollable and irreversible change in the quantum state of the entity and transforms the state of the entity into a specific state that yields definite value of the measured property. We do not know why and how does it happen. Bohr was aware of the problem but sidestepped it by suggesting that quantum principles are insufficient to cover biological matter and claiming that definite outcome in measurements is the result of the classical nature of the interface of measuring device with the brain [12]. Von Neumann developed formulism for measurement using quantum measuring devices [13]. He envisaged that the observer and the observed get entangled into a composite system in any measurement. The entangling continues to occur along the chain from the observed entity all the way up to the brain and definite outcome of a measurement is a consequence of classical nature of brain and mind interface. Von Neumann ascribed quantum nature to biological matter and classical nature to brain-mind interface implying classical nature of consciousness. He separated the world into two categories, one containing all stuff including the brain and others containing minds. It resembles the separation into *dṛśya* and *dṛṣṭā* propounded in Vedānta. One problem that remains unresolved is the reason of the classical nature of mind. Henry Stapp is developing a solution that has strong leaning to Vedānta. He is postulating quantum nature of all entities in the chain starting from the observed object to the mind of the observer, so that every measuring process should get entangled in all entities in the measuring chain. Measurement yields some outcomes, and not all possible outcomes (or eigenvalues), because of the prior training of the mind. Prior training is imparted to a mind that continuously repeats the same measurement on a system. The first measurement gives one of the eigenvalues as outcome and changes the observed system into the state that is eigenstate for the observed eigenvalue. The repetitions of measurement yield the same outcome and do not change the state. There remains only one loose end to tie, how to obtain different outcome in the measurement made after a time gap. Stapp tied the loose end in a toy model and showed how to change the observed value and state of an object from one to another coherent state in finite time through repeated measurements [14]. Stapp proposed that an act of observation or measurement is a query put to nature, and nature responds to the query by releasing a stream of consciousness. Each stream of consciousness produces one result and one state. The variants of the Stapp model may ascribe the release of stream of consciousness to *Jagadīśa* or even *Jīva* considered as a composite system of autonomous parts. The new ingredient in the Stapp's proposal is the flickering of consciousness. Flickering of consciousness should cause flickering in the properties of every entity along the entangled chain. It opens the possibility of defining and measuring attributes of flickering consciousness. Vedānta mentions *spandan* (vibration) as an essential requirement of life, which is similar to flickering and, perhaps, not different from it.

4 Unexplained Biological Data

Biology and Physical Sciences were different disciplines, but they have been coming closer and closer. They may merge in the near future. Some portions of Biology have already merged with and have assimilated in Physical Sciences. They are the success stories of Biology, which are used for identifying measurable properties of biological systems. Olatavi and Barbasi [15] have summarized the status of current biological knowledge in a four-level pyramid. The levels in the pyramid appear to be hierarchical in character, which means that a property specified in a higher level depends upon properties specified in lower levels but is not expressible by them. The bottom level is described by various biomolecules and genes and the next level by regulatory motifs and metabolic pathways of biomolecules. We have a good understanding of these two levels, and success stories of Biology are mainly associated with them. The only blemish in the understanding is our inability to comprehend cooperative and coordinated functioning of biomolecules and genes. The remaining two levels are poorly understood. Functional modules and large-scale organization characterize these levels, and empirical information allows to attribute qualitative features to them. The top or organismic level is the most extensively investigated level, and qualitative features like health, psychological behavior, consciousness, etc. are used in its description. Many of these features are experienced and felt. Vedānta is also aware of these experienced features, particularly in human beings, but is not aware of hierarchical pyramid. Cognitive and Health Sciences study organismic features, and their studies are concentrated mainly on human beings because human beings can easily communicate their experiences to investigators. Similar experiences can usually be ordered in qualitative manner, so that severity of illness, deepness of concentration, degrees of happiness, etc. are common concepts. The ordering of a feature is a precursor to its eventual measurement by a suitable procedure and thereby its conversion from feature to property. The correct procedure has not been found, and various suggested procedures have one or another drawback. Human health is the most extensively studied orgasmic feature, and many procedures have been suggested for its quantification and efficient management. These procedures are based on two different methodologies; one employs properties of the bottom level of the pyramid and the other employs features of the top level of the pyramid. The former methodology is used in Modern Medical Science (MMS), and its procedures quantify ill-health and diseases, and not good health, by Parāmeters measured in pathological and analytical platforms. The latter methodology is used in many systems contained in the collage of Complementary and Alternative Medicines (CAM), and their procedures identify diseases and good health using questionnaires and organismic symptoms. The MMS and CAM differ in identifying health and diseases as well as in their management. They succeed partially managing health and diseases. Partial success perhaps results from the use of only one level and not the whole pyramid. The whole pyramid can be used only if the reason for its existence is known. It has been our contention that the pyramid exists because living systems are composite quantum systems, whose constituents retain their identity in different levels. The incessant and spontaneous emission of visible range photons of nearly unchanging flux by almost all living systems provides direct evidence in support of our contention. The constituents describing the two bottom levels of the pyramid cannot emit these photons. The whole pyramid must be involved in their emission. The number of photons emitted in a definite interval fluctuates, fluctuations are not random but contain information, quantum framework can extract this information, and extracted information determines organismic level properties. It has profound implications for Stapp's proposal and Vedānta. It supports quantum nature of living objects, essentiality of fluctuations, and their linking with organismic properties. Perhaps, flickering in the stream of consciousness and fluctuations in photon number to fluctuations and *Spandan* are synonyms or at least have common origin. The entanglement of photon field and material stuff implies that a live object can be investigated and manipulated by photon field or material stuff or a combination of the two. Material stuff is localized, but photon field extends everywhere and is a source of information about the organismic properties of living objects. This information can be accessed noninvasively from a distance. The photon field is a recipe for remote sensing. The entanglement is recipe for remote intervention as well; it requires the tuning of detector to specific photon property, so that its detection will imply that the entangled material stuff is in the corresponding state. This mechanism or its variants may be responsible for incidences of the remote healing. The two loosely integrated living parts in a composite structure may have the capability to access information in each other's photon field reflected by the environment. The composite structure can use this information for self-reflection and exploration of environment [16].

References

1. Sue LC. www.realityshifters.com (2016).
2. Eliot D, Rohit D. The essential Vedānta: a new source book of Advaita Vedānta. Even Gītā mentions that neither the beginning nor the end of cosmos are known. 2004
3. Amit G. www.cs.ubc.ca/~goyal/age_of_universe.php (2016).
4. Greene B. The hidden reality. New York: Viking Books; 2011.
5. The Krishna Path. Did quantum physics come from vedas? http://upliftconnect.com/quantum-physics-vedas/ (2016).
6. Tye M. Ten problems of consciousness : a representational theory of the phenomenal mind, Representation and mind. Cambridge, MA: MIT Press; 1995.
7. Bajpai RP, van Wijk EPA, Van Wijk R, van der Greef J. Attributes characterizing spontaneous ultra-weak photon signals of human subjects. J Photochem Photobiol B. 2013;129:6–16.
8. Bajpai RP. Biophonic route for understanding mind brain and the world. Cosmos Hist J Nat Soc Philos. 2015;11:189–200.
9. Vitiello G. Links. Relating different physical systems through the *common QFT algebraic structure*. In: Unruh WG, Scheutzhold R, editors. Quantum analogues: from phase transitions to black holes and cosmology. Lectures notes in Physics, Vol 718). Berlin: Springer; 2007, p. 165–205.
10. Towler M. De Broglie-Bohm pilot wave theory and the foundations of quantum mechanics. www.tcm.phy.cam.ac.uk/~mdt26/pilot_waves.html (2009).
11. Takeshi Y. Thought experiment on the reality of a matter wave. Int J Quant Found. 2016;2:127.
12. Stapp HP. Mindful universe: quantum mechanics and the participating observer. Berlin: Springer; 2007.
13. von Neumann J. Mathematical foundations of quantum mechanics. Princeton: Princeton University Press; 1996.
14. Oltavi ZN, Barbas AL. Systems biology: life's complexity pyramid. Science. 2002;298 (5594):763–4.
15. Bajpai RP. Tinkering with the unbearable lightness of being: meditation, mind-body medicine and placebo in the quantum biology age. J Nonlocality. 2013;II(2), December 2013 in Panel discussion.
16. Bajpai RP, Rastogi A, Popp A. Quantum response of human skin to hydrogen peroxide stimulation. J Nonlocality. 2015;IV(Supplement), April 2015.

Chapter 12
Quantum Vacuum and Beyond: An Exegesis in the Light of the Concept of *Śūnya*

Rajeshwar Mukherjee

Abstract The quantum field theory (QFT) has unmasked a nature of Reality which is radically different from general human perception. According to QFT, the fundamental particles, which are the basic building blocks of the physical universe, are not the primary reality. These particles are simply the excitations of their respective underlying quantum fields, which constitute the primary reality. The ground state of a quantum field is the vacuum. The concept of void or vacuum is a cardinal concept in Indian tradition. The void or emptiness is denoted by the term śūnya. The word *śūnya* has been derived from "śūna" which means "to grow" or "to swell." Though it generally denotes void, yet, in some places, it is also used as a synonym of "pūrṇa"; and therefore, it has also been identified with the concept of infinity. As the theory of *śūnya* evolved as a universal philosophical doctrine, it can comfortably be applied to shed new light on the unexplained issues of the quantum vacuum, which is considered to be the primary reality of the physical universe. The paper explores some novel dimensions of quantum reality in the light of the theory of śūnya. But Reality is not exhausted within the purview of physics, and hence, there is dire need to discover other dimensions of Reality transcending the physical vacuum. The philosophy of *śūnya* is a wonderful theory which provides us a framework of knowledge useful to have an import of the Absolute Reality and its manifestations as different modes of existence. The paper also aims at utilizing the theory of śūnya to inquire into the Reality beyond the physical vacuum.

Keywords Quantum vacuum · Śūnya · Veda · Upaniṣad · Cosmology · Microcosm · Macrocosm · Fluctuation · Consciousness

---

R. Mukherjee (✉)
School of Indian Heritage, Ramakrishna Mission Vivekananda Educational and Research Institute, Belur, West Bengal, India

S. R. Bhatt (ed.), *Quantum Reality and Theory of Śūnya*,
https://doi.org/10.1007/978-981-13-1957-0_12

1 Introduction

The idea of *śūnya* is one of the cardinal concepts of Indian tradition. Though the concept is popularly attributed to Buddhist thinkers, its root can be traced in the Vedas and in the later schools of knowledge based on Vedic wisdom. It will be superfluous to mention that the concept got accentuated and evolved as a rational system of thought by dint of the philosophical exercises of the Buddhist scholars. Nevertheless, the concept of *śūnya*, as found in the Vedic lineage, is of immense importance which deserves proper exploration and comprehensive interpretation. It is true that some of the early texts do not bear any explicit reference to the word *śūnya*, yet the concept of *śūnya* can easily be extracted through an in-depth study of the Vedic canons. The concept of *śūnya* is found to be embedded in the *Nāsadīya Sūkta*, which is contained within the corpus of *Ṛgveda*. Although The Principal Upaniṣads very rarely refer to the word *śūnya*, they dwell on the concept of *pūrṇa*, which is also considered to be the synonym of *śūnya*. The minor *Upaniṣad* called the *Tejobindu Upaniṣad* mentions the concept of *śūnya* and its philosophical implications. The *Yoga-sūtra of Patañjali* along with its enlightening commentary by Vyāsa expounds *śūnya* in connection with the real nature of the self. Gauḍapāda's *Kārikā* of *Māṇḍūkya Upaniṣad* which is said to be influenced by the teachings of Buddha also suggestively expatiates on the concept of *śūnya*. However, it is an undeniable fact that the concept has been fine-tuned by the Buddhists seers giving rise to a concrete philosophy. The Mādhyamika school of Buddhism has contributed significantly to the development of the philosophy of *śūnya*, which is known as *śūnyavāda*. Nāgārjuna, the exponent of the Mādhyamika school, has extensively worked on the theory of *śūnya*. Arguably his theory appears to be the restatement of the doctrines of the *Upaniṣad*s within the shadow of the Buddhistic views [1]. On the other end, Ācārya Śaṁkara chose to write his magnificent commentary on the verses of *Māṇḍūkya Kārikā* bringing about the grand syncretism of Buddhist thought and Vedic wisdom. Thus the concept of *śūnya*, which is found as a germ in the Vedas, gradually nourished itself deriving its vital sap from the subsequent works of the Vedic and Buddhist scholars to flower into an exquisite philosophy.

The philosophy of *śūnya* is an epoch-making discovery which has been gifted to the world by the Indian thinkers ushering in a thought revolution in different fields of study. The legacy of the philosophy of *śūnya* appeared as the concept of zero in the mathematical system which is considered as one of the most significant contributions of Indian mathematicians to the discipline of mathematics. Amazingly, the theory of quantum vacuum, a fundamental concept in modern physics, has wonderful parallels with that of the concept of *śūnya*, which springs from the Indian philosophical tradition. In-depth study reveals that the philosophy of *śūnya* is not only capable of adding novel insights to the theory of quantum vacuum but transcends it to shed new light on the nature of the Reality.

2 *Śūnya* in Indian Tradition

The word *śūnya* means void, devoid of, empty, zero, null, and absence of something. It has been derived from *śūna*, which Pāṇini recognizes as the past participle of "śvi". *Śvi* means "to go," "to grow," or "to swell." The *Amarakośa* records three different synonyms of *śūnya*: empty, worthless, and absent [2]. In the *Ṛgveda*, the word has also been used in the sense of "absence" or "deficiency." Therefore, the word *śūnya* commonly connotes the absence of objects. But surprisingly the word *pūrṇa*, which means fullness, is very often used as synonymous to *śūnya* in Indian tradition. The *Maitreyī Upaniṣad* reads:

antaḥ pūrṇo bahiḥ pūrṇaḥ pūrṇakumbha ivārṇave/
antaḥ śūnyo bahiḥ śūnyaḥ śūnyakumbha ivāmbare//[3].

[It is internally full (infinite), externally full (infinite) and exists as a filled pitcher in the sea. It is internally void, externally void and exists as an empty pitcher in space.]

The word *pūrṇa* has been derived from the verbal root *pṛ* which denotes sustenance, fullness, self-sufficiency, and self-containedness. So the *pūrṇa* means fullness. Therefore, *śūnya*, which popularly means void, also suggests fullness; and it is treated as equivalent to *pūrṇa*, infinity.

Void=Emptiness=Śūnya=Pūrṇa=Fullness=Infinity

Indian tradition has harnessed this profound philosophical concept to express the nature of the Reality. The *Upaniṣad* profoundly expresses:

om pūrṇam adaḥ pūrṇam idaṁ pūrṇāt pūrṇam udacyate/
pūrṇasya pūrṇam ādāya pūrṇam evāvaśiṣyate//[4].

The Absolute is infinity (void), the relative is infinity (void). The infinity (void) emerges out of the infinity (void) leaving behind infinity (void).

The *Nāsadīya Sūkta* of the *Ṛgveda* depicts the nature of the Ultimate Reality through the concept of void. The Vedic *sūkta* reads:

nāsad āsīn no sad āsīt tadānīṁ nāsīd rajo no vyomā paro yat/
kim āvarīvaḥ kuha kasya śarmann ambhaḥ kim āsīd gahanaṁ gabhīram//
na mṛtyur āsīd amṛtaṁ na tarhi na rātryā ahna āsīt praketaḥ/
ānīd avātaṁ svadhayā tad ekaṁ tasmāddhānyan na paraḥ kiṁ canāsa//[5].

[Existence was not then, nor non-existence,
The world was not, the sky beyond was neither.
What covered the mist? Of whom was that?
What was in the depths of darkness thick?
Death was not then, nor immortality,
The night was neither separate from day,
But motionless did That vibrate
Alone, with Its own glory one—
Beyond That nothing did exist.] [6].

The seer of *Nāsadīya Sūkta* conceived of the concept of void while inquiring into the Absolute Reality. The Absolute is not only devoid of all material entities but is the negation of all affirmative statements about it. It is beyond all the conjectures about the relative existence and nonexistence. The relative terms like day and night, death and immortality, etc. are meaningless in regard to the Absolute, as it is free from all sorts of relativities. According to Sāyaṇa, the Absolute is beyond space, time, and relations. Mind and life are also nonexistent with reference to it. Though it is void, it is the generator of the worlds of space, time, and matter. It is the source of the entire relative universe with its constituents. If this void is the source of all, then the question arises: "How can something come out of nothing?" The *sūkta* asserts that beyond the negation of all relativities, there is an affirmation of the nondual self. The *sūkta* records this affirmation of absolute existence with the idiom *tad ekam*. The Ultimate Reality is *śūnya*, void, from the viewpoint of the phenomenal existence; but it is *pūrṇa*, infinity with reference to the Absolute.

The *Māṇḍūkya Upaniṣad* and its commentary by Ācārya Śaṁkara profoundly bring out the essence of these two aspects of the Ultimate Reality. The *Upaniṣad* records the Ultimate Reality as: *adṛṣṭam avyavahāryam agrāhyam alakṣaṇam acintyam avyapadeśyam ekātmapratyayasāraṁ prapañcopaśamaṁ śāntaṁ śivam advaitaṁ caturthaṁ manyante sa ātmā sa vijñeyaḥ*/[7].

[...The Ultimate Reality is unseen, unrelated, incomprehensible, uninferable, unthinkable, indescribable, essentially of the nature of consciousness constituting of Self alone, negation of the world of manifestations, tranquil, bliss and nondual...]

Prapañca means the world of manifestations. The seer of the *Upaniṣad* suggests that the Ultimate Reality is the negation of the world of manifestations and, at the same time, it is the affirmation of nondual consciousness. *Upaniṣad*s being *rahasya-vidyā* often speak in the language of paradox, and, as such, the Ultimate Reality has been expressed as void as well as infinity. The *Chāndogya Upaniṣad* also considers the Absolute Reality Brahman as bliss and void [8].

The concept of *śūnya* is also equally relevant in obtaining the import of the Reality in the microcosmic domain. In Indian tradition, an individual living being is considered as microcosm. Through the praxis of yoga, the individual being transcends all the worldly perceptions to discover its real nature, which is essentially of the nature of consciousness. The state of highest realization in yoga tradition is called the Kaivalya, the total aloofness from the objects of the relative world. The *Yoga-sūtra* observes *puruṣārhaśūnyānāṁ guṇānāṁ pratiprasavaḥ kaivalyaṁ svarūpapratiṣṭhā vā citiśaktir iti*//[9].

The *guṇa*s, which are the basic building blocks of this phenomenal universe, disappear into their ultimate cause and cease to be the objects of puruṣa anymore. The puruṣa, as the observer, observes the reality as a void *śūnya*. From an alternative perspective, the puruṣa abides in its own real nature and experiences its identity with the absolute consciousness, which is essentially infinite, *pūrṇa*. Endorsing these two alternative perspectives of looking at the Reality, theseer of *Tejobindu Upaniṣad* exclaims:

sarvaśūnyasvarūpo 'haṁ sakalāgamagocaraḥ/
mukto 'haṁ mokṣarūpo 'haṁ nirvāṇasukharūpavān//[10]

[I am of the nature of absolute void which is beyond the ken of scriptures. I am the liberation incarnate, free and the epitome of transcendence.]

vāṅmano 'gocaraś cāhaṁ sarvatra sukhavān aham/
sarvatra pūrṇarūpo 'haṁ bhūmānandamayo 'smy aham//[11].

[I am beyond the ken of mind and speech and I am omnipresent as bliss. Everywhere I am the essence of Infinity and I exist as the bliss which is the self.]

The philosophy of *śūnya* attained its acme with the speculations of the Mādhyamika school of Buddhism championed by Nāgārjuna, who is credited of reviving the philosophy of *Prajñāpāramitāsūtra*s. The popular epithet given to his philosophy is "śūnyatā." According to him, *śūnya* is the permanent principle which is transcendental, indefinable, and immanent. It is the basis of all the worldly manifestations.

The relative universe of forms having *śūnya* as its basis may also be equated with *śūnya*. The famous *Heart Sūtra* records that form is in essence emptiness and emptiness is form:

rūpaṁ śūnyatā śūnyatāiva rūpaṁ
rūpān na pṛthak śūnyatā
śūnyatāyāḥ na pṛthag rūpam//[12].

The philosophy of Nāgārjuna has a close association with the conclusions of Advaita Vedānta. Though the term *śūnya* means nothingness, yet it hints at a transcendent and permanent principle which is beyond any definition. The essence of *śūnyatā* has been captured in the cadence of *Sarvadarśana-saṁgraha*:

astināsti ubhaya anubhaya iti catuṣkoṭivinirmuktaṁśūnyatvam//[13].

Śūnyatā is that which cannot be called void or not void, or both or neither, but in order to indicate it is called void, because absolute negation is an utter impossibility. In essence, the concept of *śūnya*, as revealed in the philosophy of Nāgārjuna, epitomizes the Absolute Reality which is devoid of all objects and phenomena of the relative world [14]. Vedanta adds that the negation of all relative existence is tantamount to affirmation of the Absolute. Elucidating further Swami Vivekananda asserts: "Buddhism said, 'Realize all this as illusion', while Hinduism said, 'Realize that within the illusion is the Real'." [15]

3 Quantum Vacuum in QFT

The wonderful parallel of the concept of *śūnya* is available in the theories of physics in terms of quantum vacuum. Vacuum is defined in quantum physics as the ground state of a quantum field. It is the minimum energy state corresponding to zero particles. Vacuum is obtained when all the matter or the particles corresponding to all the known fields are removed from a region of space. So apparently vacuum is an empty space. Heisenberg's uncertainty principle does not allow the vacuum to be totally empty. It contains the vacuum fluctuations. In fact, quantum vacuum is a complex entity, which, though appears to be empty, is the source of all potentiality. It is filled with potential particles, space-time networks, unit pairs of virtual matter-antimatter, and potential properties at the quantum level. The quantum vacuum is an important concept in the theoretical framework of quantum field theory (QFT) propounded by Schwinger, Feynman, and Tomonaga. In QFT, quantum vacuum is considered as the ground state in the Hilbert space. The state is almost analogous to the ground state of a harmonic oscillator in the case of noninteracting quantum fields. Quantum fluctuations, as allowed by the Heisenberg's uncertainty principle, are the temporary emergence of the energetic particles coming out of vacuum. The quantum fluctuations are sometimes attributed to the "appearance and disappearance of two (or more) virtual particles." Virtual particle refers to a disturbance in the field and is not particle at all. A real particle is actually a regular ripple in the field. There is a great debate in the world of physics as to whether matter is composed of atoms or an underlying continuum. The field theory has given a satisfactory answer to this question. According to field theory, field is a continuum which pervades the space and yet displays a discontinuously granular structure in its particle aspect. Wave-like and particle-like aspects of radiations are the different manifestations of the same reality, the quantum field. Thus the two different aspects may be unified as the two different aspects of the same reality. Quantum field is the reality which sustains both these aspects. The ground state of a quantum field is the vacuum. Here all interactions take place through the exchange of particles. The stronger is the interaction, the stronger is the resulting force. As the resulting force is stronger, the probability of such exchange processes becomes higher giving rise to more frequent interactions. But the role of virtual particles is not only restricted to interactions. A nucleon is itself capable of emitting and reabsorbing virtual particles provided the created meson disappears within the time allowed by the uncertainty principle. Due to strong interaction, the probability of such self-interaction process is appreciably high for nucleons. These nucleons are the centers of continuous activity surrounded by the cloud of virtual particles which continue to live for an exceedingly short duration. When two nucleons come very near to each other, their cloud of meson overlaps. Some virtual particles sometimes are not absorbed by the nucleon from which they have come into being but jump across to be absorbed by the other nucleon. So the forces between particles are determined by the composition of the virtual clouds. The range of the interaction also depends on the extension of the virtual clouds. The virtual particles can under special circumstances become real particles. However, virtual particles can come out of nothing and can again merge into nothingness even without the presence of any nucleon or any strongly interacting particles. Thus the theory of fields compels to abandon the distinction between matter and void. Quantum field theory and Einstein's theory fields suggest that particles cannot be separated from their surrounding space. Einstein's general theory of relativity suggests that space-time and matter form a continuum and the presence of matter leads to the curvature of space. They cannot be separated from one another. They are conceived of as the condensation of a continuous field, and hence they cannot be treated as isolated entities. The field is the substratum of all particles and their mutual interactions. The ground state of this field is the vacuum. The virtual particles and their interactions appear out of the void and disappear in it. Thus the quantum vacuum lies beneath the existential level of real matter, although it teems with innumerable structures. Ontologically it is the seed of plenty of possibilities [16]. Therefore the concept of void resembles to a great extent with the concept of *śūnya* as conceived in Indian philosophical tradition, though it will not be apt to claim the exact equivalence of *śūnya* with the quantum vacuum. But it is beyond question that the concept of quantum vacuum echoes the concept of *śūnya* as projected in Indian philosophies. Focusing on the similarities of quantum vacuum and *śūnya*, physicist Fritjof Capra writes: "Like the Eastern void, 'the physical vacuum'— as it is called in the field theory—is not state of mere nothingness, but contains the potentiality of all forms of the particle world. These forms in turn are not the independent physical entities but merely transient manifestations of the underlying void"[17].

4 Quantum Vacuum and Cosmology

The concept of vacuum is also relevant to the events at the beginning of the universe. The Hawking-Penrose singularity theorem, based on Einstein's general theory of relativity (GTR), inflationary cosmic models of the early universe, and some other theories, leads to the notion of the origin of the universe from the big bang singularity. The Hawking- Penrose singularity theorem states that the Friedman-like universe has started with a big bang singularity. Basically, singularity is a point in space-time where the space-time curvature becomes infinite and the radius of curvature is zero. The validity of this theorem depends on the validity of the GTR. It is almost recognized among the physicists that the GTR fails to apply in the physical world where the quantum mechanical interactions predominate. The predomination of quantum mechanics is found at temperature at and above 10^{32}K and at and above the density 10^{94}gm/cm^3, and the radius of curvature comes to the order of 10^{-33}. Such conditions are found during the Planck era, that is, at first 10^{-43}s after the singularity. The big bang theory based on the GTR cannot be used as standard theory to reconstruct the physical processes happening during this time, as laws of quantum mechanics call into play in this regime and a standard theory of quantum gravity is hitherto unavailable. To avoid such big bang singularity, a host of theories known as vacuum fluctuation models of the universe came into being. The quantum cosmological models explain the beginning of the universe largely from the fluctuations of vacuum. The models developed by Tryon, Brout, Englert, Gunzig, Atkatz, Pagels, Gott, and others pictured the universe as coming out of empty background space. These vacuum fluctuation models are governed by the uncertainty relations. The model of Vilenkin depicts the physical universe to come out of void without cause. Thus there are strong opinions about the emergence of the macroscopic physical universe from the fluctuations of vacuum [18].

5 Microcosm and Macrocosm

Quantum vacuum and its fluctuations are the fundamental reality both in the microscopic and in the macroscopic domains. Though both the domains are different and are governed by different set of laws, the fundamental principle in either of the domains appears to be similar in nature. Einstein's search for a unified field, popularly known as the holy grail of physics, was though an unfulfilled pursuit to reach the unity at the fundamental level, yet the weaving of the three of the four fundamental interactions into one is a significant contribution toward validation of the grand unification of four fundamental interactions. Finally, at the dawn of the new millennium, the superstring theory emerged as vindication of the "unified field theory" conceived by Einstein with the unification of the fundamental interactions. According to the new vision, the macro- and the microworlds are governed by the same grand principle. The same master equation would govern both the micro and the macro aspects of the universe.

While it would be too naive to attempt a justification of the unification scheme of modern physics in the light of the conclusions of Indian philosophy, it is nonetheless true that the unity of macrocosm and microcosm as worked out in Indian tradition provides some insights in investigating the happenings of the physical world. However, it requires mention here that the micro-macro concept in physics differs from the microcosm-macrocosm as discussed in Indian philosophy. The micro-macro in physics is confined to the physical world outside, while in Indian philosophy, microcosm and macrocosm denote individual being, *piṇḍa*, and universe, *brahmāṇḍa*, respectively. It is observed that the concept of *śūnya* has been used in the Indian systems in connection with both microcosm and the macrocosm, as it is found that the concept of vacuum in physics is common to both micro- and macro-systems. Therefore, the theory of *śūnya* seems to arouse a speculative question: Is quantum field the primary reality in this physical world which is equally consistent in explaining the dynamics of both micro-systems and macro-systems?

6 Quantum Vacuum vis-à-vis *Śūnya*

The study of quantum vacuum engenders several pertinent questions. To list a few:

1. Can quantum vacuum be identified with the *śūnya* of Indian tradition?
2. Is this vacuum a consciousness entity?
3. Is vacuum the primordial entity of which everything is made? [19]
4. What is the material which makes the quantum vacuum? [20]

Śūnya is regarded as the Absolute Reality in Indian tradition. On one hand, it is the absolute negation of all the phenomena, and on the other hand, it is the infinite consciousness. According to Vedānta, the infinite consciousness is one without a second; and it manifests as the relative existence by dint of *māyā*, which is the substratum of space-time and causality. The Absolute Reality manifests as the relative world order of names, forms, and manifoldness. The Absolute, conceptualized as *śūnya*, is the essence of all the manifestations; and yet it is beyond all manifestations. It is devoid of names, forms, attributes, events, concepts, and ideas, and hence it is a void. Quantum vacuum, on the other hand, accounts for the manifest world, which is the relative expression of the Absolute. It is a defined concept capable of explaining all physical phenomena. Though the concept is abstract and generalized, yet it cannot be identified with the *śūnya*, the Absolute Reality. However, the concept of quantum vacuum is the essence of the reality of all the events in this physical universe. It is the fundamental principle representing the physical universe as a whole.

As the manifest world, according to Vedānta, is a superimposition of matter (*jaḍa*) on consciousness (*cit*), therefore the world of relativity must have the trace of consciousness in it. Naturally the quantum vacuum being the fundamental principle of physical universe is expected to have some sort of consciousness manifesting through it. Experimental evidence shows that vacuum fields are characterized with unpredictable and spontaneous activity. No predictions can be made as to the occurrence of a particular event. The fields are considered to be fluctuating with a tremendous speed with a typical time period of 10^{-24}s or less. Besides these wild fluctuations, the fields have remained exactly the same on average since the beginning of time and throughout the entire universe containing regions, which are too far to have any sort of communication even at the speed of light. Though the universe has undergone a massive change, the fields remained the same across the universe in each element of space-time [21]. With this observation, scientists wonder: "Does it not suggest the existence of some sort of self-referral scheme that is responsible for maintaining the fidelity of the quantum fields in spite of their frenetic fluctuations, their prodigious dynamism, their spontaneity and unpredictability?" [22]. Such self-referral scheme is attributed to non-Abelian fields which have a feature of strong self-interaction. Scientists have started believing that such self-interaction, self-organization, or self-referral is the characteristic of awareness. Moreover, the dynamic nature of the vacuum contains memory seeds which not only create the structure of the physical universe but also the intelligent beings; and therefore, the vacuum may have some elements of consciousness within it. Moreover the Wigner-Von Neumann interpretation in the context of quantum measurement problem postulates consciousness to be necessary to complete the process of quantum measurement. Neumann in his *Mathematical Foundations of Quantum Mechanics* theorized that quantum mechanics allows the collapse of the wave function to be placed at any position in the causal chain which extends from the measurement device to the "subjective perception" of human observer. Wigner further proposed that the consciousness of the human observer is the line of demarcation which brings about the collapse of wave function [23]. Henry Stapp, elucidating the Copenhagen interpretation, writes: "The complete description of nature at the atomic level was given by probability functions that referred not to the underlying space-time realities but rather to the macroscopic objects of sense experience" [24]. Penrose-Hameroff furnishes a logical scheme which suggests that individual awareness is linked with the cosmic awareness. Thus the fundamental cosmic reality, according to the modern scientists, has some sort of consciousness associated in it. But unfortunately the domain of physics is restricted within the matter, and it does not have access to the world of consciousness. Physics need not be apologetic for such inaccessibility, because every branch of science has its own jurisdiction. The jurisdiction of physics is matter and not consciousness. Physicists peer into the world of matter ranging from the queer play of subatomic particles to the grand dynamics of the clusters and galaxies and rejoice in discovering the rules of the ever-changing phenomena of the relative universe. However, in pursuing the search in the realm of matter gets the glimpse of "an unknown content" which is of the nature of consciousness. The great mathematical physicist Sir Arthur Eddington writes: "All through the physical world runs that unknown content, which must surely be the stuff of our consciousness. Here is a hint of aspects deep within the world of physics, and yet unattainable by the methods of physics." [25]

Though modern researches on cognitive sciences are devoted to build a bridge between matter and consciousness, the real nature of consciousness is still undiscovered.

The question of primordiality of quantum vacuum is a debatable issue in the field of physics. There is no consensus over the issue as to which is fundamental among vacuum and space-time. A group of scientists claim that space and time are the emergent properties of the field, while the others are unwilling to accept this. Further, if this vacuum is accepted as the substratum of all physical phenomena, the question arises: Does it also account for the events occurring in the mental frame? Can the emergence and dissolution of thought waves be substantiated on the basis of quantum vacuum? The answers to these questions will blow up another set of puzzling questions: Is mind fundamental to matter? Is it an epiphenomenon? Or is it complementary to matter?

Although some physicists are in the quest of the discovering the material that makes the vacuum, others are rest satisfied with the mathematical abstraction. Still the mystery is unsolved. Some think that geometry is fundamental, while others conceive it as the property of empty space emerging from a deeper structure [26]. Therefore reality of quantum vacuum is still shrouded with the mist of incertitude. Looking at it from the philosophical perspective, some physicists consider it as a network of potentialities which are continuous in nature. The actualization of these potentialities is discrete by nature. The vacuum, from the pragmatic point of view, is a space of possibilities from which actualities arise [27]. In the precise language of physicist, "Quantum physics allows for the vacuum state to have an intrinsic structure that provides the "possibility grid" for events, or for entities we can call real with full confidence. Potentialities are thus actualized as properties when the vacuum is disturbed or measured in specific ways." [28]

7 Conclusion

The concept of quantum vacuum as a set of potentialities, which actualizes as the objects and forms of the physical world, often reminds of the wonderful abstraction of *avyakta* conceived by the seers of *Upaniṣads* [29]. This philosophical concept of *avyakta* was later illustrated in the most scientific terms by the seers of Sāṁkhya philosophy. The *avyakta* is a continuum which is formless, limitless, undifferentiated, ubiquitous, indestructible, undeceiving, ungrounded, uncontrolled, indeterminate, and infinite [30]. It is the cosmic potentiality which gives rise to the elements, structures, properties, and patterns when its equilibrium is disturbed. It not only justifies the formation of the physical universe but also accounts for all forms and phenomena in the mental world. It is a profound concept which can explain every event occurring in the relative realm. Though a school of Sāṁkhya philosophy considers *avyakta* as an inanimate principle, the Vedantins on logical ground attributes consciousness to this all-pervading principle. The concept of *avyakta* is also the closest approximation of *śūnya* in the world of manifestations. An inquiry into the nature of *avyakta* may provide a deeper insight in comprehending the holistic view of quantum vacuum.

References

1. *Cf.* Sarvepalli Radhakrishnan (2008, Vol.1, p. 555).
2. *śūnyam tu vasikaṁ tucchariktake*: *Amarakośa* III.1.56.
3. *Maitreyī Upaniṣad* II.27.
4. *Īśopaniṣad* Invocation.
5. *Ṛgveda* X.129.
6. Swami Vivekananda (1921, Vol. VI, p. 152).
7. *Māṇḍūkya Upaniṣad* VII.
8. *Cf. Chāndogya Upaniṣad* IV.10.4.
9. *Yogasūtra* IV.34.
10. *Tejobindu Upaniṣad* III.40.
11. *Tejobindu Upaniṣad* III.38.
12. *Prajñāpāramitāhṛdaya* II.
13. *Sarvadarśanasaṁgraha, Bauddha Darśana.*
14. Sarvepalli Radhakrishnan (2008, Vol. I, pp. 570–572).
15. Swami Vivekananda (1971, Vol. VIII p. 273).
16. G.S. Paraoanu (2014, p. 1).
17. Fritjof Capra (p. 247).
18. *Cf.* Boi Luciano (2011, pp. 10–15).
19. G.S. Paraoanu (2014, p. 7).
20. Loc.cit.
21. *Cf.* Mani Bhowmik (2014, pp. 248).
22. Loc.cit.
23. https://en.wikipedia.org/wiki/Von_Neumann%E2%80%93Wigner_interpretation
24. Henry Pierce Stapp, (1972) "The Copenhagen Interpretation". *American Journal of Physics* 40 (8):1098–1116.

25. Arthur Eddington (2009, p. 182).
26. G.S. Paraoanu (2014, p. 7).
27. Loc.cit.
28. *Ibid.* p 8.
29. *Cf.* *Kaṭhopaniṣad.* 3.11.
30. *Cf.* Brajendra Nath Seal (1958, pp. 2-3).

Bibliography

Bhowmik M. Is the source of awareness present in the quantum vacuum? In: Menon S, editor. Interdisciplinary perspectives on consciousness and the self. New Delhi: Springer India; 2014.

Fritjof C. The tao of physics. Great Britain: Flamingo; 1991.

Arthur E. Space, time & gravitation (ebook). London: Cambridge University Press; 2009.

Luciano B. The quantum vacuum. Baltimore: The John Hopkins University Press; 2011.

Paraoanu, G.S. *The quantum vacuum*, arXiv:1402.1087v2[quant-ph]; 12 Dec 2014.

Radhakrishnan S. Indian philosophy. New Delhi: Oxford University Press; 2008.

Seal BN. The positive sciences of the ancient Hindus. Delhi: MLBD; 1958.

Stapp HP. The astrophysical journal, vol. 40., Copenhagen Interpretation; 1972.

Vivekananda S. The complete works of swami Vivekananda, vol. VI. Advaita Ashrama: Kolkata; 1921.

Vivekananda S. The complete works of swami Vivekananda, vol. VIII. Advaita Ashrama: Kolkata; 1921.

Chapter 13
How to Approach Śūnyatā as the Quantum Reality Through Biological Consciousness?

Bal Ram Singh

Abstract Although there is a dispute as to who discovered the concept of śūnyatā, there is an agreement that the symbol of śūnya was available in India in the fifth century AD. The dot symbol and its subsequent form of the open circle indicate not only nothingness as we know the meaning of śūnya, but it symbolizes infinity as well as there is no beginning or end in the symbol. Therein lies a major symbolic meaning on how the śūnya explains infinite from nothing. Mathematically also the śūnya represents the whole as it cannot be divided by anything. The concept of quantum was basically developed to explain observed reality that was not possible to examine by material science laws that were created in the nineteenth cent ury. It ended up describing the whatness of the reality, without addressing the whyness. It has nevertheless become a major concept in physics today, and its mathematical expressions are being employed to many observations, from origin of the universe to functioning of atoms. However, it is worth noting that the quantum concept is a system developed by admitting the failure to comprehend and making that as the basis to attempt explaining all the physical phenomena. It has obviously succeeded in explaining and predictive verification of many processes to a certain extent. However, it has also become limited when applied more broadly. In other words, it has failed to develop a unified principle for even all things physical.

The basic narrative of physical science, or even science for that matter, is observational, although theoretical concepts (both philosophical and mathematical) do play some role. In the observational world, sensory reality is perceived first.

Keywords Śūnyatā · Quantum Biology · Consciousness · Time · Space · Self-organizations

B. R. Singh (✉)
School of Indic Studies, Institute of Advanced Sciences, Dartmouth, MA, USA
e-mail: bsingh@inads.org

S. R. Bhatt (ed.), *Quantum Reality and Theory of Śūnya*,
https://doi.org/10.1007/978-981-13-1957-0_13

1 Introduction

Although there is a dispute as to who discovered the concept of śūnyatā, there is an agreement that a symbol of śūnya was given in India in the fifth century AD. The dot symbol and its subsequent form of the open circle indicate not only nothingness as we know the meaning of śūnya, but it symbolizes infinity as well in its symbol as there is no beginning or end in the symbol. Therein lies a major symbolic meaning on how the śūnya explains infinite from nothing. Mathematically also śūnya represents the whole as it cannot be divided by anything.

The concept of quantum was basically developed to explain observed reality that was not possible to examine by material science laws that were created in the nineteenth century. It ended up describing the whatness of the reality, without addressing the whyness. It has nevertheless become a major concept in physics today, and its mathematical expressions are being employed to examine many observations, from origin of the universe to functioning of atoms. However, it is worth noting that the quantum concept is a system developed by admitting the failure to comprehend and making that as the basis to attempt explaining all the physical phenomena. It has obviously succeeded in explaining and predictive verification of many processes to certain extent. However, it has also become limited when applied more broadly. In other words, it has failed to develop a unified principle for even all the things physical.

The basic narrative of physical science, or even science for that matter, is observational, although theoretical concepts (both philosophical and mathematical) do play some role. In the observational world, sensory reality is perceived first. If that reality is observed with repeated frequency, the real becomes regular, and when that regular observation is made by others as well, it rises to the level of common. A commonly observed phenomenon over a period of time makes it normal and becomes an element of value judgment. The normal if not challenged in due course leads to setting up of standard and invites enforcement. The standards that are not broken or disrupted with rationality or alternative observations become ideal. Many of the current scientific ideas, including quantum, thermodynamics, genetics, evolution, equilibrium, etc., have reached the state of ideal to explain reality, using the sequence of real->regular->common->normal->standard->ideal. However, in another domain when one conceives the idea of reality beyond perception, the ideal is the no attachment, no connection, or nonexistence, which in fact could be the śūnya or emptiness, or all consciousness.

The consciousness in its most expanded and extensive form encompasses all, recognizing nothing separate, and thus simply is just being. It may in fact be opposite of the concept of the quantum, which primarily separates even the energy into packets or quanta. What if that is not the feature of the physical world, rather a feature of degenerated consciousness? If mind were the quantized consciousness, then the physical reality is just a reflection of that perception. Erwin Schrodinger has alluded to this possibility for the creation of genetic material.

To approach the reality from the ideal of none, a different kind of scientists would be needed, those who at least attempt to practice techniques to get a peek into the consciousness by self-realization. This presentation will explore means and ways to expose this concept for the future of science and society.

2 Quantum Reality

The two main features of quantum physics are that (1) the matter is just a state of energy and (2) matter, at least the tiny forms (electrons, other particles) of it, is nonlocalized. The state of the energy allows the matter to be considered as a wave, while the nonlocalization precludes following the particles in a classical way.

Quantum physics is largely a mathematical expression about the existence of the energy states which can be defined by waves and a statistical analysis for the probability of finding a given particle in a specified space. The reality as we perceive is thus theorized.

Quantum chemistry in fact provides the first glance of the reality, albeit indirect, by measurements of the spectral characteristics of real matter which is made of atomic particles known to be described via quantum mechanics in contrast to the classical mechanics. Modern synthetic chemistry particularly is helpful to construct molecules of different bonding (single, double, conjugate bonds of varying length, etc.) and shape (linear vs. cyclic), thus providing variants where the concept of quantum mechanics can be experimentally tested. For example, a linear carbon structures, such as hexatriene and butadiene, provide a good model for restricting an electron or a group of electrons to the given length of the conjugated double bonds. The bond length is then linked to the space an electron can be found, and a correlation is possible based on the Schrodinger wave equation to predict energy state of the molecule. The longer the length of the conjugated double bonds, the lower the energy state which can be monitored with spectral characteristics showing transition at lower wavelengths of the light. This exercise provides the first glimpse of matter that is otherwise not perceived at minutest level of its construction.

Quantum biology is indeed the field that deals with direct perception in living system, developed as a complex structure function system from the basic elements identified in physics and chemistry (Fig. 13.1). So far, quantum coherence and incoherence have been used to explain efficient harvesting of solar energy by pigments in green bacteria and plants, to explain magnetic geopositioning in birds by radical pair spin of proteins, and to explain sensory perception in olfactory systems and vision by quantum tunneling of the excited energy states [8]. In contrast to quantum physics and quantum chemistry, quantum biology is an observational phenomenon where a variety of biological systems are operating in state that cannot be explained entirely by classical mechanics or physics. Curiously these are all sensory phenomena and encompass from lower organisms such as green bacteria to mammals. In other words, biology is full of processes which can be approached through what we call quantum mechanics.

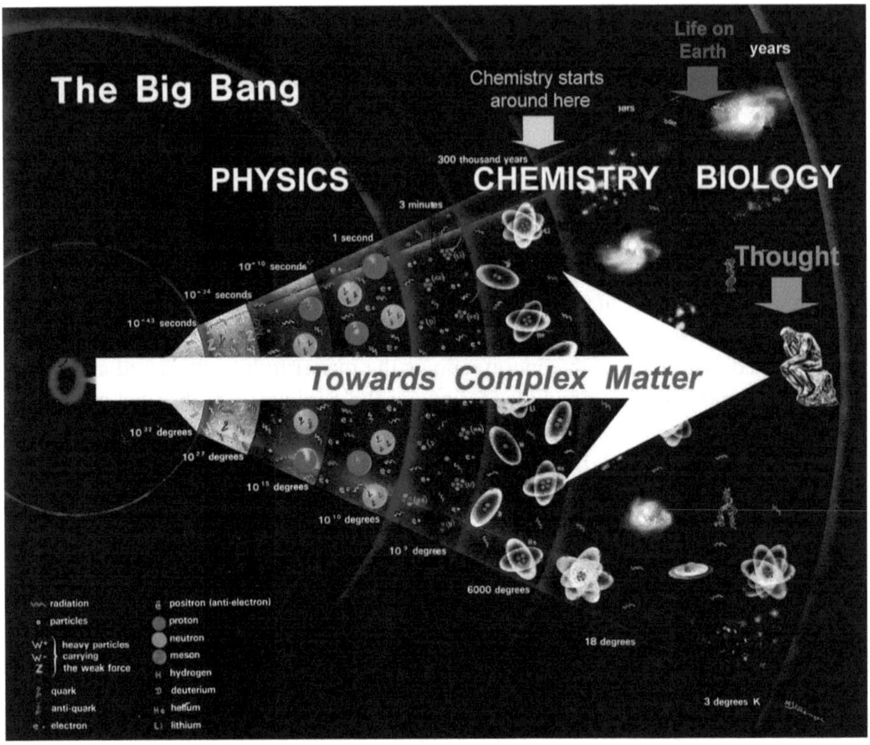

Fig. 13.1 Evolution of the universe toward increasingly complex (conscious) forms of matter: from particles to atoms, to molecules, to life, and to the thinking organism. (Taken from Ref. [11])

The question is whether the quantum concept is a mathematical operation to understand the external physical reality or it is the reflection of an internal biological process in general and sensory process in particular. If the latter, then whether the external reality, including its description through quantum mechanics, is simply a reflection of the internal phenomenon. Is the incomplete nature of the quantum description of the external world simply a reflection of the lack of complete comprehension of the internal process? What physical and philosophical tools may be available to examine the internal quantum-related process? Is it possible that the quantum reality is in fact a negation of the total reality that is reflected in the concept of śūnya to mean nothing separated or all included?

2.1 The Concept of *Śūnya*

It is interesting to note that the śūnya meaning the zero first initially was represented as a dot, which was later denoted as the a circle. In numerical sense it means nothing or existence of nothing. For the numbers to exist, the zero represents both the negative and positive numbers. In a way this then can be referred to as the origin of the numbers, i.e., zero is the basis of all the numbers, and thus the quantitative world, which is the root basis of reality. While the zero acts as the basis of the world reality, zero by itself dissolves it into nothing. It dissolves the duality, yet by its capacity to spring out the positive and the negative in equal proportion, it is the origin of the infinity.

By representing it as a dot of infinitesimally small in size, it can refer to the infinity of the smallest proportion, whereas in its current form of the circle, it reflects the endless and beginningless structure, again representing the infinity. Mathematically, it provides a place value that is never ending, and it when multiplied into reduces every other number to zero, and when divided into any number, it increases it to infinity. Perception of both has not been possible either for any instrument or for ordinary human faculties. Nevertheless these concepts have not only been in the human perception but also used for deriving useful mathematical equations. Human perception being the origin of such ideas needs to be examined further. Human perception with sensory faculties is different than the perception through consciousness, subconsciousness, and even unconsciousness. There are reports of the role of unconsciousness in human decision process [5], which provides a practical value to the absence of consciousness that is indeed a presence of zero playing a role in the decision process. Consciousness is therefore a key element to be examined to understand the limitless nature of śūnya and even the basis of the origin of quantum concepts.

2.2 Consciousness

A critical feature of consciousness is the existence of time and space, which in fact connects it to the quantum description of the matter and energy. Consciousness is generally defined as wakefulness, alertness, perception of surrounding and response, etc. However, these definitions are generally limited at the human level and also at the animal level. When considering plants, bacteria, and virus, one has to resort to basic biological processes at the microscale, such as signal transduction and membrane transport, where one can find ample examples of reception of signals and responses which could run in a time scale of seconds to milliseconds, with processes involved at molecular and atomic levels going further down to picosecond levels or lower.

Consciousness considered from atomic to human levels thus can be given a general term of reception and response process of matter, which is related to scale of time involved in the processes which reflect "consciousness." In contrast to sub-nanosecond phenomena, there are mega and cosmic levels of consciousness which involve thousands to billions of years and include origin of the universe, origin of life, evolution of life, etc. The time scale used here is important to emphasize the relationship between these processes, human perceptions, and memory, all of which are objects of consciousness.

2.2.1 Consciousness as the Concept of Time

Information processing and awareness can be explained through neurobiology. According to Crick and Koch [3], Crick [4], Koch [9], and others, certain 35–75 hertz neural oscillations in the cerebral cortex seem to be correlated with awareness in a number of different modalities, and a mechanism of binding (synchronization of separately represented pieces of information) has been hypothesized. In our model, the "consciousness coefficient" [17] can be used to "control" the time needed to resonate cortical oscillations with a chosen object of awareness. Thus, the self chooses what objects enter into the "frame of awareness" and how long they stay; therefore, a framework independent of physical time can be developed. However, it is logical that "multitasking" or "continuous partial attention" is antithetical to this process and concentration or focus will be needed to manipulate neurobiology. Davidson's work with Tibetan Buddhists [12] has demonstrated the ability of Lamas with many years of meditative practices (>10,000 h) to generate gamma waves ($\approx$40 Hz) when asked to meditate on positive emotions.

In a similar study, Newberg et al. [13] examined meditation and the brain activity of Tibetan monks. According to the subsequent findings, gleaned from scans of each monk's brain during meditation, an increase in activity was found around the frontal region of the brain, in which attention on specific tasks is processed; on the other hand, a decrease in activity was found around the area at the back of the brain, where one's processing of orientation and spatial awareness occur. Newberg found that "During meditation, people have a loss of the sense of self and frequently experience a sense of no space and time and that was exactly what we saw." He concluded, "When someone has a mystical experience, they perceive that sense of reality to be far greater and far clearer than our everyday sense of reality. Since the sense of spiritual reality is more powerful and clear, perhaps that sense of reality is more accurate than our scientific everyday sense of reality." In REM sleep (the state where dreams occur), time dilation is reported, but this state is outside of the purview of self-control.

Thus, manipulative time dilation is only possible if we are in the waking state and can increase the gap between two thoughts/sensory perception modalities. This is possible through yoga/meditation (yogaś cittavṛitti-nirodhaḥ).

In *suṣūpti* state, deep sleep, or non-REM sleep, Ādi Śankarācārya states [Vivekacūḍāmaṇi (171); [19]]:

suṣūpti-kālé manasi pralīné
n'aiv'āsti kincit sakala-prasiddheḥ

In the time of *suṣūpti*, when the mind is absorbed in *Māyā* – when there is no mind – there is nothing. I and the world are not distinct, but this is still a level of ignorance, *avidyā*, in that the experiencing self is not aware that it is but a part of the whole. However, in *turīya* state, the level of awareness (beyond waking, dream, and deep sleep) can be characterized as the purest state of awareness in which the consciousness transcends limitations of space, time, and causation and merges with ultimate. This is not a state of knowledge but a pure experience [17].

Thus, any model to quantify consciousness must be able to encompass the above states of awareness and, more importantly, also be able to control the agency of space and time without any external input.

2.3 Concept of Space: Origin of Space

Space is a product of the big bang singularity, but it can be posited that the big bang singularity itself is a state before which everything was pure existence and pure consciousness, ā*nīdavātam*, without vibration, as described in the Hymn of Creation, the Nāsadīya Sūktam (ṚṚgveda, 10.129).

- There was neither nonexistent nor existent.
- That one thing, breathless, breathed by its own nature.

It then existed without vibration. *Prāṇa* (primal energy) existed then, but there was no motion in it. Then when the "*kalpa*" begins after an immense interval, the *ānîdavātam* commences to vibrate and blow after blow is given by *prāṇa* to *ākāśa*. The atoms become condensed, and as they are condensed, different elements are formed.

All that is power in the universe, manifesting as force or attraction – or even thought – is *prāṇa*.

$$\textit{Prāṇa} + \textit{Ākāśa} = \text{Universe}$$

Every manifestation of power is *prāṇa*
Every material manifestation is *ākāśa*.

This vibration then becomes the first source of time and space, as we understand the gap between two events or gap between two points, respectively. Quantitative measure of these terms runs from zero to infinity, a concept well examined in Vedas. An article by D. R. Sharma [15] explains that although mathematics regards zero and infinity as the two extremes of the number system, they are very different from ordinary numbers. By their very definition, numbers are quantitative. They have meaning only in relation to some objects. By themselves they mean nothing; five has nothing five-like, and it is only a word. Zero and infinity have no quantitative aspect to them; they can be referred to only qualitatively. Each is an indivisible whole. They do not depend on other objects for their meaning. In other words, they are absolute, not relative. Any number multiplied by zero or infinity loses its identity; it merges into the multiplier. Any number divided by zero yields infinity but divided by infinity yields zero. Zero and infinity exist side by side. The same idea is expressed in Vedānta from a different perspective.

Sharma continues as follows: "In Vedānta zero and infinity are regarded unknowable and therefore explaining or defining them is out of question. Even in mathematics zero and infinity cannot be arrived at; a variable can only tend to either.

Looking for the smallest number greater than zero or the largest number smaller than infinity is futile. For any unknowable, we cannot say what it is; it can be described [is] only in terms of what it is not. That is why in answer to the question 'what is the Ultimate Reality *(Brahman)*' the Vedas say 'not this, not this (*neti neti*)'."

Sharma points out that the Sanskrit word for zero and all its equivalents in other languages are *śūnya*. "It is far more comprehensive in meaning than zero; it means zero only in the context of numbers. Its general meaning is void or emptiness, a concept that the mind cannot grasp unless it is itself empty (the initial premise of yoga philosophy). The same is true for infinity. In Sanskrit there is no single word for infinity. Several words like *anādi* (without beginning), *ananta* (without end), etc. are used."

According to Sharma, "Upaniṣads characterize Brahman as 'smaller than the smallest and larger than the largest' [Śvetāśvatara Upaniṣad, 3-20; Kaṭhopaniṣad, 1-2.20.]. In the usual sense in which the words small and large are used, this statement may seem to be a contradiction in terms. This is because our concept of small and large is tied to the perception of space and time. Small and large are a pair of opposites, which according to Vedānta are two faces of the same coin. Small is contained in the large and large is contained in the small in a latent form." As Kabir, one of the great Indian mystics, said: "The drop merges in the ocean, where do you search for the drop? The ocean merges in the drop, where do you search for the ocean? The end of the search is the realization that each is in the other [16]. A singularity brings zero and infinity together."

The cosmological singularity combines zero space-time and infinite energy [15]. "In Vedānta the primordial state of existence of the Reality is an infinite emptiness (śūnyatā). When the mind is completely empty and one transcends space-time in deep meditation, one is able to realize this state. However, this emptiness is not nothingness or non-existence. It is pervaded by energy and consciousness. One arrives at the same concept coming from quantum physics; the quantum void is not empty. There are myriads of virtual particles appearing and disappearing constantly and it is filled with an all-pervading cosmic energy. This state is the substratum of all existence and is called the 'ground' by mystics as well as physicists [10]. The dictionary meaning of singularity is 'the state of being unique or only one of its kind'. In that sense we can regard the Ultimate Reality as the cosmic singularity. In quantum physics one cannot see or probe 'naked singularity'; it is enveloped by the event horizon. From Vedānta's point of view we cannot 'see' the Reality because it is veiled by māyā. In cosmology space-time goes to zero at the singularity but the concept is still wrapped up in space-time; it is defined as a point singularity. In case of the Ultimate Reality the singularity is devoid of any geometrical connotation. It is beyond the grasp of the rational mind."

Existence of anything is only represented in space-time. Anything that has name and form is therefore limited in space-time. How one can transcend space-time? It is physically impossible, but physical body is not the true identity of oneself. There are subtle bodies – ethereal, astral, mental, and spiritual – which can transcend the space-time limitation to which physical body is bound. According to Sharma [15], "space-time implies separation - distance between two points or two events. When every object is viewed as an event and the object vanishes, space-time also vanishes. The vanishing of the subject-object duality essentially means that the person has merged with the universe. In this state one has transcended space-time and zero and infinity have come together. The Upaniṣads describe this as the state in which a person sees everything in himself and himself in everything [Īśopaniṣad, 5, 6]."

Concept of zero and infinity is inseparable and is consistent with the Vedāntic idea of undivided wholeness which expresses the omnipresence of supreme or *Brahman*. Īśopaniṣad says: "That *(Brahman)* moves and remains still at the same time; That is near as well as far; That is inside as well as outside" [Īśopaniṣad, 5, 6]. Everything in this universe and beyond is in fact the manifestation of the *Brahman*; thus nothing is beyond it. One of the great sayings (*mahāvākyas*) of Vedānta is *"aham Brahmāsmi"* (I am Brahma; *Bṛhādaraṇyaka Upaniṣad* 1.4.10 of the *Yajur Veda*). "I" refers here not to the body or name but the all-pervading power of Brahman. If "I" is everywhere at the same time, then for "I," there is no space and there is no time. Still there is movement because Brahman is consciousness (*prajñānam brahma* – "Consciousness is *Brahman*" *Aitareya Upaniṣad* 3.3 of the *ṚṚgveda*), which is dynamic [15]. However, now the concept of movement is totally different; space and time are not involved in it. The concept of nonlocality in modern quantum physics comes close to this viewpoint.

According to Vedānta, *Brahman* is pure consciousness, which is in every living and nonliving being. So everything that we perceive or know is through that Brahman, which is the ultimate knower. The subject-object duality precludes the possibility of being conscious of our own consciousness in the same way as we are conscious of an object [15]. How can the knower be known is a famous exclamation in Vedānta (*Bṛhadāraṇyaka Upaniṣad*, 2–4.14).

Two questions thus arise in deriving quantitative measures of observations made:

1. Can objective and subjective observations be examined scientifically? This is an important question, as according to Vedic literature, the whole cannot be known under dual nature of the observer and the observant.
2. Can there be a quantitative measure of the consciousness itself? This is an important issue if we assume answer to the question 1 above is a "yes." Since it is the attainment of infinity with the consciousness that brings about zeroness to the vibrations, there is a possibility that intermediate stages between the beginning and the end of this attainment can be estimated.

2.4 Modern Scientific Approach to Consciousness and Śūnya

Modern science is akin to western knowledge system which in large part is influenced by the western philosophy; even certain elements are borrowed from the eastern world. In order to apply modern scientific methods and fields of learning to understand ancient concepts like consciousness or *śūnya*, especially when there may be major cultural and philosophical differences between the society where these scientific tools are developed and the society of the ancient knowledge, it is important to examine the basic premise in the western philosophy or science and compare them with some of the fundamental assumptions in the Indian philosophical systems.

Western thoughts [1] hold that (1) man cannot know metaphysical truths by direct experience; (2) even if they can be known, so far no man has known them; and (3) being pure speculations, various schools of Indian thought, like the speculative system of the West, must be mutually contradictory and that if one is true, then all the others must be false. Presuppositions for Indian thoughts are (1) man can know metaphysical truths directly; (2) there have been such men, and there may still be seers with such knowledge; (3) seers or *ṛṣis* teach metaphysical truths after knowing them directly; and (4) while all the *ṛṣis* know the same truths, they teach these truths in different standards represented by the texts of the schools [1].

It has been realized that human sensual faculties are limited. Therefore, scientists have adopted models to examine and understand the matter as a substance. These models have been developed, and are acceptable [2] because they (a) describe the new in terms of the old, (b) express unfamiliar in terms of the familiar, (c) relate the abstract to the concrete, or (d) integrate several previous models in an intuitively satisfying way.

It is notable that with the application of these models, we acknowledge as scientists that we are not even dealing with reality, just a simple perception of it. At best, these objects in the solid world around us are really only images or approximations around us [18]. Our scientific theories are just theories; their relationship to reality again depends on the perception. Even the instrumental measurements are based on a perceived reality. Echoing these sentiments, Stephen Hawking said in a 1994 lecture: "I don't demand that a theory correspond to reality because I don't know what it is" [7] (Fig. 13.2).

Given the above limitations, it is still possible to model reality in terms of the modern scientific languages. Chemistry certainly represents a language of science that can be used to model more than the two other fields of natural sciences, biology and physics, primarily due to their complexity and abstractness, respectively.

3 Hypothesis

Consciousness is a general concept of existence of any and all the things known and unknown and is universal. The consciousness in its ideal and complete form is experienced equally by everything, including human beings. Awareness of the consciousness occurs with limiting the consciousness to one or more of the infinite dimensions of the consciousness, including space and time.

Any one of these limitations is reflected in mind which acts as the relativist of a human being. Mind is a form of energy that changes frequently with time and space and any other changes in the universe by being an integral part of the consciousness. The relationship of mind with time and space opens the possibility of its examination

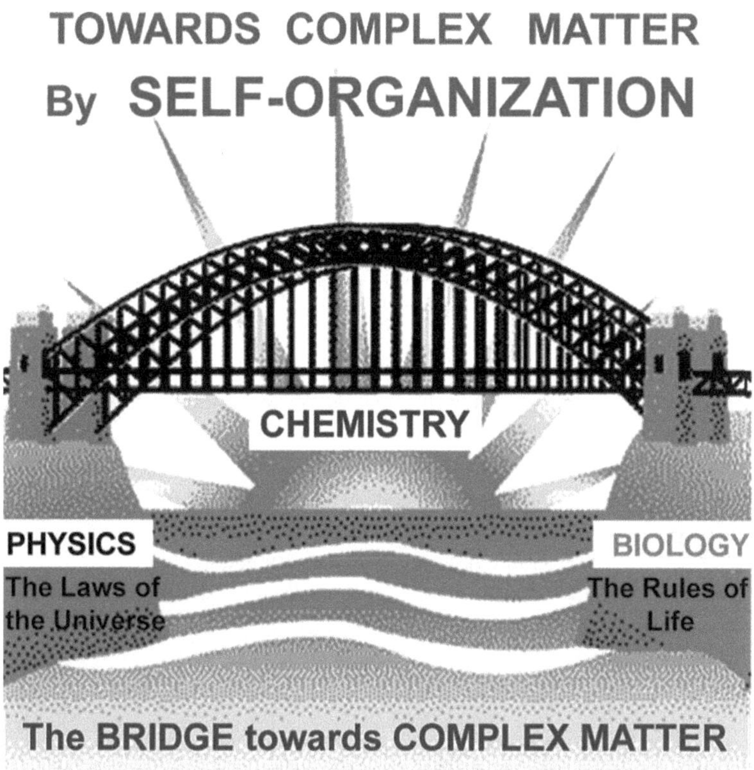

Fig. 13.2 Chemistry, the bridge between the abstract laws of the universe and their expression in a highly complex feature (life), thereby unraveling the generation of complex matter by (conscious?) self-organization, the driving force of the evolution of the universe. (Taken from Ref. [11])

by the tools of modern science, which is primarily the study of matter as a function of time and space.

In order to examine the consciousness, one has to create a model for consciousness which could be based on intuitive assumptions yet lead to measurable parameters to validate the model and create an understanding of the consciousness in a scientific manner. We make a model of the consciousness by assuming the following.

1. Consciousness is all pervasive.
2. Expression of consciousness can be represented by vibrations.
3. Different things may have different frequencies of vibration.
4. Vibrational frequency can be defined as periodic change in signal in any of the infinite dimensions of consciousness with respect to another dimension of the consciousness (e.g., space vs. time, energy vs. time, space vs. energy, etc.).
5. Vibrations are natural to all matter and non-matter components of the universe and para-universe.

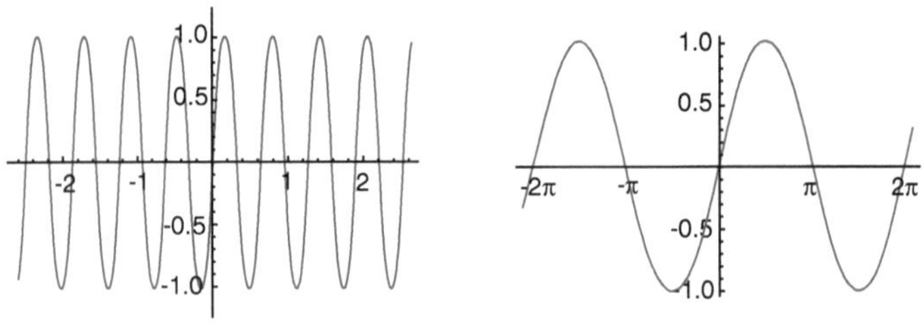

Fig. 13.3 Signal changes with high (left panel) and low (right panel) frequencies, maintaining constant amplitude irrespective of the frequency

- Assuming the consciousness to be represented by space vs. time, a signal in space will increase in frequency with decrease in time (Fig. 13.3), and as the time reaches zero, the signal will become all pervasive and will represent reaching of the consciousness. Similarly, with increase in time (reaching infinity), wavelength of the signal will increase to become all pervasive (Fig. 13.3, right panel), again reaching the consciousness. In either cases, the amplitude of the signal remains constant irrespective of variation in time. Contraction and expansion of time is thus a defining factor to attain consciousness.

3.1 Time in a Circle as a Representation of Consciousness and Mind

The correct geometric model for time has been argued over for many years by various philosophers and thinkers, most seeing time as either linear or circular. While the theory presented here does not by any means end this debate, it does present a new look at time in terms of how it is perceived and presents new ideas as to why time may travel in circular, rather than linear, motion.

The main purpose of this model of time is to show how the mind's perception of time changes with respect to one's consciousness. Its basis lies in the fact that a certain amount of physical time may be perceived by us to be much longer or shorter than it actually is. For example, Einstein once conducted an experiment dealing with this very issue. For the first part of his experiment, he sat on a hot plate for what he thought was an hour; the actual amount of time he spent sitting on the hot plate was closer to a minute and a half. Afterward, he sat with a pretty girl for what he felt was a minute, in actuality sitting with her for 56 minutes. Thus, our mind's perception of time can often be vastly different than the physical measurement of time.

Einstein's experiment and our observations kindled in us an inspiration to create a model that could map this phenomenon. The circle, which represents time as a whole, is the basis for the model (Fig. 13.4). The angle measure of any piece taken of the circle represents a physical amount of time; so, for example, if the entire circle

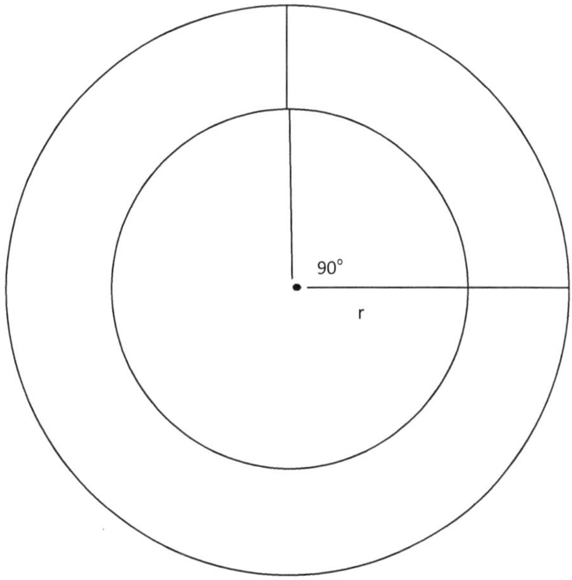

Fig. 13.4 The circle of time showing an inner circle and an outer circle. According to our model, the 90° angle shows two perimeter arcs representing two different levels of consciousness indicated by the respective radii

was to represent 1 h, the 90 degree portion shown above would be the equivalent of 15 physical minutes.

The radius of the circle that one uses depends upon the consciousness of the mind. A mind completely focused or engrossed in an activity will be represented by a circle closer to the center, while one less focused, distracted, or unengaged will be represented by a much larger circle, being placed farther from the center.

The mind's perception of time corresponds to the length of the arc created on that mind's specific circle. The less engrossed one's mind is, the longer its perception of time becomes. When one's mind is less engrossed, it is represented by a circle farther away from the center, and while the angle and actual measure of time stay the same, the mind's perception of time and the length of the arc both increase.

This model created to map our perception of time also fits the Hindu view of time and consciousness. First of all, Hindu philosophy sees time as circular and cyclical, such as the cycle of *yugas* and the cycle of birth and rebirth. Thus the circular model presented fits well.

Further, in Hindu philosophy, it is believed that the further one gets from this physical world and the closer one gets to total consciousness or God, the less time and space matter or even exist. This also suggests that as time and space matter less, one is closer to total consciousness. This also means that the center of the circle is in fact total consciousness or Brahman.

Using the center of a circle to represent *Brahman* or total consciousness further fits Hindu philosophy, which says that while *Brahman* is infinite and everywhere, he is also simultaneously intangible, unable to be seen, and thus to the human eye existing as a result of faith in his existence. The center of a circle is in similar situation: it cannot actually be found, but it is known to exist. The moment any physical point is picked as the center of a circle, that point takes up physical space, and thus it itself becomes another circle, whose center must be found. In this way, the center, being infinite yet not able to be found, easily fits with the Hindu description of total consciousness.

In addition, interestingly Hinduism sees physical or what we might call "real" time to actually not be real at all. In Hindu philosophy, all worldly dimensions, including that of time, are part of "Māyā." In this way, a mind less bound by the restrictions of worldly time is also closer to the Hindu definition of reality and thus at a higher level of consciousness.

Altogether, using the relationships this model creates between consciousness and the mind's perception of time, it is possible to use the model in order to take a measurement of one's level of consciousness and its relationship specifically to time.

Mind and Einstein's Theory of Relativity: Linking Time to Temperature

The easiest way to make a connection of time and temperature with mind is the observation Einstein made in 1938 [6]:

- "When a man sits with a pretty girl for an hour, it seems like a minute. But let him sit on a hot stove for a minute and it's longer than any hour. That is relativity" The important thing to note here is the interconnectedness of temperature with the observation of time. This is equivalent to the level of mental tolerance of adverse conditions by the people, something that can be empirically measured. This also allows the individualization of the observation which is subjective. It would, however, be measurable and comprehensively by others.

3.2 *Measurement of Consciousness: Time in Terms of Temperature*

Once a connection is established that it is the mind that is responsible for the time-temperature perception, one can derive the time in terms of temperature, allowing a direct measurement of mind (which is the limited version of consciousness) by making assumptions about the instrument of mind, which is brain, and the lifetime experience of human being:

- Kinetic theory of gases
- $V_{rms} = (3RT/M)^{1/2}$;

  V_{rms} – root mean square velocity
  R – universal gas constant
  T – temperature, K
  M – molar mass

- Defining Vrms as d/t (d, distance in meters; t, time in seconds) and rearranging the equation, one can get:
  - $t = (d^2M/3RT)^{1/2}$

- Assuming d and M to represent constants, the expression can result in:
- $t = C/T^{1/2}$, where C is consciousness coefficient that can be estimated for each individual or entity experimentally.
- $C = (d^2M/3R)^{1/2}$

- For humans, M would be 1.4 Kg, and d would be 7.0956×10^{17} meters (average life span, 75 years, in seconds × speed of light, 3×10^8 meters/second). R = 8.3145 J/K.mol
- An unevolved human, defined totally on material basis, would have a consciousness coefficient, C, of 1.68×10^{17} m.K.mol/J.Kg, where, m is meter, K is Kelvin, mol is moles, J is Joules, and Kg is kilogram.

The C could become a powerful tool to examine individuals for the evolution of their mind, and it could perhaps be useful to their medical and social care. This being so practical, more data can be collected for evaluating the scientific nature of this subjective science and compare it with the objective science being practiced in the modern world.

3.3 Human Technology to Approach Śūnya

While there is really no physical tool that can measure either zero or infinity, there are human tools that can be developed to realize these elements of the universe. An experiment conducted by Paul Ekman (http://mindful.org/the-science/neuroscience/a-monk-in-an-mri) elucidates the importance and linkage of time, consciousness, and perception. His experiment focuses on the most neurologically evolved sensory system: the visual system. It utilizes a videotape in which a series of faces show a variety of expressions very briefly. The challenge is to identify whether you've just seen the facial signs, for instance, of contempt or anger or fear. Each expression stays on the screen for just one-fifth of a second in one version and for one 30th of a second in another – so fast that you would miss it if you blinked. Each time the participant must select which of the seven given emotions he or she has just seen. The ability to recognize fleeting expressions shows an unusual capacity for accurate empathy. Such expressions of emotion – called micro-expressions – happen outside the awareness of both the person who displays them and the person observing. Because they occur unwittingly, these ultrarapid displays of emotion are completely uncensored and so reveal – if only for a short moment – how the person truly feels.

Ekman found that a Tibetan monk and another advanced western meditator were two standard deviations above the norm in recognizing these super-quick facial signals of emotion, albeit the two subjects differed in the emotions they were best at perceiving. They both scored far higher than any of the 5000 other people tested. "They do better than policemen, lawyers, psychiatrists, customs officials, judges—even Secret Service agents," the group that had previously distinguished itself as most accurate.

Thus, understanding the perception of time itself is therefore a valid subject of inquiry to bring consciousness into the quantitative domain of science and technology.

To perceive, understand, and comprehend processes and phenomena beyond open perceptual range (subseconds for processes and years for memory), a human being uses cognitive and imaginative processes which are still subject to intrinsic and sensory processes of brain, as recognized by modern science, and mind which still remains elusive to most mainstream scientific fields. The three levels of human state of consciousness (wakeful, REM sleep, and deep sleep states) provide the basic framework of examination for developing a conceptual basis of further elaboration of time in relation to mind.

However, there are perceptions beyond physical realm which still need to be considered. Access to such perceptions, or more appropriately experience, is beyond the duality of observations and is referred to as the turiya state of the consciousness. Is it possible for a modern scientist to approach that state with mathematics, technology, and theories? Would scientists of modern times agree to be trained with the art of experiencing the universal knowledge or the universe itself? Would the scientists still follow the description of universe in time-space framework? How would that knowledge be communicated in general?

As proposed in the scientific model of the consciousness, it is all encompassing, reaching infinity or zeroness with time dilation. One reaches the singularity, negating the idea of diversity which is embodied by the concept of entropy, defined as kInW, where k is Boltzmann constant and W is different ways or diversity. Under the condition of singularity, $W = 1$, and $\text{In } 1 = 0$. In other words, when one reaches the singularity with the university, it is mathematically the state of śūnya or zero. This infinity/śūnya is the treasure trove of all that is described as the universe. At least theoretically, it is shown that matter and energy originate from the empty or śūnya [20].

The larger question is whether the quantum phenomenon is a reality or only a perception. Is the quantum perception derived from the consciousness precipitating into mental thoughts and matter? This is the same concept that Schrodinger addressed in his thoughts of how the genetic material is formed as the precipitation of the consciousness to perform routine work reproducibly [14].

References

1. Agrawal MM. Six systems of Indian philosophy. Delhi: Chaukhamba Sanskrit Pratishthan; 2001.
2. Barton A. Chapter 1, States of mind states of matter. Bristol: Institute of Physics Publishing; 1997.
3. Crick F, Koch C. Toward a neurobiological theory of consciousness. Semin Neurosci. 1990;2:263–75.
4. Crick F. The astonishing hypothesis: the scientific search for the soul. New York: Scribners; 1994.

5. Dijksterhuis A, Bos MW, Nordgren LF, van Baaren RB. On making the right choice: the deliberation-without-attention effect. Science. 2006;311:1005–7.
6. Einstein A. On the effects of external sensory input on time dilation. J Exothermic Sci Technol. 1938;1(9):1938.
7. Hawking SW, Penrose R. The nature of space and time. Sci Am. 1996;275:44–9.
8. Lambert N, Chen YN, Cheng YC, Li CM, Chen GY, Nori F. Quantum biology. Nat Phys. 2012;9:10–8.
9. Koch C. The quest for consciousness: a neurobiological approach. Roberts and Company: Englewood; 2004.
10. Krishnamurti J, Bohm D. The ending of time. San Francisco: Harper and Row Publisher; 1985. p. 96.
11. Lehn J-M. Perspectives in chemistry—steps towards complex matter. Angew Chem Int Ed. 2013;52:2836–50.
12. Lutz A, Greischar L, Rawlings NB, Ricard M, Davidson RJ. Long-term meditators self-induce high-amplitude synchrony during mental practice. Proc Nat Acad Sci. 2004;101:16369–73.
13. Newberg A, D'Aquili E, Rause V. Why God won't go away: brain science and the biology of belief. New York: Ballantine; 2001.
14. Schrodinger E. What is life? Cambridge: Cambridge University Press; 1964.
15. Sharma DR. Zero infinity and God. 2003. Sulekha.com (http://dharambir-rai-sharma.sulekha.com/blog/post/2003/03/zero-infinity-and-god.htm), March 2, 2003.
16. Sharma RK, editor. Kabir Granthavali (Complete Works of Kabir). Allahabad: Oxford University Press; 2001. p. 169–70.
17. Singh BR, Singh RB, Sengupta S. A proposed model to quantitatively assess consciousness employing time and temperature relationship to mind. In: Singh BR, editor. Exploring science in ancient Indian texts. New Delhi: DKPW Publishers; 2012. p. 285–310.
18. Stapp HP. Mind matter and quantum mechanics. Found Phys. 1982;2:363–99.
19. Turiyananda S. VIVEKACUDAMANI of Sri Sankaracarya (Sanskrit Text with Transliteration, Translation and Index), Ramakrishna Math; 2003.
20. Wilson CM, Johansson G, Pourkabirian A, Simoen M, Johansson JR, Duty T, Nori F, Delsing P. Observation of the dynamical Casimir effect in a superconducting circuit. Nature. 2011;479:376–9.

Chapter 14
Quantum Physics, Consciousness and Śūnya

S. Rammohan

Abstract '*Śūnya*' is the Indic symbol that bridges the *Brahman* of Vedānta and the material universe. *Śūnya* denotes void as well as infinite space. Aryabhata employs 'Śūnya' to denote number zero. '*Śūnya*' or 'nothing' is something that cannot be defined. While Vedanta denotes *Brahman* as formless energy or consciousness, quantum scientist defines the universe as an inseparable web of vibrating energy patterns in which no single particle has reality independent of the entirety. The entirety includes the observer too.

Our consciousness has emerged from cosmic consciousness and becomes a part of the universal whole, however separated in space. This is epitomized in the very first verse of *Kenopaniṣad*. At the quantum level of reality, the demarcation between the realms of phenomena seems to blur. There is obviously only one alternative, viz. the unification of mind and consciousness. Quantum vacuum (QV), *Brahman* and *Śūnya* (void) are all indescribable entities. Quantum scientists tell us that the only universal language is that which can be at the level of gauge bosons. Can human mind be trained to transmit and receive at the level of QV? Probably the answer lies in the development of our consciousness through the practice of Yoga and meditation or the practice of inward journey, advised by Sri Ramana Maharshi.

Keywords Consciousness · Wave-particle duality · Bell's theorems · Quantum field theory · Supersymmetry · Turīya

1 Introduction

'Quantum model of universe' is based on the observations of quantum (subatomic) physics. To the quantum physicist, the universe is an inseparable web of vibrating energy patterns in which no single particle has reality independent of the entirety; and included in the entirety is the observer. The particle behaviour is so strange that

S. Rammohan (✉)
Sarasvati Research Centre, Chennai, India

S. R. Bhatt (ed.), *Quantum Reality and Theory of Śūnya*,
https://doi.org/10.1007/978-981-13-1957-0_14

one gets a doubt as to whether two particles are communicating and co-operating with each other by telepathy. The underlying fields of the particles in the standard model are classified as matter fields and force fields. The matter fields package energy to produce the building blocks of the matter, the fermions. The force fields, through their carrier particles – the bosons – govern the inner action of matter, producing the basis for physical activity.

At the foundation of things, elementary particles exist in a kind of reality that is different from ordinary reality; they have rudimentary mind-like properties, they can exert instantaneous, long-distance influences on each other, and they can exist in superpositions of states which are not quite real but, as Heisenberg wrote,'between the idea of a thing and real thing'. Like a mind, quantum systems can react to changes in information and act spontaneously. The visible order of reality is determined by non-material probability fields whose nature is close to that of a 'thought' than a 'thing'. The existence of the periodic table of the elements is the result of the symmetry of the wave functions of elementary particles. Physical reality is not what it looks like, and elementary particles are not just miniaturized editions of the ordinary objects of our conscious experience, but they are different in essence.

'Modern atomic theory', Heisenberg wrote, 'is essentially different from that of antiquity in that it no longer allows any reinterpretation or elaboration to make it fit into a naïve materialistic concept of the universe'.

Newton's physics is materialism, i.e. the doctrine that 'to be is to be material' [1]. In classical mechanics, all reality was reduced to the motion of particles obeying Newton's laws. The universe was clockwork and closed, and nature was a machine. There was no room for the spiritual and mental, and the nature of mind, non-material and not subject to determinism, was a violation of its premises. In Newtonian physics, there was no room for the spiritual and mental, and the nature of mind, non-material and not subject to determinism, was a violation of its premises.

In contrast, quantum physics has opened a window to a different kind of reality similar to Indian philosophy. Elementary particles are not real in the same way as the things they form. The three aspects of quantum physics can be highlighted as follows:

1. The basis of the material world is non-material.
2. Quantum systems can exert instantaneous, long-distance influences on each other.
3. Quantum systems have mind-like properties.

The wave properties of elementary particles are a manifestation of the wave-particle duality; when observed, elementary entities always appear as particles; when not observed, they act like waves. In atoms and molecules, for example, electrons are standing waves, wave patterns or mathematical forms. Close analysis has shown that the nature of these waves is that of probability waves or probability fields, because they provide probabilities for finding an electron in space. Since probabilities are dimensionless numbers, probability waves are empty (*Śūnya*) and carry no mass or energy, just information on numerical relations. Yet all visible order in the universe is determined by the rules of their interference. The waves of molecules determine the intermolecular interactions which are at the roots of chemistry. In contrast, quantum physics has opened a window to a different kind of reality. A crack has evolved in the materialist world of classical physics, and elementary particles are not real in the same way as the things that they form.

This is how quantum physics shows that we cannot decompose the world into independently existing elementary units. 'As we shift our attention from macroscopic objects to atoms and subatomic parcels, nature does not show us any isolated building-blocks, but rather appears a complex web of relationship between the various parts of a Unified Whole'. [2]

The various Hymns of Creation in Vedic writings dwell on this *Brahman-Jīvadichotomy* in a variety of ways – it is like a matrix of metaphysical concepts. The authors tiptoe through ontological and conceptual difficulties with the help of myths and metaphors. Dr. Paul Deussen gives an extensive review of the *Vedic literature* and says 'The motive of the conception that dominates all these passages may be described to be the recognition of the first principle of the universe as embodies in nature is a whole, but especially and most of all in the soul (the universal and the individual). Hence the idea arose that the primeval being created the universe, and then as the first born of the creation entered into it'. [3] The world is because of the individual consciousness 'I am' there to observe it.

So long as big and small are merely relative concepts, it is no help to explain the big in terms of the small. 'At this stage it becomes important to remember that science is concerned only with observable things and that we can observe an object only by letting it interact with some outside influence...... The concepts of big and small are then purely relative and refer to the gentleness of our means of observations as well as to the object being described. In order to give an absolute meaning to size, such as is required for any theory of the ultimate structure of matter, we have to assume that there is a limit to the fineness of our powers of observation – a limit which is inherent in the nature of things and can never be surpassed by improved technique or increased skill on the part of the scientist'. [4]

The wave-particle duality means that reality is created by two processes. In the first process, between observations, elementary entities constantly evolve into superpositions of possibilities or tendencies for actual events to occur. This is the wave-like state of reality. In the second process, the transition from the possible to the actual take place when an observation is made. One of the states contained in the superpositions of possibilities becomes the actual one. This is the particle-like state of reality. Heisenberg compared the superpositions of possibilities to Aristotelian potential. In superposition states, the quantum entities are not quite real but have the potential of becoming real.

In contrast, in the quantum world, influences can act instantaneously over arbitrarily long distances. Experiments involving Bell's theorem have shown that, under certain conditions, decision made by an experimenter in one laboratory can have an instantaneous effect on the results of an experiment in another laboratory a long distance away. In experiments of this kind, two particles which at one time interact and then move away from one another are found to stay connected, as though they are one thing, no matter how far apart in space they are. This is the non-locality of the quantum world. Quantum non-locality implies a holographic universe that necessitates a holistic approach [5]. Since our consciousness has emerged from the universal wholeness and is part of it, it is possible to infer that there is a cosmic consciousness.

At the quantum level of reality, the line of demarcation is blurred between realms of phenomena, which are characteristically disconnected at the level of consciousness, such as the natural and the supernatural. The two realms are seen to merge, like the physical and the metaphysical, the mental and the material and the mechanistic and the teleological. At the foundations of physical reality, the nature of material things reveals itself as non-material. Deterministic process alternates with expressions of choices in creating the visible order. Entities with mind-like properties are found. Till Einstein, everybody thought 'absolute space' is the arena (mathematically, a 'manifold') on which things change with 'absolute time'. In special theory of relativity (STR), he redefined the manifold to be flat space-time by making both space and time relative with respect to inertial observers but keeping space-time absolute. In general theory of relativity (GTR), he propounded this manifold to be curved space-time that can act on matter unlike the flat space-time. Then the quantum theory (QT) pointed out this manifold to be an abstract mathematical space called Hilbert space, since the space-time description of quantum processes is not available. Quantum field theory (QFT) that originated in a successful merge of QT and STR requires this manifold to be the quantum vacuum (QV). Unlike the ordinary vacuum, QV contains infinite number of 'virtual' particles that give rise to all matter and interactions. With the advent of quantum mechanics, these seemingly contradictory differences between particle and wave lost their sharpness in the quantum world. A quantum object can simultaneously 'be' a particle and a wave, until a measurement is made on it. According to QFT all fundamental entities are quantum fields (not material in the conventional sense) that are neither particle nor wave in the classical sense.

The concept of symmetry (invariance) with its rigorous mathematical formulation and generalization has guided us to know the most fundamental of physical laws. Symmetry as a concept has helped mankind not only to define 'beauty' but also to express the 'truth'. Physical laws try to quantify the truth that appears to be 'transient' at the level of phenomena, but symmetry promotes that truth to the level of 'eternity'.

'One overarching lesson we have learnt in the last 50 years is that all forces are associated with nature's symmetry principles', as observed by Prof. Greene [6]. Forces exhibit their separate identity when nature's symmetry is spontaneously disturbed. Manifestation of one more symmetry of the loss of nature, known as supersymmetry, is indicated, even though circumstantially. Supersymmetry is surmised to facilitate unification of bosons and fermions at the Planck scale, where the fields behave as just different aspects of one field, the super unified field.

According to QFT, the basic matter-fields interact by exchanging messenger fields (technically called gauge fields) that define the most fundamental level of communication in nature. Both matter fields and gauge fields originate in the fluctuations of QV, and in this sense everything in universe, including consciousness, is, in principle, reducible to QV and its fluctuations.

Seeing the tremendous success of physical sciences in the external world, one would think its methodology to work for understanding the inner world. It is not obvious 'a priori'; why should not QT work in this third ontology when it has worked so successfully with two different ontologies? A physical space-time description of mind is not possible because thoughts that constitute the mind are acausal. T-space (thought-space) is identical with mind, but it need not be so if mind can exist in a thoughtless but awake state, often called '*turīya*'.

Unlike the Cartesian dichotomy of mind and body, some schools of Indian Philosophy like *Yoga* treat both mind and body in a unified manner. Since (western) science is based on Cartesian paradigm, it cannot synthesize mind and body unless it takes the clue from oriental philosophies and then blend it with its own rigorous methodology.

A great conceptual step taken by Indian philosophy in this regard is to introduce a fourth state of consciousness called '*turīya*' that is defined to be none of the above but a combination of all of the above states. This state is claimed to be the superconscious state where one transcends the limitations of perceptions constrained by space-time. *Patanjali* has provided very scientific and step-by-step instructions to reach this fourth state through *samyama*; *dhāraṇā*, *dhyāna* and *Samādhi* are different levels of *samyama*. Sometimes the possibility of having a universal language (like *Sāndhyabhāṣā*, etc.) to communicate with everything in the universe is also mentioned. Modern physics tells us that the only universal language is at the level of gauge bosons and QV. Is there any connection between these two? Can a (human) mind be trained to transmit and receive at the level of QV?

Nature manifests itself not only at the gross level of phenomena (accessible to direct senses) but also at the subtle level of natural laws (accessible to 'refined' senses). Consciousness is the ability to access nature at both these levels. Hence everything in nature is conscious.

According to quantum theory, the quantum vacuum contains neither matter nor energy, but it does contain fluctuations, transitions between 'something' and 'nothing' in which potential existence can be transformed into real existence by the addition of energy. (Energy and matter are equivalent, since all matter ultimately consists of packets of energy.) Thus, the vacuum's totally empty space is actually a seething turmoil of creation and annihilation, which to the ordinary world appears calm because the scale of fluctuations in the vacuum is tiny and the fluctuations tend to cancel each other out. Hindu philosophy states, in a similar vein, *Sṛṣṭi*, *Sthithi* and *Pralaya* are cyclic and metaphorically, time is a curvature, not linear.

A solid scientific foundation supports the basic insight of quantum field theory that the primary elements of reality are the underlying fields, which permeate all space and time. It is reasonable to assume that all these fields come from a common source. It is indicated in the very first verse of *Īśāvāsya Upaniṣad*, which states that all the things, moving and nonmoving, are permeated by the cosmic consciousness. Recent researches into the isolation of God particle in the CERN laboratories are steps in this direction. At the very substratum of space, the Planck length, all the underlying, fluctuating quantum fields of nature are surmised to be unified today in a virtual quantum way, even in the absence of great heat, typical of the beginning of universe.

The quantum vacuum is different from the classical vacuum because it contains the fluctuations even after you have turned the field down to zero. The quantum theory asserts that a vacuum, even the most perfect vacuum devoid of any matter, is not really empty. Rather the quantum vacuum can be depicted as a sea of continuously appearing and disappearing (pairs of) particles that manifest themselves in the apparent jostling of particles that is quite distinct from their thermal motions. These particles are 'virtual', as opposed to real, particles. At any given instant, the vacuum is full of such virtual pairs, which leave their signature behind, by affecting the energy levels of atoms.

What is called the vacuum state or the quantum vacuum, it is by no means a 'simple empty space', and again it is a mistake to think of any physical vacuum as some absolutely empty void. According to quantum mechanics, the vacuum state is not truly empty but instead contains fleeting electromagnetic waves and particles that pop into and out of existence.

'*Śūnya*' is the Indic symbol that bridges the spiritual *Brahman* and the materialist universe. Mathematically it is supposedly used by *Āryabhaṭa* to denote the numerical zero. It is useful to note that *Āryabhaṭa* never used numerical forms in his works but used words. '*Śūnya*' is something which cannot be defined, yet it cannot be denied. It is like empty space. Its emptiness cannot be determined and its existence cannot be questioned. It is like infinite time. Its finiteness cannot be calculated and its presence cannot be questioned. Together they form '*Śūnya*'. *Śūnya* and quantum theory are powerful expression to comprehend the universe and unravel the mysteries of science and spirituality.

As Prof. Srikantan had observed, when a physicist looks at quantum reality or relative reality, he is not looking at things in themselves – noumenon – a direct and non-mediated reality. Rather the physicist is looking at nothing but a set of abstract equations, not a reality itself but a mathematical symbol of reality. Einstein has observed that matter when we perceive is merely nothing but a great concentration of energy in very small regions. We may therefore regard matter as being constituted of space in which the field is extremely intense. Field is the only reality. Reality according to Einstein is just field. The field itself cannot be accurately delineated; we can only give it a mathematical definition. There is no longer physical entity which we can identify as field. As defined by Prof. Dirac, all matter is created out of some imperceptible substratum, which is nothingness according to him. Nothingness is ultimately quantum mechanical vacuum. Reality is only quantum mechanical vacuum and fluctuations on it. Oriental philosophy, Buddhism and Hinduism hold identical views. The Hindus call the ultimate reality as the *Adhiṣthānam* and the Mahāyāna Buddhist call it *Śūnyatā.* The *Yogācāra* Buddhists give a new term to this as *Ālaya Vijñāna* or store-house consciousness. Reality is either spontaneous fluctuations or something brought about by deposition of energy. Nothingness according to Prof. Dirac is unimaginable and undetectable. But it's a peculiar form of nothingness out of which all matter is created. As Weinberg says, elementary quantum fields are governed by symmetries. They are not objects that we are familiar with but as things that we have to only postulate.

David Bohm has stated that, once they were in a plasma, electrons stopped behaving like individuals and started behaving as if they were part of a large and interconnected whole. Here we can recall the telling comments of Heisenberg that it is important to realize that the behaviour of smallest particles cannot be described unambiguously in ordinary language. It is interesting to see Francis Crick who wrote in the initial portion of his famous book *The Astonishing Hypothesis* that there is nothing else other than the activity of the neurons to explain everything and ultimately comes to the very strange conclusion that his hypothesis may be right or it may be totally wrong; there may be some other religious or other explanation.

Nature controls everything, which Einstein calls as 'cosmic religion'. He comes to the conclusion that just by randomness and chaos operating, all this cannot happen. Does God play dice with the universe? He concludes that there is a cosmic intelligence. But there is a difference between personal god and cosmic intelligence.

Bohm described all phenomena in terms of order. A simple example of order is the description of a straight line as an ordered array of short line segments of equal length laid end to end, with all of the successive segments having the same orientation and difference in position. According to Bell's theorem on quantum field of the universe, the apparently separate parts could be ultimately covered at a deep and fundamental level. David Bohm asserts that the most fundamental level is an 'unbroken wholeness' which he described as 'that-which is'. Indian Ṛṣis called it 'TAT' and the Buddhists '*Tathathā*'. This is also posited by Gary Zukav in *The Dancing Wu li Masters*, where he states that there already exists an instrument of thought based upon an 'unbroken wholeness'. There exist a number of sophisticated philosophies distilled from 1000 years of practice and introspection whose whole purpose has been to sharpen though instrument. The movement of self-realization of the yogis, through this process, is similar to a 'quantum jump' by the quantum physicists.

Indic scriptures begin with the notion of '*Ātman*' and take the approach of quantum physics to comprehend the universe. However, the Indic scriptures boldly postulate that the ultimate or the supreme *Brahman* is the unchanging reality amidst and beyond the world and which cannot be exactly defined. *Brahman* is derived from the Sanskrit root 'bṛh' which means 'to grow' and connotes 'immensity'. *Brahman* is the indescribable, inexhaustible, incorporeal, omniscient, omnipresent, original, first, eternal, both transcendent and immanent, absolute infinite existence and the ultimate principle who is without a beginning and without an end, who is hidden in all and who is the cause, source, material and effect of all creation known, unknown and yet to happen in the entire universe, as eloquently described in the *Upaniṣads*.

Ancient India's gift to the world is the concept of zero. This was developed from the '*Śūnya*' or 'Nil'. This numeral was used in computation and it was indicated by a dot. The first Indians to evolve the concept of zero appear to be *Pāṇiṇi*, the great grammarian, and *Piṅgala Brahmagupta* in his *Brahma- Sphūta Siddhānta* mentions about the operations of zero. The concept of zero in mathematics opened up a new era of negative numbers, a concept unknown to the western thinkers at that time. The representation of '*Śūnya*' or zero by '0' made the representation of high-value number with limited characters and calculations very simple. Indian mathematicians, 'the heroes behind the zero', had a clear conception of the abstract numbers as distinct from the numerical quantity of objects or spatial extension, thus taking mathematics to a different plane.

Erwin Schrodinger, the renowned quantum physicist, considered the complex number as an illustration. The concrete aspect is represented in a complex number, by a real number, and the indeterminate portion of the complex number 'Z' is written as $z = a + ib$ where 'a' is the real part and 'b' is the imaginary part. 'Z' is just one variable, but it carries with it two concepts simultaneously, the real 'a' and imaginary 'b'. This is what Schrodinger used to integrate the duality of quantum mechanics. Schrodinger, a great admirer of *Upaniṣads*, has written in his book *Mind and Matter*, 'The reason why over-sentient percipient and thinking ego is met nowhere within our scientific world –picture can easily be indicated in words because it is itself the world picture. There is obviously only one alternative namely the unification of mind and consciousness. Their multiplicity is only apparent in truth. This is the doctrine of the *Upanishads*'.

Consciousness, the window through which we eventually connect scientific knowledge, is an integral part of the universal reality. 'The principle argument is that thought process and consciousness are the primary concepts, that our knowledge of our external world is the content of our consciousness and that consciousness, therefore, cannot be denied', as observed by Eugene Wigner [7]. Again, Prof. Wigner says 'It will remain remarkable in whatever way our future concepts may develop – that the very study of external world led to the conclusion that the content of consciousness is an ultimate reality'[1]. Roger Penrose adds to this, 'this is only the phenomenon of consciousness that can conjure a putative theoretical universe into actual existence'. [8] Again in the emphatic words of Prof. Penrose, 'A world-view which does not profoundly come to terms with the problem of conscious minds, can have no serious pretension of completeness. Consciousness is part of our universe, so any physical theory, which makes no proper place for it, falls fundamentally short of providing a genuine description of the worlds' [9].

The 'nothing' part of the vacuum state as a theoretical object is that the average value of a series of measurements of the field will be zero. The 'something' part of the vacuum state is that the value of any single measurement will in general not be zero. When we cannot predict single measurement results and how they will vary over time, we often find that we can predict average values and how the average values will vary over time.

If we take 'nothing' to be the same as 'zero' and 'something' to be the same as 'not-zero', the vacuum state is both 'nothing' and 'something'. The famous '*Antaḥpūrṇo bahiḥ pūrṇaḥ pūrṇakumbha ivārṇave* and *Antaḥśūnyo bahiḥ śūnyaḥ śūnyakumbha ivāmbare*' bring to focus the concept that *Śūnya* and *Pūrṇaḥ* are really identical. The Buddhist philosophy, especially that of the Mādhyamika, places great

[1]Greene (1999, p. 169).

emphasis on the concept of *Śūnya*. According to *Mādhyamika* Buddhism, there is no reality or nonreality. All is *Śūnya*, void and empty, and all realities are disclaimed. Thus 'whatever is' is not describable by any concept. Being devoid of any phenomenal characteristics 'void' or '*Śūnya*' is the real nature of things. For them, reality lies in the middle (*madhyama*) and not in any extremes of realism or idealism. Ācārya Nāgārjuna rejected the alternative standpoints of 'is' and 'is not' and their conjunction and disjunction. Very interestingly, the Śaivitelupagarma *Devikalottaram* verse 26 mentions about the efficacy of destroying all assumptions and latencies with the weapon of '*Śūnya Bhāvanā*' or contemplation of 'void'; verse 42 says that those who are established in the 'Sarva Śūnya Padam' will escape the cycle of birth and death.

Dr. George Gheverghese Joseph refers in *The Crest of the Peacock: Non-European Roots of Mathematics* to the Buddhist doctrine of 'emptiness' or '*Śūnyatā*' that may possibly have given rise to the zero digit. The zero digit is widely believed to have arisen on the subcontinent somewhere between 200 BCE and 500 CE.

Whereas all other signs refer to concrete presence or aspects of duality, zero may be said to act as tangible reference to the intangible or the non-dual realm. In that sense zero acts as 'gateway' between the dual and non-dual and may be seen as 'vestigial birthmark' of the ineffable – raison d'etre of India's Vedic tradition, which eventually gave rise via the versatile '*Śūnya/pūrṇa*' concept to fruitful application in various fields including religion, philosophy, literature, art and eventually mathematics. It is seldom explicitly recognized that the present-day world of science and technology and all it has wrought would simply not have been possible, if, for example, mathematics in the West had remained hobbled, due to the doctrinal resistance to the introduction of the Hindu numerals that remained fierce for centuries based on deep-seated Greek and later Christian aversion to the concept of 'emptiness' – literally seen as the domain of the devil – by the unwieldy Roman or Greek numerals that were only displaced by the Hindu numerals in the Renaissance.

In its empirical study of 'emptiness' [i.e. the quantum vacuum], modern physics arrives at a comparable worldview as did the ancients in India, who studied emptiness intuitively. But, whereas the West adopted the zero only functionally, for its practical utility, its deeper meaning has as yet not been fully grasped. For example, the '0-based' philosophy '*Śūnyatā*' or the neologist 'nonism' is therefore still missing in the West, where only monism, dualism and pluralism are considered, consequently trapping quantum physics in its ontological quandary, unable to comprehend, as its laboratory experiments and indeed quantum theory itself indicate, that subatomic phenomena arise and decay within a vacuum presumed to be empty.

Eddington has emphatically declared 'Universe is of the nature of a thought or sensation in a Universal Mind. . .to put the conclusion crudely, the stuff of the world is mind-stuff. As is often the way with the crude statement, I shall have to explain that by "mind" I do not exactly mean mind and by 'stuff' I do not at all mean stuff, till this is about as near as we can get to the idea in a simple phrase. The mind-stuff of the world is of course something more general than our conscious minds; but we may think of its nature as not all together foreign to feelings in our consciousness.

Having granted this, the mental activity of the part of the world constituting ourselves occasions no surprise [10]. As in Immanuel Kant's ontology, there is the world of appearances and that of the things in themselves. All our understanding is of the appearance of things, not of the real things. Kant called the things in themselves noumenon; it denotes mind-like.

We can recall the theory of evolution by the French palaeontologist, Teilhard de Chardin, which shares with quantum physics the view that reality is wholeness and the elements of consciousness are active at all levels of reality. He posits that matter and consciousness are not 'two substances' or 'two different modes of existence' but two aspects of the same cosmic stuff [11].

'Before on this planet, the physical-chemical conditions allow the birth of organic life, the universe was either not yet anything in itself, or it had already formed a nebula of consciousness' [12]. If in the singular point of convergence of matter, the biological evolution has revealed itself as an 'expansion of consciousness,' [13].[2]

The famed Prof. Stephen Hawking, while taking about the anthropic principle, which proposes that the physical constants of nature are precisely tailored in this universe for unfolding of consciousness, states that 'many scientist dislike the Anthropic principle because it seems rather vague and does not appear to have much predictive power. But the Anthropic principle can be given a precise formulation and seems to be essential when dealing with origin of universe'. [14]

As brought out in this brief account of development in quantum physics in relation to the concept of *Śūnya* and the self or consciousness, it seems to hit a road block in an attempt to explain comprehensively 'mental' or 'thought' processes, as extensions of 'consciousness' present in all forms of biological life. It's averred that it should be possible to overcome the road block by introducing the parameters of cognitive systems in the *Vedic* traditions. It has also been hinted that *Śūnyatā* as a concept can be explored further to bridge the dialectics and knowledge systems.

In conclusion, we can recollect the words of Prof. J Wheeler in this context, 'Quantum mechanics has led us to take seriously and explore. . .that the observer is as essential to the creation of the universe as the universe is to the creation of the observer'. [15] This is a beautiful echo of the philosophy of the renowned Indian mystic, J Krishnamurty. In the event that awareness is an inseparable aspect of all-pervading quantum fields or their common source, it should permeate through all space and time in the universe. This aspect has to be explored in the further researches.

References

1. Margenau H. Open vistas. Wood Bridge: Oxbow Press; 1983. p. 47.
2. Capra F. Tao of physics. Boston: Shambhala publications Inc.; 1999. p. 9.

[2]de Chardin (1956, p. 172).

3. Deussen P. The philosophy of upanishads. Edinburgh: T & T Clark Publications; 1993. p. 183.
4. Diarac PA. *Principle of quantum mechanics*. Oxford: Oxford University Press; 1947. p. 3.
5. Bohm D, Hiley B. *The undivided universe*. London: Routledge.
6. Greene B. *The elegant universe*. New York: WW Norton & Co; 1999. p. 357–8.
7. Wigner E. *Quantum theory and measurement*. Princeton: Princeton University Press; 1983. p. 173–4.
8. Penrose R. *Shadows of the mind*. New York: Oxford University Press; 1989. p. 448.
9. Penrose R. *Shadows of the mind*. New York: Oxford University Press; 1989. p. 8.
10. Eddington AS. Philosophy of physical science. New York: Mc-Millan Co.; 1939. p. 151.
11. de Chardin T. *Heart of matter*. London: Collins; 1978, p. 25.
12. de Chardin T. *Mon universe*. Paris: Editions du Seuil; 1965. p. 74.
13. de Chardin T. *La Place de l-hommeDans la Nature*. Paris: Editions du Seuil; 1956. p. 48.
14. Hawking S. *The universe in a nutshell*. New York: Bantam Book; 2001. p. 86–7.
15. Wheeler J. *At home in the universe*. New York: Springer-Verlag; 1996. p. 44.

Select Bibliography

1. Bag AK, Sarma SR. The concept of Śūnya. New Delhi: Aryan Book International.
2. Datta B, Singh AN. The history of Hindu mathematics. New Delhi: Bharatiya Kalaprakashan; 2004.
3. Joseph G.G., The crest of the peacock: Non-European roots of mathematics, Princeton University Press, Princeton.
4. Kaplan R, Lane A. The nothing that is: a natural history of zero. London: The Penguin Press; 1990.
5. Plofkin K. Mathematics in India. Princeton: Princeton University Press; 2008.
6. Radhakrishnan S. Indian philosophy. London Ltd.: George Allen & Unwin; 1948.
7. Rotman B. Signifying nothing, the semiotics of Zen. London, Palgrave, Mac Milan; 1987.
8. Seife C. Zero: the biography of a dangerous idea. London: Souvenir Press; 2000.
9. Dawkins R. The blind watchmaker. New York: Noon & Comp; 1986.
10. Eddington AS. The nature of the physical world. New York: Macmillan Comp; 1929.
11. Hiesenberg W. Philosophical problems of quantum physics. Woodbridge: Ox Bow Press; 1952.
12. Hiesenberg W. Physics and philosophy. New York: HarperTorch books; 1962.
13. Kafatos M, Nadeau R. The conscious universe. New York: Springer Verlag; 1990.
14. Shedrake R. The presence of the past. Rochester: Park Street Press; 1988.

Chapter 15
The Concept of Reality in Quantum Mechanics and Śūnyavāda: From the Perspective of Yoga

Geo Lyong Lee

Abstract There is a considerable similarity between the physical concept of reality implied by quantum physics and the philosophical concept of reality articulated by Nāgārjuna. For neither is there a fundamental core to reality, rather reality consists of systems of complementary and interacting objects (*dharmas*). From the perspective of sādhanā (practice), the Mādhyamaka and the Yoga have something in common in that both of them head toward nothingness of "I-ness" (*ahaṃkāra*). As long as there is "the I," no-self (*anattā*) cannot come out, and "non-duality" cannot be realized. The goal of yoga is essentially to cause the mind to become like zero. When we look at the complementarity and interaction of the particles with the observer, we find interesting correspondences between quantum physics and Mādhyamika philosophy. This observation or mental involvement is actually also apparent in the area of yoga and meditation. In fact, one of the most important aspects of yoga practice, even in haṭha-yoga, is this mental involvement. There is a maxim in the field of yoga: "*Cakras* are fed with observation of yogi." This means that the act of observation turns potentiality of *cakras* into actuality.

Keywords Mādhyamika · Anātman · Śūnyavāda · Pratītyasamutpāda · Awaking of *Cakras* · Darśana

1 Introduction

Quantum physics is enchanting because of its potential to transform our understanding of reality and also its striking parallelism with Buddhist ideas, as well as with some other Indian traditions that involve meditation. There are considerable commonalities between the metaphysical concept of reality enunciated by Buddhism and the physical concept of reality implicated by quantum mechanics. "Quantum physics is indeed very much in agreement with Buddhistic emptiness. For neither is there a

G. L. Lee (✉)
Graduate School of Integrative Medicine, Sun Moon University, Asan, South Korea

S. R. Bhatt (ed.), *Quantum Reality and Theory of Śūnya*,
https://doi.org/10.1007/978-981-13-1957-0_15

substantial kernel to reality, rather reality consists of systems of interdependent and complementary objects (*dharmas*).

Buddhism is based on the idea of *anātman* or "all things being empty of intrinsic existence and nature." It stresses not only that no permanent, substantial self (*ātman*) can be found to exist but that mental processes and physical objects of the world including human beings are devoid of any inherent and separate nature (*svabhāva*). None has its exclusive and individual existence but only an existence in connection with another object interacting as a reference point.

One of the interesting aspects of quantum mechanics from the viewpoint of Buddhism is that particles of classical physics are regarded as processes consisting of continuously changing wave functions. These processes reveal themselves as segregate and localized particles at the moment they are observed. So particles are continually changing and lack any inherent and separate existence independent of the act of observation. Consequently, everything composed of particles in quantum mechanics is also impermanent and ever changing, and there is no static basis for its existence. Therefore, at a generalized level, the understanding of reality in quantum mechanics is converged with the fundamental insight of the Buddha.

2 The Concept of Reality in Quantum Mechanics and *Śūnyavāda*

The idea of dependent origination (*pratītyasamutpāda*) is one of the most central theories in all of Buddhism. Nāgārjuna, the most famous exponent of the Mādhyamika school of Buddhism, developed a rigorous and sharp method of philosophical analysis which penetrated into the ultimate nature of reality. He exploited a rather unusual terminology for the status of all things. Since they are interdependent, he says, and since depending on something else is to have no nature of one's own (*svabhāva*), they must be without any nature, that is to say "void" (*śūnya*). His philosophy is frequently called the doctrine of the emptiness (*śūnyavāda*).[1]

Mādhyamika emptiness (*śūnyatā*), the core view which is logically induced from the law of dependent origination (*pratītyasamutpāda*), is the insight that there is nothing in the world which exists as an independent entity in its own right. According to Nāgārjuna, when the Buddha said that everything was interdependent, he did not mean that some things depended on other things which were themselves independent. He said that things do not arise substantially either out of themselves or out of something, or out of both, or without a cause. Therefore nothing exists as a totally independent feature of reality as was thought by Western philosophy and science until quantum mechanics came on the stage.

The concept of reality in quantum mechanics can be expressed by the concepts of complementarity and interaction. Niels Bohr, a leading founder of quantum mechanics, first introduced the concept of complementarity into quantum mechanics in his

1927 lecture in Como. According to John Wheeler, "Bohr's principle of complementarity is the most revolutionary scientific concept of this century and the heart of his 50-year search for the full significance of the quantum idea." According to this concept, the particle and wave aspects of physical objects are not two separate forms but complementary phenomena. The two forms do not contradict and exclude each other but are mutually complementary forms. They only together can provide a complete description of physical manifestations.

The term "*śūnya*" used by the negative dialectic of Nāgārjuna means *pratītyasamutpāda* (dependent origination). However, according to Nāgārjuna, the concept of "dependence" in *pratītyasamutpāda* does not assume any positive reality upon which other things depend but which does not depend on anything else. The law of *pratītyasamutpāda* does not allow any kind of fixed state, and it can be said that the Mādhyamika śūnya or emptiness implies the duality of nothingness and tendency of escape from state of nothingness. The duality of emptiness can be compared with the wave-particle duality of quantum mechanics.

According to Niels Bohr, "a complete knowledge of phenomena on atomic dimensions requires a description of both wave and particle properties." Based on experimentation, he said that such phenomena as light and electrons exhibit both particle-like and wavelike behavior; i.e., such things have a wave-particle duality. Consequently, in quantum world, a particle cannot have both a definite position and momentum. Therefore, he argued that in quantum mechanics there are no independent quantum objects because they all are in an interactive and interdependent relationship with each other. For Niels Bohr the fundamental physical reality consists of interacting and complementary quantum objects. There stability of the entities and events of the world is based on the unstable interactions of their component parts. Indeed, quantum mechanics implies an essential interconnectedness of nature.

3 Emptiness or Zero Point as the Goal of Yoga

From the perspective of practice (*sādhanā*), the Madhyamaka and the Yoga have something in common in that both of them head toward nothingness of "I-ness" (*ahaṃkāra*). As long as there is "the I," no-self (*anattā*) cannot come out, and "non-duality" cannot be realized. It is not until the sense of I-ness turns into zero that "*tat tvamasi*" (Thou art That) in the Upaniṣads can become a reality. As long as "I" exists, "He" does not exist; when the I becomes zero, "He" turns up. For example, if A and B are in the relationship of non-duality, it means there is neither A nor B. Even when only one of the two exists, the relationship of non-duality cannot come into existence. That is because "non-duality" is not different from "unity of being" or "pratītyasamutpāda." Emptiness, as a logical consequence of pratītyasamutpāda, is not nothing; it has its own fullness.

In terms of their orientation for the zero point, the Madhyamaka and the Yoga are the same. We can describe as "nothingness" all the efforts made to reach this point.

The goal of the practice of yoga is to condition the mind to become like zero and in so doing, to establish *kaivalya* (solitariness) by *vivekakhyāti* (discriminative knowledge). The classical yoga lifestyle strives toward the goals of *abhyāsa* (practice) and *vairāgya* (detachment) as mentioned in the *Yoga sūtras* and the *Bhagavadgītā*, which seeks to zero-out all desires, attachments, emotions, and ego clinging. The goal of yoga is essentially to cause the mind to become like zero. In fact, the goal of meditation (the central feature of the yoga lifestyle) is to zero-out thoughts, to zero-out the mind, and to realize the true condition of reality which was suggested by the Mādhyamika and quantum mechanics.

> Let him meditate on his own reflection in the sky as beyond the Cosmic Egg in the manner previously described. Through that let him contemplate on the Great Void unceasingly. The Great Void having no beginning, middle, or end, has the brilliance of tens of millions of suns, and coolness of tens of millions of moons. By meditating continually on this, one obtains success.[2]

The Mādhyamika agrees with the yoga: It is only through the understanding and realizing of emptiness that *mokṣa* or *nirvāṇa* from saṃsāra is attained. Insight into emptiness is therefore called "the gateway to liberation." It is reasonable to see the *aṣṭāṅga* of Patañjali yoga system as paths leading to a zero point in the end. That's because through the practices and self-denial, *ahaṃkāra* (egoity) is getting weaker and purified, and it eventually gets to nothingness and then returns to the purified original *buddhi*; out of the *buddhi* emerges *vivekakhyāti* by which *kaivalya* is attained. Therefore, the entire process of yoga practice can be said to be "nothingness" itself.

The first of six *pāramitā* (six perfections), highlighted in Mahāyāna Buddhism, is Buddhist alms, which is a straightforward example of Buddhism aiming for the zero point. When the core meaning of anatta is considered as the notion of "I-ness," the idea of "mine," the easiest and most direct way to realize selflessness is to give yours away. By giving alms nothingness can be realized and therefore selflessness can be reached.

Like in coordinate geometrical figures, the zero point is "nothing" and, at the same time, "the origin" where the X axis and the Y axis meet each other on the coordinates. As long as the X axis or the Y axis has some value on the coordinates, they cannot meet at the origin. *Ahaṃkāra* (egoity) constantly wants to have some value of the X axis and the Y axis. Fake self tries to come away from the origin as far as possible. This is the nature of the fake self. To the contrary, practice is an effort to go back to the zero point. At this point, the Mādhyamika and the Yoga are all the same. The closer fake self is to the zero point, the more likely it is to go glimmering. As long as it comes away from the zero point farther, the true self never comes into light, because it is not until the fake self dies out that the true self makes its appearance. Like this, the Mādhyamika and the Yoga have in common in heading for nothingness.

4 Observer Effect of Quantum Mechanics and Awakening the *Cakras*

When we look at the complementarity and interaction of the particles with the observer argued by Niels Bohr, we find interesting correspondences between quantum mechanics and Madhyamika philosophy. The theoretical implication of quantum mechanics shows that in quantum world there would be no such things that are completely independent from observation. What is being observed is not independent of the observer. According to Niels Bohr, "isolated material particles are abstractions, their properties being definable and observable only through their interaction with other systems." In the same manner, Wheeler said "no phenomenon is a phenomenon until it is an observed phenomenon." According to quantum mechanics, the observer is part of the system.

The Mahāyāna Buddhist philosophers claim the mind is a fundamental aspect of reality, which is not reducible to a physical basis. For them the freakiness at the smallest scale of mechanics is the result of our realization of our mind's involvement in producing reality, that is, the observer is part of the system. The concept of observer effect in quantum mechanics indicates that at the quantum level, being and knowing, perception and reality are inextricably entangled; perception changes the ontological structure of reality.

In fact, one of the most important aspects of yoga practice, even in *haṭha-yoga*, is this mental involvement. From the perspective of Indian philosophy, one of the chief elements of a possible connection between quantum mechanics and yoga and any other India traditions of sādhana (practice) is found in the observer effect of quantum mechanics. In yoga practice, by the act of observation, yogi actually affects the state of the object being observed such as his body, breath, *cakras*, or mind.

Even in *haṭha-yoga*, which is usually regarded as a form of physical exercise, the involvement of the observer's mind becomes inescapably obvious. When he/she tries to practice an *āsana*, if he/she does not observe a *cakra* corresponding to the *āsana*, his/her physical posture is not *āsana* itself but a mere physical exercise. There is a maxim in the field of yoga: "*Cakras* are fed with observation of yogi." This means that the act of observation turns potentiality of *cakras* into actuality. Observation resolves the question of what the *cakra* actually "is" through a combination of the *cakra's* inherent potentials and the manner in which it is observed.

Quantum mechanics confirms the very basis of meditation practice, the last three limbs of *aṣṭāṅgayoga* outlined by Patañjali in his *Yoga sūtras*. Through the very act of observation, the yogi (observer) changes his mind which is being observed and makes it calm. The double-slit experiment proves how an electron that usually behaves like a wave, instead, behaves like a particle when under observation. Just as the presence of the observer can affect the behavior of an electron, our observing conscious mind in meditation can change our psychological state.

It is not easy to feel the change in our mind and body after short-term meditation. Only through a long-term and regular practice of meditation we can expect a meaningful result. According to the *Yoga sūtras* (1.13), "That practice when continued for a long time without break and with devotion becomes firm in foundation." The higher the intensity of observation, the greater the effect that the observer has on the observed. Objective inward observation through the practice of meditation leads to heightened discernment of the mind-matter phenomenon at its subtlest level.

5 Hindu Way of Life as *Darśana*

Quantum mechanics raises profound questions not only for physicists but also philosophers today, including the relationship between the observer effect and our understanding of consciousness. Insight into these questions can be found in the practice of yoga and meditation. In particular, observer effects of quantum mechanics have important implications for yoga training and meditation.

The premise of the observer effect is the interrelationship of mind and body which is demonstrated by quantum mechanics. Indeed, this has already been known for a long time in the philosophical traditions of India. According to the Sāṃkhya, mind is also a product of *prakṛti* in the same way of the physical body, where mind turns out to have the material quality. The difference between the two is only relative; the former is a rough substance, whereas the latter is subtle substance. Because both of them are the products of *prakṛti* (matter), we can straighten our bent mind through meditation just as we can bend a steel bar with the power of our arms.[3]

From the Indian viewpoint, the word "philosophy" suggests "observing" the existence. In Sanskrit, the philosophy is referred to as "*darśana*" which has its root in the word "*dṛṣ*" that means "to see." "Seeing" the reality and "observing" the facts of experience form the basis of philosophy. Seeing is not simply a sensory activity but contemplation with a conscious mind. Seeing is a perceptual observation. It may also concern the conceptual knowledge or an intuitional comprehension. Thus the term *darśana* suggests a whole view revealed to the inner self.

Why does Indian philosophy and sādhanā traditions so much emphasize the act of observation? The answer of this question is related to the observer effect of quantum mechanics. A fundamental conclusion of quantum mechanics acknowledges that the observer creates the reality. As observers, we are personally involved with the creation of our own reality. The act of observation turns potentiality into actuality. Observation resolves the question of what the particle actually "is" through a combination of the particle's inherent potentials and the manner in which it is observed.

Even the way the Indian people overcome the pain, the secret of the observer effect is hidden. Let me give you an example. While sitting to meditate, you should not move your body first. But what if you feel an itch on your face while sitting? Will you scratch the itch with your hand? No. If you scratch the itch with your hand, your meditation will be disturbed. Then what will you do? You just have to observe it for a while with a relaxed mind. Then your itch disappears quickly. How do you see such a result? Your activity of observation changes the itch into comfort.

Yoga and meditation are a way of observing our lives as they are. Just as when we observe the itch itching disappears, if we can observe our painful life as they are, we will be able to cross the river of saṃsāra. It is only through observing our lives as they are that changing saṃsāra into *nirvāṇa* or *mokṣa*. I understand that Indian people prefer a slower way of life than active and dynamic life, because they know that the ultimate goal of life is attained by observing their lives as they are. If we want to observe our lives as they are, we need to choose a slower way of life than a bustling life.

Notes

1. Baggott ([1], p. 97).
2. *Śiva-saṃhitā*, 160–161.
3. Lee ([4], p. 136).

References

1. Baggott J. The quantum story: a history in 40 moments. Oxford: Oxford University Press; 2011.
2. Bohr N. Atomic physics and the description of nature. Cambridge: Cambridge University Press; 1934.
3. Encyclopædia Britannica.
4. Lee GL. The salvational meaning of the Śarīra (body) represented in Āyurveda, yoga and Tantra: focusing on the parallel structure of five elements (Pañca-mahābhūta), the multi-layer structure of five covers (Pañca-kośa) and the bipolar structure of Śiva-Śakti. J Korean Soc Indian Philos. 2013;39:135–70.
5. Hinterberger T, von Stillfried N. The concept of complementality and its role in quantum entanglement and generalized entanglement. Axiomathes. 2012; 22(1).
6. Nayak GC. Mādhyamika Śūyatā - A Reappraisal. Indian Council of Philosophical Research: New Delhi; 2001.
7. The Śiva-saṃhitā, – A Critical Edition & English Translation by James Mallinson; 2007. YogaVidyam.com.
8. Vedral V. Decoding reality: the universe as quantum information. Oxford: Oxford University Press; 2010.
9. Williams P. Mahayana Buddhism: the doctrinal foundations. New York: Routledge; 2008.
10. Wheeler JA. No fugitive and cloistered virtue"—a tribute to Niels Bohr". Physics Today, January 1963.
11. Wheeler JA. The 'past' and the 'delayed choice' double-slit experiment. In: Dolling LM, Gianelli AF, Statile GN, editors. The tests of time: readings in the development of physical theory. Princeton: Princeton University Press; 2003.

Chapter 16
Quantum Mechanics Shows the Limit of Naïve Realism

Bhakti Vijnana Muni

Abstract The ancient Indic philosophy of Vedānta harmonizes the truth as the original Organic Whole (*pūrṇa*). The non-dual truth of reality is seen by different sages as *Brahman*, *Paramātmā*, *and Bhagavān* according to the development of their inner spiritual life. Thus, the personal conception of reality is the higher vision. Modern science has also considered a more subjective conception of reality due to the progress in quantum physics. Thus science is undergoing a process of self-critique due to its own progress. The role of the observer is found to have a significance that was not found in the classical physics. Although QM is not a science of life, it has given some hints that we cannot consider reality in the sense of naïve realism. The subject contributes to the object as much as the object contributes to the subject. The objects cannot be thought of as existing independently of the observer. Therefore the focus of science should be to study this relation between the subject and the object in everything. In this sense QM is advancement over conventional science. Further life process is essentially an organism and the parts are never independent of the whole. The parts are the inseparable members of the whole. The logic of life presents us a higher category in nature than physics and chemistry. Logic of machines, physics, and chemistry does not lead to living processes.

Keywords Naïve realism · Vacuum polarization · Dependent co-organization · Brahman · Kant · Observer · Life

B. V. Muni (✉)
Sri Chaitanya Saraswat Institute, Bengaluru, Karnataka, India
e-mail: bvm@scsiscs.org

S. R. Bhatt (ed.), *Quantum Reality and Theory of Śūnya*,
https://doi.org/10.1007/978-981-13-1957-0_16

1 Introduction

The ancient Indic philosophy of Vedānta harmonizes truth as the original Organic Whole (*pūrṇa*). The Vedāntic thought and the Buddhist thought have different conclusions in terms of the causal principle. The Vedāntic concept of origin explains that *Brahman* is the source of everything (*sarva-kāraṇa-kāraṇam*), and it is *nirguṇa* (no material quality). The Buddhist thought is centered around the idea of *dependent co-originotion*. The concept of vacuum polarization of the relativistic quantum field theory has encouraged some of these dialogues as the vacuum potential is both the negation of the particle-antiparticle pairs and the source of these. Thus the quantum vacuum is not an absolute zero but is the very source of observed quantas. The other aspect of quantum physics is that of complementarity which has been extended by some philosophers like Pauli and Jung to include both the mental and physical domains as something to be considered as mutually exclusive epistemic domains and yet both necessary for a comprehensive description of reality.

In the twenty-first century, the wholistic thought is again becoming more acceptable. According to Hegel the truth is the whole. The whole constitutes the basic underlying dynamic truth or reality. This is also the Vedāntic understanding which describes the *Brahman* or spirit as absolute truth. *Brahman* is complete in itself or full existence, full knowledge, and full bliss. It is the true infinity. In Vedānta absolute is sentient, and therefore the personal realization of the absolute truth is its topmost realization. There can be many wholes, but in order to comprehend the absolute truth, we have to comprehend the original whole, which is the original complete, the Original Person, or the *ādi puruṣa*.

2 Vacuum Polarization

The concept of vacuum polarization comes out of the problems in nuclear and subnuclear physics. For example, Julian Schwinger could explain the anomaly in the decay of fluorine. Oppenheimer imagined that there could be some kind of new interactions, and yet Julian Schwinger explained it through creation of electron-positron pairs or the concept of vacuum polarization. This had given Schwinger a head start over Feynman who for long did not believe in vacuum polarization [1].

The development of the subject of quantum mechanics (QM) has changed some of the views about nature of reality. The idea of complimentarily is one of the important outcomes of QM. Physics developed from classical mechanics to relativistic physics and relativistic quantum field theory, and it has brought new understanding of the concepts of space, time, and fields. Relativistic QM includes the concepts of quantized fields. The quantities which were considered as having continuous values now have discreet values. After the acceptance of one-particle, two-particle, as well as many particle systems, QM also recognized a no particle state called the vacuum state. Particles are thought of as something that are created out of the vacuum, which is a potential where particles have not yet been actualized. At present the idea of quantized fields is our deepest level understanding of the microscopic phenomenon. With the development of higher and higher energy machines, we have discovered an array of subatomic particles, some of which are stable and some which are not stable, and they decay into each other in all possible conceivable ways. In the words of Julian Schwinger, "We send electrons, protons into the various kinds of nucleons, where we perform experiments in which these particles enter a certain very small region. We make no attempt to describe what goes on there and simply try to finally characterize what emerges when the particles are separated again. Is the purpose of theoretical physics to be no more than a cataloging of all the things that can happen when particles interact with each other and separate? Or is it to be an understanding at a deeper level in which there are things that are not directly observable (as the underlying quantized fields are) but in terms of which we shall have a more fundamental understanding?" [2].

3 The Philosophical Study of Reality

The philosophical study of reality begins with the study of the triad consisting of the world of objects, the knower of them, and the process that leads to knowledge of them. The analysis of this triad is called *trikuṭi vinyāsa* in Vedānta. In Vedānta these tenets of the triad have been established on the grounds of the perceptual, rational, scriptural, and transcendental realization of the sages as well as the revelation from the absolute source. When the subject becomes the object of his own study, then spiritual life begins. The difference between life and nonlife is that only living beings exhibit three qualities of thinking, feeling, and willing. Therefore for the study of life, we must include the study of consciousness and free will. In this regard James A. Shapiro mentions from the work of Barbara McClintock, a Nobel Laureate in Biology, that "Her work and line of thought underlined a departure from the usual mechanistic thinking in two important ways. (i) She drew attention to cognition and purposeful action by the cells, and (ii) Her work highlights the sophistication and richness of inter and intracellular communication and control in the cell." [3]. All cells are sentient and life displays cognition at all levels [4]. Biology must be conceptually rich enough to include concepts like sensory response, intelligence, self-identity, internal teleology, and consciousness. Life is a fundamental reality as the Vedas have been revealing from time immemorial. The logic of life corresponds to a higher ontological category of holistic understanding. Life is not an additive process. Life develops from an already existing individual whole of the species by the process of reproduction.

The mechanistic thought is an idealization, which is based on the possibility of reducing a complex behavior to a simpler explanation on the basis of the parts. The physical sciences and the chemical sciences are based upon such idealizations. However philosophy is concerned with being and nature. Being implies existence and nature means the determinate qualities. For example, being of sodium means the existence of sodium. But we talk about sodium only in terms of its determinate qualities. For example, the nature of sodium is that it is yellow in color, it is an alkali group metal, etc. But to study the being of sodium, we require a more philosophical approach in terms of ontological questions. Naïve realism has relied on the ability of science to explain an objective world independent of any observation. Concepts of substance, causality, and logic have been understood mainly in classical sense for a long time. But with the advent of quantum mechanics, there arose a large number of philosophical problems in the foundations of science in the twentieth century. But we would like to point out that quantum mechanics is not really a theory about life. There are many interpretations of QM which rely on the existence of consciousness (observer), Yet how consciousness is deduced is not yet answered within its framework. QM has mainly shown that naïve realism cannot be an answer to the concepts of substance, objectifiability, causality, and logic. David Bohm, for example, argued for a wholistic understanding of the phenomenon, by interpreting the pilot wave theory. He explains, "Not only is undivided wholeness implied in the content of physics (notably relativity and quantum theory) but also in the manner of working in physics." [5].

4 Buddhist Principle of Causality: Dependent Co-originotion

The Buddhist understanding of origin of observed reality is based upon the concept of dependent co-origination. Dependent co-origination is the literal translation of the term *praṭicca samuppāda*.[6] This idea contains the basic Buddhist insight into the nature of reality and working of nature. As a principle it is understood to be universally applicable in Buddhism. One of the characteristics of Buddhist principle is that the nature is objective. Thus reality is not a creation of Buddha but Buddha only reveals the nature of reality. In this way *dependent co-originotion* is not a subjective explanation but is a statement of objective reality. The second characteristic is *necessity* which means there are necessary conditions in the presence of which a given effect must happen. The third characteristic of reality is *invariability* which means there is a constant relation between cause and effect. The fourth characteristic, conditionality, means that there is a group of conditions that come together to result in a particular effect. Buddhist idea in this way is an objective view of reality. In this scheme consciousness seems to be a result of conditions arising out of a sequence of previous conditions beginning with beginning-less ignorance. Buddhism does not believe in concepts of eternalism, like soul. Causality in Buddhism means a complex of events which is interdependent and interrelated. We cannot speak of any unconnected phenomenon. Human beings are understood as an important aspect of reality. The nature of human beings and their relation to reality are essential aspects of dependent co-origination. The different layers of reality acceptable to Buddhism are therefore the physical inorganic, the physical organic, the sphere of mental activity, the moral sphere of moral action, and finally the sphere of higher spiritual life. The mental sphere should not be misunderstood as being caused by an agent like the soul. Foundation of the concept of soul and universal soul as found in the Vedic tradition is not found in Buddhism. All these different spheres are understood as causally interdependent web of conditions. In this way nature, with its all aspects, including human beings, is considered to be a phenomenon governed by one universal principle, namely, dependent arising, co-existence, and dissolution [6].

Everything including human being is characterized by the triple qualities of impermanence, unsatisfactoriness, and non-substantiality.[1] In Buddhism everything animate as well as inanimate shares certain fundamental features. It is mainly used to explain how suffering arises and how it ceases. This depends on various causes and conditions in the individual. This doctrine has also been used to explain the problems that we collectively suffer in the world. This principle of dependent co-origination denies any central or supreme position to any particular entity.[2] Specially, in Mādhyamika emptiness and intrinsic existence have been considered as incompatible rivals [7]. The *Abhidharma* scholars developed the concept of intrinsic nature or *svabhāva*. According to Nāgārjuna, "When *svabhāva* exists the universe will be unborn, non-ceased, remaining immutable and devoid of variegated states. If empty is not seen then reaching what has not been reached, the act of terminating suffering as well as the relinquishing of all defilements also will not be seen.[3]" In this way the concept of *śūnyatā* becomes important in the Buddhist doctrine.

The position of Buddhism is to critique the position of other traditions when they state that something exists due to its *svabhāva*, i.e., it has an intrinsic or inherent existence. The Buddhist methodology is dependent on arguments. The essence of this *prasaṅga* method, as set out in the Mādhyamika, sets out to refute the position of the opponent, who has posited something as real existence, through reasoning. The stress is on the egotistical grasp that is bound to appear in the arguments of the opponent.[4] Nāgārjuna's thesis is the denial of any existentially intrinsic reality. For him all *dharmas* lack intrinsic existence and so they are not universal truths. Any other position thus has unresolvable paradoxes. In this way Buddhism comes to the notion of emptiness. It is the notion of dependent co-origination that Nāgārjuna calls as emptiness (*śūnyatā*). Entities do not exist from their own side but are also imputed by the mind. Emptiness is not a vague absence. It is still less an absolute reality. It is the absence of *svabhāva* or intrinsic existence of the object being examined. Emptiness is not even a *dṛṣṭi* or a point of view. What is true about an object is that it does not have any real fundamental existence. In this way as everything lacks any intrinsic existence, even emptiness lacks any intrinsic existence. In this way we

---

[1]Tilakaratne, [6], pp. 5.

[2]ibid., [6], pp. 9.

[3]ibid., [6], pp. 8.

[4]Williams, [7], pp. 71.

come to the emptiness of the existence of the emptiness itself, and this is called *śūnyatāśūnyatā*.[5]

Another Buddhist view is called as *Yogācāra*, and it pertains to a formulation of universal idealism.[6] According to this all things can be subsumed under three natures. The first nature is called constructed or conceptualized nature. This is experienced everyday by the unenlightened. It is how things appear to us, i.e., the realm of subject-object duality. But in reality it is not how things are. The second nature is the dependent nature of the origin of *dharmas*, and it is the causal flow. It is that which appears in opposition to the way in which it appears (first nature). Thus all things of experience as well as the experience itself are just a flow of perceptions.[7] We do not know if something is out there. Only due to the beginning-less ignorance are these perceptions made out into enduring subjects and objects confronting each other. This flow of perceptions forms the basis of our mistaken construction and is the dependent nature, and we should let it all go. The final nature observed in deep meditation is called the perfected nature. It is said to be a complete absence in dependent nature of objects (of conceptualized nature). It lacks the enduring fixity of objects. It is expressed as a nonduality where there is neither the subject nor the object. There is only a single flow. It is also emptiness for it is empty of another. It is the flow of perceptions – the dependent nature – which is empty of enduring entities. The perfected nature here is hence the actual absence, the absence that occurs when one thing that does exist is lacking something that does not exist there. It is not explained as the absence of intrinsic existence. Rather, it is the absence that is the perfected nature – that is, the complete absence, in the dependent nature, of the conceptualized nature.[8]

Different schools have their own different concepts of error. From these theories of error, they establish their concepts. Thus Buddhism describes the phenomenal reality in terms of the practical efficiency. According to the Mādhyamika school, the error lies in holding the nonexistent as the existent. But the school fails to acknowledge that this error really arises due to sensory contact under certain conditions. The Vedāntic school considers that this is really an error due to misappropriation of one real object for another real object [8].

The Buddhist Yogācāra school holds that the error of the Nyāya school is in projecting the subjective ideas as objective extramental facts. But Jayatīrtha of the Madhva school shows that this view of error is not supported by the sublating consciousness. The theory of internal reality of objects represented in illusions (as held in Yogācāra view) would lead to absurd results. It would imply that there would be an actual burning sensation inside the body when there is an erroneous superimposition of fire on a heap. The *Asat-Khyāti* of the Mādhyamika school of Buddhism holds that error lies in the manifestation of the nonexistent as existent. But

[5]ibid., [7].

[6]ibid., [7], pp. 89.

[7]ibid., [7], pp. 90.

[8]ibid., [7], pp. 89–92.

it fails to recognize that this illusion is due to the contact of the sense organ with the shell as the *adhiṣṭhāna* and under certain conditions. According to the Madhva school, a proper realistic theory of knowledge should also include the possibility of errors and illusions. The reality of errors and illusions, however, does not upset the foundations of realistic knowledge. The Vedāntic theory provides a place for dreams, errors, and illusions in life. Life holds both truth and error. The errors and illusions are rather a proof of the limitations of the individual [9].

The Buddhist scheme merely takes the question of errors and illusions to the extreme. Rather the proper view according to Vedānta is that the possibility of error implies that there is a possibility of true knowledge as well. Madhva defines illusions as the contrary appearance of an unreal, nonexistent object as real and existent and vice versa [8]. The root of this theory is the presentation of some nonexistent entity, fact, or relation as existent and real in a given substratum, with which the sense organ is in contact and when other conditions propitious for the erroneous cognition are also there. This element of false presentation results in illusions. The *Asatkhyātivādin* is therefore correct in holding that the object of illusions is utterly unreal. But he is not justified in denying any kind of reality to any of the constituents of that experience, including the *adhiṣṭhāna*, etc., and dismissing the entire fabric of that experience as a baseless fancy. Madhva's detailed critique of Buddhist and the Nyāya view of error are given in his doctrine of *Abhinava-anyathākhyāti* [8].

Error is the reverse of valid knowledge. It is an experience of an object cognized as something else having certain characteristics that lie outside of its being. Examples are the cognition of silver as the shell and a rope as the snake examples. This explanotion of errors is a very important aspect of philosophy. What is the source of error? Does it lie in the subject or in the object or both? The Indian schools that deal with the theory of error include the Buddhist Idealistic school (Yogācāra), Buddhist Nihilistic school (Mādhyamika), Mīmāmsakas, Naiyāyikas, and the Vedāntins of different affiliations. Thus *Advaitin* school, the Rāmānuja school, the Madhva school, as well as the *Acintyabhedābheda* school all have developed theories of error. According to Rāmānuja, the object represented in error is always real. Strictly speaking there is no invalid cognition. It can be said that the cognition of silver in the shell is true in a relative sense and is not false in any absolute sense. Rāmānuja says that every object in the world has some elements of the other objects, in its constitution in varying proportions. This is the understanding that comes out from the Upaniṣads. The perception of silver in the shell is to be taken true to the extent that there are the silver elements in its constitution. Only it is a partial experience of the silver in the shell. We have only a partial experience of silver in what is shell silver. The error is not in ascribing reality to something unreal. The perception is no doubt wrong. It is not because it is cognition of the unreal or what is not in fact there. Rather the error is due to the partial view or an imperfect cognition of the real. Thus the Buddhist view blunders about the subtle nature of the truths [8].

The *vaiṣṇava* Vedāntic schools hold that the world of perception is real. It is the basic foundation of theism. Srila B R Srīdhar Swāmī, one of the foremost scholars in the Gaudiya tradition of Vaiṣhṇavism of *Acintya-bhedābheda-tattva*, explains, "Proper knowledge is not possible under the philosophical systems of Buddha and

Śankarācārya. If we say that the world is false, is a suicidal position. Error or misconception has reality. The basis of material existence is the possibility of the *taṭasṭha jīva* committing a mistake and developing misconception. The soul is *aṇucetana*, atomic consciousness. And as atomic units of consciousness, our freedom is not perfect. Our defective freedom is the cause of this illusion. The soul must have freedom. Before a crime is committed, the possibility of committing a crime is present in the ordinary peace-loving subject. The possibility of disease is there, so hospitals, medicine, and special diets are all necessary. In the same way, the possibility for misconception is there in the soul because we are weak and limited." [10].

The Vedāntic literature deals with both identity and distinction and explains *Brahman* as the non-dual truth. The idea of the "other" is important to consider in understanding the nature of the non-dual truth. For example, the subject-object is a most important instance of distinct but yet an inseparable reality. By a thoughtful analysis, we find that one term implies the "other." A subject does not exist without corresponding object. If it were not so, then there would be no possibility of talking about the "subject." In the same way, an object can only be called an object because it is in relation to a subject. This type of interdependence is exhibited by all kinds of opposites. For example, light-darkness, sleep-waking, day-night, etc. have meaning only in relation to their opposition to one another, and hence opposition is essential to the existence of either term. Further unity and multiplicity are also opposites and hence must also be aspects of a single reality, i.e., the duality is contained within a unity [11].

However there is another consideration when we come to the true infinity. This has been explained by Hegel and in the words of Sripad Bhakti MadhavaPuri Maharaja, "When we try to go 'outside' of something but find the same thing there as 'inside' have we really gone outside at all? Such a situation in which we find the same thing inside as outside, in which the other is the same thing as that which it is other to is called the infinite, or literally non-finite. The infinite is non-finite because there are no boundaries, no limits, no place where it is not or where it is finished, finite or fini - ended. At the same time it does not thereby exclude negation. It is rather negation that establishes it as infinite, because it is only by negation that we apprehend that the infinite is indeed infinite or that which becomes other to itself and yet remains itself in such 'othering'." "This means that the other or negation is present in the infinite. If we think about this carefully, it also means that the finite is present in the infinite, because other implies distinction, and distinction implies definite boundaries or finitude." [11].

In Vedānta, the origin is *Brahman* and is not an absolute emptiness. *Brahman* is *pūrṇa* or complete in itself. In the Vedāntic thought, the process of negation will not lead to voidism or absolute emptiness. When we negate something, it does not produce a void. Therefore everything in nature is posited as if it is existing for itself. But by the process of negation, independent positing of entities is being negated in various ways and forms by nature. In this way everything exists for some higher and higher purposes in nature, in which everything is appearing and vanishing as transitory moments. Therefore reality is dynamic and it is never fixed. Thus everything is being negated for a higher purpose than that which is for itself. When we come to negation of all such negations, we arrive at the purpose of absolute or the *Brahman*. Therefore *Brahman* is the cause of all causes or that which constitutes the ultimate purpose. Everything exists for the satisfaction of the absolute. *Brahman* is not some finite entity that can be limited by some boundary. Hegel, the German philosopher summarized this beautifully by saying, "Reality is by itself and for itself." All causes ultimately are to be traced to be dependent on *Brahman*. But *Brahman* is by itself or is the substantial truth of all reality. Thus the entities are not only inseparable but also they are all dependent on *Brahman* for their being and becoming. This is the major difference between Vedānta's principle of *Brahman* as the cause of all causes (*sarva-kāraṇa-Kāraṇam*) and Buddhism's principle of dependent co-origination.

5 Revaluation of the Concepts of Substance, Causality, and Logic Due to Quantum Physics

QM deals with the question of indeterminism because it states that there are physical measurements whose results are not definitely determined by the state of the system prior to the measurement. The determinism of QM is opposed to the determinism of classical mechanics. It means that the result of quantum mechanics is not an unequivocal result of the previous state of the system. To circumvent the problem of indeterminism of QM, the hidden variable theories were proposed. Apart from these there is the question of locality. Nonlocal correlations have been found to be true. An instance of this is given by particles which are separated in space but whose properties are correlated in some way. Two particles in a state of a certain kind are not independent of each other. It has become clear that it is not possible to develop a theory that conserves determinism and locality in QM that could reproduce all the experimentally proved results. There are discrepancies between the results of the hidden variable theories and the predictions of quantum mechanics. Any hidden variable theory ends up predicting results that contradict quantum mechanics. This point was proved by the Bell's theorem.

The change in the wave function as a result of measurement is called wave function reduction. The problem in the measurement theory is to establish at precisely which point this reduction takes place. There is no fixed point that we could determine at which the wave function is reduced. This has resulted in some physicists like Eugene Wigner proposing some subjective theories in which this reduction takes place when the information reaches our brains. This theory assumes that our mind is of a different nature that the physical world that it is interpreting. Here physics loses its objective relevance in the usual sense. Further it does not establish how the different observers (persons) come to the same conclusions about the results of the experiments if we do not admit the existence of an objective physical world. Apart from these the subjective theories do explain the results of the quantum mechanical experiments, and some philosophers consider them to be the best explanation available at this time. In the Copenhagen interpretation, it is impossible to separate the quantum mechanical systems from their measuring devices.

The physical conditions for knowledge we encounter in atomic physics are such that not all measurable properties can be in the same way attributed to an observed object. We cannot relate all measurable observables to a well-defined object. The concept of substance, causality, and logic has changed considerably with the understanding coming from atomic physics and relativity. In the following section, the author has taken ideas from Mittelstaedt [12].

5.1 Substance

It is generally Kant to whom credit is given for making the traditional concept of a substance precise. Before him, Descartes was a substance dualist. He explained "substance" as something that is causally independent as well as a subject independent thing. The properties are the accidents or the modes of the substance. When the question is posited that what truly is a piece of wax, when it changes its properties, Descartes arrives at the conclusion that wax is different from all that is perceived w.r.t. to the senses. It is only through the mind that it could be apprehended. Thus the perceptible properties of the wax are changeable and are accidents. But wax as such is unchangeable and is not perceptible to the direct perception. Thus substance (wax) is that which persists in the course of the changes of its apparent form (the observables). Thus all properties of thing can be related to a substance in the sense that they are the accidents of the substance. Further regardless of the changes of the accidents, the substance persists in time, i.e., substance remains unchanged. Further the immutability of a substance w.r.t. time defines what a temporal change is. Thus traditionally we objectify a substance. The concept of substance is a point of view for ordering and subsuming phenomenon.[9]

But Hume regarded such a concept of substance as a misconception because nothing in our experience corresponds to it. Kant responded to this critique of Hume by saying that substance is a category and as such is one of the conditions by which things at all become manifested in experience.[10] Kant explained that the persistence of substance is a synthetic but an a priori valid proposition that "All appearances contain the permanent (substance) as the object itself, and the transitory as its mere determination, that is, as a way in which the object exists."[11]

Kant's position leads to the question: how pure concepts of understanding could be applied to appearances? Kant considered the concepts of space and time by

[9]Mittelstaedt, [12], pp. 115.

[10]ibid., [12].

[11]ibid., [12], pp. 116.

assuming them as a form of intuition, a priori. Time and space was condition for the possibility of phenomenon. It implied from here that time was one dimensional. Further different times were not simultaneous but were sequential. Space and time were not objects in the Kantian scheme but were conditions because of which experience of objects becomes possible. Thus these constituted the transcendental ideality.[12] But these concepts also have an empirical reality. Kant says, "Time is therefore a purely subjective condition of our ... intuition ... and in itself, apart from the subject, is nothing. Nevertheless, in respect of all appearances, and therefore of all the things which can enter into our experience, it is necessarily objective." [13]. According to Kant the mediating concept is called the transcendental schema. It is purely intellectual on the one hand as well as sensuous on the other. The time-order is given by the categories of relation, i.e., substance, causality, and interaction. As Kant explains in *Analogies of experience* that "Now since absolute time is not an object of perception, this [determinotion of position in time] cannot be derived from the relation of appearances to it. On the contrary, the appearances must determine for one another their position in time.[13]" In this way by the aid of the category of causality, the time ordering of phenomena occurs. Further if two phenomena could be considered as the cause and effect of a process, then the temporal sequence is established.

But with the advent of theory of relativity, a critique of such a concept of time was realized. A contradiction appears in theory of relativity if one assumes that time relations that are applicable to Kantian categories are also applicable in all events. Before Einstein there was no realization of the need to define physical concepts by stating the measurement process. Physics as a science was not considered as an observer-dependent system of knowledge. Physics could be tested experimentally, but the possibility of the observer's influence upon them was not considered. This would be true so long as influence of the measuring process on the results is negligible. But the scenario was changed when the measurement techniques in electromagnetic theory were developed. The real properties of the measuring instruments in such cases could not be neglected. A direct influence of the wave equation of light was observed on the results of measurements. Thereby the results now always contain some features that are traced to the measurement process. This became true both in relativistic physics as well as quantum physics.[14]

There is always a consideration of time dependence of empirical events in a physical theory. These principles were realized fully according to the Kantian principles in classical physics. Time was uniform and absolute and the same in all inertial frames. For the purpose of experimental realization of the same, it was assumed that there is instantaneous transmission of signals in classical physics. But in relativity it was assumed we can perform all measurements with real instruments. The most standard procedure for doing space-time measurements was of

[12]ibid., [12], pp. 27.

[13]Smith, [13], pp. 226.

[14]Mittelstaedt, [12], pp. 24.

course light. But unlike the assumptions of instantaneous signal transmission, light travels only with a finite velocity. This severely restricted the applicability of the Kantian principle of time sequence and simultaneity. These were now applicable to events only when they lie within the light cone. Due to this it is not possible to make assertions about the simultaneity of the space-like events. Therefore Einstein was now forced to make new generalizations about simultaneity by taking into account the finite velocity of light so that it is applicable to spatially separated events. This also implied a change in concept of causality. The causal relations could occur only within time-like-related events. Thus for Einstein the categories were not given a priori, but the choice of the understanding of concepts must be such that it applied to reality. This shows that finding propositions about experience which are a priori valid is not restricted. Thus it is not possible to apply a uniform time for all events due to relativity. But for events within the light cone, the old Kantian concepts are still valid.[15]

In quantum physics the observations of the atomic systems have become possible only with measuring devices that occur naturally, being themselves constructed of elementary particles, like photons. These have an inherent influence on the observational results. These influences are determined by the physical laws, i.e., the quantum laws that govern these measuring instruments.[16]

We began with a discussion about the function of the Kantian concept of substance. It is in ordering the objects of perception into the objects of experience, and this concept was applicable to all objects of our experience in classical physics. A "thing" was regarded as the completely determined content of experience, i.e., all possible properties and their contradictions. In classical mechanics, the state is uniquely determined by any canonically conjugate pair of variable like position and momentum, *x and p*. In QM we can relate such properties only in a restricted sense to the state$|\varphi\rangle$. This state lives in the Hilbert space. Thus the attribution of observables to substance consists of relating the properties to the state $|\varphi\rangle$. This state symbolizes substance in QM. A quantum mechanical object means that not all but only the objective properties can be considered by the state. A consolidation of these objective properties with the aid of quantum mechanical concept of substance does not lead to "things" in the classical sense. The quantum mechanical concept of substance is only in the sense of quantum mechanical objects. Therefore these quantum mechanical objects cannot be completely determined. This is quite unlike the concept of substance in the classical sense which was derived from Kantian concepts. It is only in the classical limiting sense when Planck's constant becomes equal to zero, when all properties become objective, and the quantum mechanical and the classical substance coincide. In quantum theory it is not possible to relate all properties in some sense to a substance. It is so because it is impossible experimentally to determine all properties without changing the system in the act of measurement. Hence it becomes impossible to interpret the results of the respective

[15]ibid., [12], pp. 29–33.
[16]ibid., [12], pp. 93–97.

measurements as accidents of a substance. Additionally over and above this, there are logical and probability theoretical contradictions that do not allow us to relate all properties to an object even by hypothetical considerations. As there are non-objectifiable properties, the quantum mechanical objects can be regarded only as fictional objects in the classical sense (in the Kantian sense). Thus we cannot regard the physical systems as carriers of all properties.[17]

5.2 *Causality*

The principle of causality was given by Aristotle in the form of four aspects of cause. They were *causa materialis* (material cause), *causa efficiens* (efficient cause), *causa formalis* (formal cause), *and causa finalis* (final cause or purpose). But in modern science, only *causa efficiens* and *causa materialis* are accepted. Causal law relates to this particular *causa efficiens*. It implies the principle that everything must have a cause from which it follows as an effect. Earlier Laplace took the assumption that these causal laws were the laws of Hamiltonian mechanics. From this he declared the fiction of a demon that (i) if it is sufficiently informed to know about the present state of the world completely, and (ii) if its capacity to make computations is sufficiently great, then the demon could know all the past and future world events. This implied that there is a unity of mathematics and nature. We can often find this conviction in many earlier scientists like Kepler, Galilei, and Descartes. But this conviction (validity of the causal law) was first doubted by Hume. He denies the necessity of the causal law. So Kant arrived at the conclusion that the principle of causal law must be a synthetic judgment. Further Kant said that there are also synthetic judgments that must be thought of together with their necessity. These are therefore valid a priori. It then implies that the causal law must be valid a priori, if there are actual experiences of things in space and time.

But the causal law in the form of Hamiltonian equations becomes invalid in quantum mechanics. This is because of the fact momentum and position (p and x) are incommensurable properties. This is unlike classical physics. The measurement of the momentum will completely annihilate the previous measurement of position. Furthermore it is quite independent of the time interval between the two measurements (momentum and position). Thus it is not at all true to say that the position was some previously measured x_0 after the momentum measurement. In quantum physics such conclusions are erroneous. Thus in contrast to physical objects, the physical objects in quantum mechanical sense (when we consider all properties) cannot be individualized. This in spite of the fact that quantum mechanical objects (when we consider only the objective properties) can still be individualized in some restricted sense. The time-dependent changes of such objects are determined not by causal laws but by probabilities. In this way these objects are no longer identifiable during a

[17]ibid., [12], pp. 129–131.

change in their properties. Therefore they can no longer be objectified as physical objects. Essential features that characterize a “thing” in classical sense are absent. They are not the objects of experience. Therefore the a priori validity of causal law asserted by Kant cannot be applied.[18]

6 Logic

Effective logic is a theory that is a priori valid for all propositions of unrestricted availability. All propositions of classical physics belong to this type of propositions. In quantum mechanical systems, also all propositions about the occurrence of objective properties are also the same type of propositions. But modifications in logic are necessitated if we want to include the arbitrary measurable and therefore the incommensurable properties. Some statements of logic, even though continue to be true, lose their applicability. But still if we insist on applying the laws of logic, then we lose the knowledge of quantum theory. Thus we are forced to consider them as fictional objects (e.g., atoms are not really classical objects, but still most of the time, we treat them as if they were).This means some of the laws of logic become false. The quantum mechanical objects are not available unrestrictedly. This is even due to the physical conditions alone under which they can be proved. Thus essential restrictions become apparent in the propositions of effective logic. For example, the logical statement $A \rightarrow (B \rightarrow C)$ becomes invalid in quantum logic. Further questions relating to the concept of matter and causality are closely connected with the validity of logic in quantum physics. These questions relate to the concerns of objectifiability of arbitrary properties. This is often discussed in relation to the double-slit experiment in literature. Probability theoretical contradiction arises when we insist on referring all measurable properties to a system and hence their objectification. The causal law also loses validity when applied to incommensurable properties, which one interprets as objective properties of a quantum mechanical system. Thus objectification of incommensurable properties is not possible. This is in spite of the fact that the law of causality is not lost in its entirety.[19]

7 The Role of Observer in Quantum Mechanics

One of the most important questions that arises due to quantum physics is the extent of bringing a special role or contribution of the observer. It means that QM introduces subjective elements to an extent. The physical system in QM has its most comprehensive description in the form of its state:$|\varphi\rangle$. The expansion of $|\varphi\rangle$ in

[18]ibid., [12], pp. 134–146.
[19]ibid., [12], pp. 156–200.

terms of states $|A_i\rangle$ (and the eigenvalues) means $|\varphi\rangle = \sum_i |A_i\rangle\langle A_i|\varphi\rangle$. This contains the coefficients $\langle A_i|\varphi\rangle$. This is connected with the probabilities associated with the occurrence of A_i. Probabilities are prepositions about the knowledge of the object by the subject. $|\varphi\rangle$ is the description of the state of the system, and it is an objective property of the system as the operator $P_\varphi = |\varphi\rangle\langle\varphi|$ has the value 1. But when we consider the measurement of the non-objective quantity A with values A_i, that one obtains from $|\varphi\rangle$, the probability statements of the form, $w_\varphi(A_i) = |\langle A_i|\varphi\rangle|^2$, results, and these statements concern the knowledge of the observer. They are not statements about knowledge of the object in the state $|\varphi\rangle$. They refer to knowledge of the observer prior to the registration of the fact how the object system will be constituted after a measuring process on the quantity A. These probability statements refer to mixture w_φ, and do not refer to the pure case P_φ. This means that the statistical statement $w_\varphi$is a statement about the observer's knowledge about the object. There is a loss of information during the transition from the pure case to the mixture $P_\varphi \rightarrow w_\varphi$. This is the epistemic cut, and the loss of information caused by the cut is the reason why in quantum mechanics probabilities, or the statement about the observer's knowledge, occur. In this way the subjective element enters quantum mechanics. The difficulties caused by the separation of the object from the measuring device leads to a situation where the statistical element cannot be avoided even in principle.[20]

8 Why Quantum Mechanics Abandons Naïve Realism

Naïve realism is the concept that there exists a world of material objects, and the world of perception is largely perception independent. This also implies that we perceive the world directly as it is. However quantum physics witnesses several scientists like Alain Aspect, Anton Zeilinger, etc. whose work does not validate naïve realism. According to these, quantum mechanics has some nonlocal character in it. Many experiments in quantum physics, like double-slit experiment, EPR experiment, Stern-Gerlach experiments, etc., are all incompatible with naïve realism. The Bohmian interpretation attempted to show that both position and momentum are simultaneously well defined through hidden variables. But John Bell through his theoretical work showed that local realism is in contradiction with the predictions of quantum mechanics. With regard to closing a loophole connected with fair sampling idea of photon, Kohler et al. state that "The recent experiment makes the photon the first physical system for which all three loopholes have been closed, albeit in different experiments...Almost 50 years after the formulation of local realism, its endgame clearly has begun." [14]. Zeilinger says in an interview, "Why do you want

[20]ibid., [12], pp. 88–112.

a hidden reality that exists independent of the observation?" And I don't get satisfactory answers [15].

The standard model is often considered to be containing the most fundamental particles. But at a deeper level, the particles are thought to be the excitations of a field. Some of the particles last for only a very short duration of time. Julian Schwinger tried to develop a phenomenological model called the source theory to explain the phenomenon of creation and annihilation of the particles, especially in high-energy physics, like those involved in strong interaction. In this domain the underlying dynamic and the fundamental theory behind it are not so well understood. In this domain the immutability of the particles, which is the foundation for ordinary physics, disappears.

At the subatomic level where Planck's constant becomes significant, there is a duality observed in behavior of light as particle or as wave. Thus we know today that the wave function concept which is used in quantum formalism is not a physical wave but gives us the probability of finding a particle when an observation is made. Therefore we can't talk about an electron or a photon in itself directly. Bohr considered that the wave function represents our knowledge of the electron and not the electron in itself. We do not know what an electron is. We only know something of its appearance. We are only dealing with our knowledge of things and not things directly. Therefore when our knowledge changes, the object also changes. We can only talk about an observation in which we interpret reality as particle or wave and that determination has dependence on the design of experiments.

Many physicists opine that we can talk about particles in physics because we do observe particle behavior. But sometimes these particles exist for only a very short length of time especially in high-energy physics. The conventional operator field theory is based upon the idea of replacing the particle with three-dimensional space itself. But Kimball, who worked with Julian Schwinger notes, "But please recognize that it is a speculation. It assumes that one is indeed able to describe physical phenomena down to arbitrarily small distance, and, of course, that goes far beyond anything we know at the moment. All we are able to do experimentally as we go to higher and higher energies is to plumb to smaller and smaller distances, but never to zero distance." Kimball Milton observes that the other idea called as the S-matrix theory is also a speculation because it assumes that the particles are basic and cannot be analyzed.

Julian Schwinger therefore ventured to give a phenomenological account of these phenomena in terms of his source theory so that involvements in speculations of the operator field theory could be avoided. He observed that a common characteristic of the particles in high-energy physics is that they must be created. It is only through this act of creation that they must exist and we can define what we mean by a particle. And these are created out of collisions.

Kimball Milton then observes that we can say these as particles rather than random bumps on the excitation curve only because its properties are reproducible. For example, we may vary the experimental parameters like energy, angles, and type of reaction, and yet the particle remains the same, i.e., the same mass, the same spin, and the same charge. This may however last for only 10^{-24} seconds and decay even before it exits the nucleus, and yet it has all the properties that we associate with the concept of a particle. Only in this limited sense we talk about particles as something elementary. But when we understand that even these are being created by nuclear processes that are governed by processes that we do not understand so well and which we have only modeled in terms of quantum concepts, then we can clearly see the limitation of the particulate concepts. In this regard Kimball A. Milton writes, "If the immutable particle has ceased to exist as the fundamental concept in terms of which a situation can be described, what do we replace it with?" [16].

Thus positivistic concepts are insufficient to explain even modern physics and what to say of life which is a cognitive phenomenon. Therefore in summary, we can ask on what rational basis do we think that mere chemical concepts can explain the phenomenon of life when there is no evidence that life could be created from any chemical process? The category of life is that of self-determination and free activity, and therefore life (organism) must be a higher category than mere chemistry. It cannot but be.

9 Vedāntic Concept of Reality

The Vedāntic concept is that the truth is an Organic Whole. By only analyzing the parts, we cannot come to the original whole. The various mechanistic concepts are merely a result of the finite egocentric thoughts of man, but it does not apply to reality. Mechanism is an idealization. Consciousness is not a result of mechanistic processes. The Original Truth is an Organic Whole and is complete in itself. In the Vedāntic aphorism, *Athāto brahma Jijñāsā*, a thoughtful person who has a good association of a self-realized sage, is asked to inquire after the spirit. If we want to know about life, and truth, we should not go in the direction of matter. Rather we must inquire after the spirit. Only material energy cannot be the source of myriad existences. Therefore to help mankind, Vedānta offers two perfect axiomatic truths that can be empirically verified. They are (i) life comes from life and (ii) matter comes from life. Absolute truth is sentient and not impersonal. The impersonal is only one facet of the absolute. The cosmic creation is not directly a transformation of the *Brahman* or the absolute truth. Rather the cosmic creation is a transformation of the potency of the absolute. That is why in Gaudiya Vaiṣhṇavism, Srila Bhakti Vinoda Thakura has explained that Vedāntic philosophy is that of *śakti pariṇāmavāda* and not *brahmapariṇāmavāda*. The absolute truth is a fully independent Original Person. All other sentient entities are dependent on the Original Person or the *ādi puruṣa* for their being. The material life of living entities begins with the misuse of their free will. Therefore the *ādi puruṣa* is of the nature of *Sat* (eternity or existence), *chit* (full knowledge) and *Ānanda* (transcendental bliss). Thus the *ādi puruṣa* is not only the substantial truth of reality but also is the original subject and hence the Original Person. The German philosopher Hegel expressed it beautifully in his expression that "Truth is the whole." Srila B.R Sridhar Dev Goswami Maharaja, a great Vedāntic authority in the line of Sri Chaitanya Mahaprabhu and

Srila Rupa Goswami, accepted two most important concepts given by Hegel. They are (i) die to live and (ii) reality is by itself and for itself. This means it is necessary for a thoughtful being who has a good association that he must die in the entire plane of misconception. The absolute truth is the plane that is above our self. Hence only by *śaraṇāgati* and by service can we get entrance in that plane. *Sevā* or service presupposes *śaraṇāgati* or surrender, and all of these presuppose that we have good saintly association. Without the saintly association, such realization is impossible. This is evident in the realizations of great scientists who are finding that all our endeavors to know the truth become limited after their limit. Thus sometimes great scientists become humble. An example can be found in the statements of Newton, Einstein, and others.

The Vedāntic epistemology has higher epistemological categories than that utilized by science. The five epistemological levels are (1) *Pratyakṣa* (faith in knowledge from direct sense perception), (2) *parokṣa* (faith in knowledge from indirect sense perception), (3) *aparokṣa* (faith in knowledge from the self. It is not a knowledge from the domain of senses), (4) *adhokṣaja* (faith in knowledge from above the plane of self, i.e., the transcendental plane), and (5) *aprākṛta* (life in continuous transcendental nature, beyond material nature). Thus the Vedāntic epistemology explains three higher categories than that understood by modern science. These five levels of epistemology can be interpenetrating, and therefore the transcendental truth can appear before the senses of the living entities by higher grace. The experiences of the great teacher of the fine theistic traditions of Vedānta are recorded in their writings, and in their school, it can be realized as a living truth. Thus the development of scientific truth can lead to the realization of the absolute when the scientists can cooperate with the wisdom of Vedānta and apply the principles in their own practical life.

According to Vedānta, matter came from life. Or matter is a symbol of undeveloped consciousness. All our science is a science of material transformations. But it does not apply to life. We have no proof that chemicals ever combined to give life. But we have proof that life comes from life. As Pasteur's experiment proved that life cannot arise spontaneously in the conditions that prevail upon the earth today, the abiogenesis theory was put forward as a hypothesis that since the primeval conditions in the earth were assuredly different than today, life might have arisen in those conditions. But there is no demonstration of that in any lab experiments in any part of the world.

Hegel clearly distinguished the logic of life from the logic of mechanism and logic of chemism. Hegel did not hold a vitalistic view of life nor did he think that nature was alive in all its parts (https://www.academia.edu/233237/Hegels_Idea_of_Organic_Life). The structure or the logic of life is dependent on life's own inner conceptual necessity. It is not a natural process which is comparable to the chemical processes. Chemical processes cannot give rise to life during the course of evolution of nature. Living beings have their individuality. A living being comes to be and also passes away. But their essence, i.e., their substantial form which is their existence as a determinate genus process or species, exhibits the categories of universal (e.g., animal or human), particular (particular species), and individual (a given instantiation of the species, an individual member, say this person or this animal).

So many species of life seem to coexist, and natural philosophers like Darwin may have conceived that there is a kind of evolution from one species to another. But the problem is that such statement about the history of the organisms is not observed in nature. Even the fossil record and the geological data are contradictory to the concept of gradual evolution, and there is no link that connects the species to some universal ancestral species. Evidence presents the species variety as a web of all life in which all species participate and exist simultaneously, by coexisting and co-depending on each other for their survival and needs. Hegel writes, "Life is essentially organism. In the organism the form is this unity, and at the same time these parts of the form are not parts but members [Glieder], they are ideal."[21]

We cannot call the chemical parts of a chemical unity as members. But in case of life, the limbs are the members of the unity of the living form. The form of the living organism is always instantiated strictly within the species process of nature or organic life as an individual strictly within the boundaries of the species. The chemical parts can be separated and reunited by appropriate chemical processes. For example, we can separate Na^+ and Cl^- ions in NaCl by the process of electrolysis. But if we hold a chemical view of living organism and do the same to them (separate the chemical constituents), then it will lead to death. We will not be able to come back to the organic unity of the living process, which is lost in the reductive process. Hence the organic unity of life is based upon its life process or its genus process, which is governed by its inner conceptual necessity. And the study of this inner necessity must be the objective of the study of science of life.

Therefore Hegel used the terminology member rather than parts for the different limbs and organelles of living organism. These members are produced by the inner conceptual necessity of the living process itself, and therefore the chemical process will never be able to produce even a single blade of grass. Kant had expressed this quite succinctly when he said, "There will never be a Newton for the blade of grass." Therefore even the insignificant blade of grass is teaching the scientists the lesson of humility.

The members of the living organism (say, the root, stem, leaves, fruit, and seed) are mutually co-dependent and co-producing members of the one whole. We will not be able to make the logic any simpler. For example, acorn is necessary to produce an oak tree, and an oak tree is necessary to produce acorn, the seed. In modern cellular biology, this circularity in causation is observed and substantiated. For example, DNA requires RNA and the cellular extra-genomic material for producing the necessary functions and structures. Therefore the information in the cellular process is contained in the whole cell itself, and when DNA is abstracted, there is no more any information. The reductionistic thinking finds its limit in modern cellular biology. The structure of biological causality is circular. The effect is the cause of the cause. This logic process of the living phenomenon necessitates proper ontologically

[21]*Vorl.*16, 139.31.

wholistic thinking in nature. Therefore the parts (members) are mutually related as co-dependent and co-producing moments of the whole. The cause of the integration of the members is therefore to be located in the unity of their common purpose which is coming from the life principle wholistically, and it is not a mechanism or mere chemistry. This proves the Vedāntic Axiom for the Origin of Life, i.e., *life comes from life*. The Organic Whole thinking of Hegel finds a parallel in the Vedāntic thought as well. The whole comes from a whole. Or the complete comes from the complete. This is also the foundation of the Vedāntic thinking about life, universe, and reality. However this unity of the living process is negative or it is a self-related negative unity. Therefore consciousness is negative. It is not the positive like rocks.

10 Conclusions

Naïve realism is being found a limited concept. Quantum physics disproves the theories of local realism. Objects cannot be thought of as existing independently of the observer. This is also the conclusion of Vedānta. The Vedāntic understanding is that matter comes from life. This principle should be experimentally verified, and it will open a new light on the question of origin of life and origin of matter. Life is a higher category and cannot be produced from matter. A practical approach to understanding the Vedāntic concept will be to take the principles in the form of the two empirically verifiable axioms, viz., (1) life comes from life and (2) matter comes from life. This summarizes the entire Vedāntic teachings in a scientific form. Thus Vedāntic truth is in agreement with those scientists who from the development of quantum physics think that there is a subjective element to the objective reality.

References

1. Milton KA. Julian Schwinger (1918–1994), Homer L. Dodge Department of Physics and Astronomy, University of Oklahoma, Norman, OK 73019. Available at https://arxiv.org/abs/physics/0606153v1.
2. Schwinger J. In: Englert B-G, editor. Quantum mechanics: symbolism of atomic measurements. Berlin/Heidelberg/New York/Barcelona/Hong Kong/London/Milan/Paris/Singapore/Tokyo: Springer; 2001. p. 24–5.
3. Shapiro JA. Barbara McClintock, genome self-repair and cell cognition: a revolutionary vision for the future of biology. Huffington post. http://www.huffingtonpost.in/entry/barbara-mcclintock_b_1223618.
4. Shapiro JA. Evolution: a view from the 21st Century. Upper Saddle River: FT Press Science; 2011. p. 123.
5. Bohm D. Wholeness and the implicate order. New York: Routledge Classics; 1980. p. 181.
6. Tilakaratne A. Director, Postgraduate Institute of Pali and Buddhist Studies, 133/19, Nawala Road, Narahenpita, Colombo 05 Sri Lanka. refer to the article: 'Dependent co-originotion: the Buddhist approach to reality', http://www.beyondthenet.net/slabs/articles/Dependent.pdf.

7. Williams P. Mahayana Buddhism: the doctrinal foundations. 2nd ed. New York: Routledge; 2008. p. 68.
8. Sharma BNK. Philosophy of Sri Madhvacharya. Bombay: Bharatiya Vidya Bhavan; 1962. p. 116–30.
9. Williams P. Mahayana buddhism: the doctrinal foundations. 2nd ed. London: Routledge; 1996. p. 68.
10. Sridhar BR. Subjective evolution of consciousness: the play of the sweet absolute, Bangalore: Sri Chaitanya Saraswat Math; 1989. p. 91.
11. Puri BM. Introduction to Hegel's phenomenology. Available at www.gwfhegel.org.
12. Mittelstaedt P. Philosophical problems of modern physics. Boston, MA: D. Reidel publishing company; 1976.
13. Smith NK. Immanuel Kant's critique of pure reason. London: Macmillan; 1929. p. 33–77.
14. Kofler J, Ramelow S, Giustina M, Ursin R, Zeilinger A. Quantum experiment preludes the endgame for local realism – photonic Bell violation closes the fair-sampling loophole. http://www2.mpq.mpg.de/~jkofler/Files/Publications/Kofler,%20Ramelow,%20Giustina,%20Ursin,%20Zeilinger%20(2013)%20-%20Quantum%20experiment%20preludes%20the%20endgame%20for%20local%20realism.pdf.
15. Zeilinger A. Discover interview: Anton Zeilinger dangled from windows, teleported photons, and taught the Dalai Lama, naive realism and awakening. http://www.naiverealismandawakening.com/?page_id=1164. interview by Eric Powell, August 29, 2011.
16. Milton KA. Julian Schwinger: nuclear physics, the radiation laboratory, renormalized QED, source theory, and beyond. http://arXiv.org/abs/physics/0610054v1, Feb, 2008.

Chapter 17
Subjective Evolution of Consciousness in Modern Science and Vedāntic Philosophy: Particulate Concept to Quantum Mechanics in Modern Science and *Śūnyavāda* to *Acintya-Bhedābheda-Tattva* in Vedānta

Bhakti Niskama Shanta

Abstract How the universe came to be what it is now is a key philosophical question. The hypothesis that it came from nothing or śūnya (as proposed by Stephen Hawking, among others) proves to be dissembling, since the quantum vacuum can hardly be considered a void (śūnya). In modern science, it is generally assumed that matter existed before the universe came to be. Modern science hypothesizes that the manifestation of life on earth is nothing but a mere increment in the complexity of matter – and hence is an outcome of evolution of matter (chemical evolution) following the Big Bang. After the manifestation of life, modern science believed that chemical evolution transformed itself into biological evolution, which then had caused the entire biodiversity on our planet. In the framework of materialism, the major attention is to find general organizational laws stimulated by physical sciences, ignoring the uniqueness of life. The main goal of materialism is to reduce consciousness to natural processes, which in turn can be translated into the language of math, physics, and chemistry. Following this approach, scientists have made several attempts to deny the living organism of its veracity as an immortal soul, in favor of genes, molecules, atoms, and so on. However, advancement in various fields of biology has repeatedly given rise to questions against such a denial and has supplied more and more evidence against the completely misleading ideological imposition that living entities are particular states of matter. In the recent past, however, the realization has arisen that cognitive nature of life at all levels has begun presenting significant challenges to the views of materialism in biology and has created a more receptive environment for the soul hypothesis (Shanta BN. Commun Integr Biol 8(5):e1085138, 2015). Therefore, instead of adjudicating different aprioristic claims, the development of an authentic theory of origin of life

B. N. Shanta (✉)
Sri Chaitanya Saraswat Institute, Electronic City, Bengaluru, India
e-mail: bns@scsiscs.org

S. R. Bhatt (ed.), *Quantum Reality and Theory of Śūnya*,
https://doi.org/10.1007/978-981-13-1957-0_17

and universe needs both proper scientific knowledge and the appropriate tools of philosophical analysis of life. This talk will highlight the uniqueness of biological systems that offers a considerable challenge to the mainstream materialism and proposes the Vedāntic philosophical view as a viable alternative for development of a theory worthy of origin of life and universe.

Keywords Consciousness · Reductionism · Organic whole · Determinism · Vedānta · Acintya-bhedābheda-tattva

1 Introduction

Reductionism is the dominant approach practiced among the physicalists and is grounded in the belief that all phenomena are based on material processes that are ultimately reducible to the natural laws. Reductionists claim that theories and laws in different fields of science are nothing but special cases of theories and laws formulated in some other, more basic branch of science, in particular of physical science. Adopting this ideology, most biologists are also reductionists and naturally presume that they can explain a biological system in principle as soon as the system is reduced to its smallest components. That is why they are busy in completing the inventory of the components within a biological system and the functions of each one of them. They are under the impression that after accomplishing that task, it would be very easy to explain everything observed at the higher levels of organization. Such approaches have progressed from gross anatomy to microscopy and from organ physiology to cellular physiology. However, all these attempts only establish their inapplicability when it comes to comprehending biological systems using the principles that are basic to the physical sciences [2]. To understand living organisms, one may dissect the body of a living organism into muscles, organs, bones, nerves, molecules, and atoms. The practitioner of that type of analysis may only gain certain useful new information, but those smallest parts will never provide all the answers that are necessary for understanding life as a whole. Even after acquiring a complete catalog of all the molecules of which it is composed, it is not possible to infer structure and function of the biological systems like the cell, liver, heart, lung, brain, and so on.

Leaving aside biological systems, reductionism is even unable to explain the nature and teleological function of artifacts. For example, to understand the nature and function of an earthen pot, reductionists may apply appropriate natural laws and also determine what kind of soil the pot is made from, and then they can study the structure of that soil under the microscope and carry on downward through chemistry to the basic molecules, atoms, and elementary particles of which the soil is composed. Such an approach cannot contribute anything toward understanding the properties of a pot as a pot. A sentient subject may use the same pot for many different purposes, and thus the purpose of the pot has an external teleological dependence (subject is outside the system) on the sentient subject. Different pots may be made of many different substances like soil, plastic, metals, and so on, and, yet, they can be used for the same function (say, storing water) by the sentient subject. Therefore, a mindless application of reductionism cannot comprehend the external teleological function of the pot, which is dependent on the sentient subject. Similarly, in a sentient living organism, a single chemical structure of a biomolecule can execute many different functions, and also one function can be produced by several different chemical structures [3]. Reductionism can at best hunt for correlations and not causal relationships between a structure and a biological function [4]. In a living cell, molecules like proteins can specifically catalyze a chemical reaction or recognize an antigen not only because their amino acids are arranged in a particular way but also because their three-dimensional structure and function are controlled by the sentient living cell. Without the existence of sentience, as in the case of a dead cell, the same proteins may be present, but they cannot do all those functions that are observed in a sentient cell. The complex functions of the body of a living organism have an internal teleological (subject is inside the system) dependence on the sentient living entity within the body. Being more complex than external teleology, it is impossible for reductionism to grasp the internal teleological functions of different chemical structures present within a sentient living organism.

A recent paper [5] also accepts that, despite a significant progression in reductionism-based cell biology, an elementary rationalization of even the simplest subcellular biological processes is missing. In this article, based on the so-called notion of theories of "active matter," the authors raised hopes on developing the physical principles of subcellular organization to help establish predictive theories of cell biology. However, the term "active matter" is grossly misleading because matter by its nature is "inactive" and it may apparently be observable as "active" only under the influence of external forces or by the subtle influence of sentient or cognitive principles. Consciousness is a force within the body, and only when it is conscious, an organism will stand up and perform its usual activities. The moment consciousness leaves, the body collapses. Therefore, by using a reductionism-based self-organization theory, biologists can never discover the natural laws that govern the actual cellular microscopic behaviors of the molecular constituents or the interactions between cytoskeleton filaments.

Aristotle's four aspects of causes [6] will be a good explanation to demolish the great brick wall that we often come up against the attempt to understand living organism from a non-reductionist viewpoint. Let us consider the "brick wall" example (which is an example for external teleology) in the context of Aristotle's four aspects of causes. If someone asks why a "brick wall" was built, then following a reductionist approach, we can only address the two causes from Aristotle's four aspects of causes: (1) the material cause – that out of which "brick wall" is made – and (2) the efficient cause, the natural laws that are important in the art of "brick wall" construction. However, the simplistic reductionist approach cannot address another two subtle causes: (1) the formal cause – the form or the shape of the "brick wall" (which was in the mind of the architect) – and (2) the final cause, the end or the purpose (external teleology) for which the "brick wall" was built. This is a major limitation of reductionist approach commonly practiced in physical sciences. It is important to note that apart from above four aspects of causes that Aristotle described, Vedānta talks about yet another cause: "the original case" or "cause of all causes." The verse 5.1 in *Sri Brahma Samhita* explains about this original cause:

īśvaraḥ paramaḥ kṛṣṇaḥ sac-cid-ānanda-vigrahaḥ
anādir ādir govindaḥ sarva-kāraṇa-kāraṇam

Translation The personification of spiritual existence, consciousness and ecstasy, Sri Krishna, who is known as Govinda, is the Supreme Lord of all Lords. He has no origin, He is the origin of all and He is the cause of all causes.

2 The Linear Logic of Physical Sciences Is Insufficient to Address Organic Whole

The commonly practiced linear causal explanations in physics and chemistry are insufficient to address the network and circular causality of an organic whole. The immensely complex organic whole does not allow reductionism to unravel all the causal relations of a functional dynamic integrated biological phenomenon [7]. Due to a misunderstanding, reductionists falsely believe that causality is a relationship between two chemicals/objects or between a structure and a function. In reality, causality is a relationship between successive events, and reductionism cannot establish a unique causal relationship between the structure and the function of a biomolecule in an organism. Therefore, a thorough knowledge of basic molecules, atoms, and elementary particles cannot explain anything about origin of life, differentiation during ontogeny, subjective experiences, and so on. An apparent proof for the same is that, despite their big claims, the overenthusiastic reductionists could not succeed in developing a purely materialistic (Cartesian) theory of biology.

The term biology is of Greek origin meaning the study of life. On the other hand, chemistry is the science of matter, which deals with matter and its properties, structure, composition, behavior, reactions, interactions, and the changes it undergoes. The theory of abiogenesis maintains that chemistry made a transition to biology in a primordial soup [8]. To keep the naturalistic "inanimate molecules to human life" evolution ideology intact, scientists must assemble billions of links to bridge the gap between the inanimate chemicals that existed in the primordial soup and anatomically modern humans. Even though the proponents of a natural origin of life express much optimism for providing their theories, presently there is a detailed compilation of information seriously questioning this doctrine [9]. This reductionistic ideology has always failed to answer two simple questions: (1) What is the minimum number of parts that are essential for a living organism to survive? (2) By what mechanism do these parts get assembled together?

Whether it is between genes and tissues, cells and other parts of the organism, and organism and its environment (which includes both living organisms and inanimate objects), a highly intricate and inseparable sentient interaction is the hallmark of biological process at all levels [10]. Due to this specific characteristic of biological systems, we must consider nature, ecosystem, social group, organs of a single organism, and so on, as organic wholes. Reductionists should understand that they have a wrong conviction that the organic wholes are mere mechanical and chemical additive sums of their parts. Unlike, mechanical or chemical systems, the parts in a biological system cannot be separated from the system, without destroying it as a working system. Therefore, they can no longer be called parts but are participants or members of a dynamic organic whole. A complete knowledge of the properties of the participating members can never provide a complete knowledge about the dynamic organic whole. Materialists must realize that, to develop proper explanations of mind and consciousness, biology needs a much more sophisticated philosophical foundation than the rather simplistic conceptual framework of the physical sciences.

3 Determinism Is One of the Stumbling Block of Egocentric Approach in Modern Science

The theories of logical positivists, physicists, and mathematicians are based on natural laws, and therefore, those theories are generally strictly deterministic. Studies in physical sciences adapted this approach throughout the history. French mathematician and physicist Laplace claimed that if we can know at one time the positions and speeds of all the particles in the universe, then we can predict their behavior at any other time, in the past or future. This framework of strict determinism does not allow any independent (of natural laws) subject to intervene and break the natural laws. In other words, everything including the living entities – subjects – is also under the complete dictum of natural laws. This is the position that is adopted by majority of scientific community.

Following Laplace, scientists blindly started presuming that by their research works and models they can predict the future/past. As the scientific research progressed further, in due course, the evidence forced the scientists to accept the fact that their capability to predict the future/past is significantly restricted by the complexity (e.g., chaos theory or butterfly effect) of the equations. In spite of these clear practical difficulties, many scientific studies are still based on the dogma of determinism.

In the nineteenth century, scientists believed that hot body (say a piece of hot iron rod) would lose energy in radio waves, infrared, visible light, ultraviolet, X-rays, and gamma rays, all at the same rate. If that is the case, then it will put all of us at the risk of exposure to the dangerous radiation, and also it will ensure that everything in the universe will maintain the same temperature. However, we all know that it is not the actual situation. German theoretical physicist Max Planck refuted this wrong notion by proposing that the radiation comes out of hot body in quanta (packets) of certain amount. As compared to infrared or visible light, the energy in the quanta is higher for ultraviolet and X-rays; hence, unless the body is extremely hot like the sun, it cannot emit (due to lack of enough energy) ultraviolet and X-rays. Even though, Planck considered the ideas of quanta as a mere mathematical trick (and did not attach it with any physical reality), scientists in due course claimed that the amount of spin (an intrinsic form of angular momentum) carried by elementary particles have a value multiple of a basic unit (discrete or quantized values). In the early twentieth century, another German theoretical physicist Werner Karl Heisenberg proposed that we cannot measure both the position and the speed of a particle exactly, and thereon a few scientists started to realize the implications of quantum behavior on the dominating notion that we find in modern science: "to proceed with a dogmatic presumption that total understanding is within our grasp."

Without light, our eyes are helpless to see things, and using this analogy, we can also understand the limitation that Heisenberg's uncertainty principle highlights. To see where a particle is, we have to shine light on it, and instead of any arbitrary amount, by following Plank's explanation, we have to use certain quantum of light for this purpose. When we impinge certain quanta of light on the particle, it will disturb the particle and thus change its speed in an unpredictable manner. To measure more accurately the position of particle, we have to use light of shorter wavelength like ultraviolet, X-rays, and gamma rays. However, as we have mentioned above, as compared to visible light, the quanta of these forms of light with shorter wavelength have higher energies, and thus they will cause more disturbance to the speed of particle. Thus, we have to sacrifice more and more the accuracy of measurement of the speed of particle when we try to increase the accuracy of the measurement of the position of particle and vice versa. We all know that the knowledge received by sensual experiences always faces problems when the objects are too small (say, an electron) or too big (say, gigantic planets). Heisenberg's uncertainty principle certainly undermines the *Laplacian* scientific determinism, which invokes the necessity of knowing the position and speed of the particle simultaneously.

Moreover, the practitioners of quantum mechanics are also unable to overcome completely the notion of determinism. As we have discussed above, according to quantum mechanics, particles do not have well-defined positions and speeds. Thus, following a deterministic view, the practitioners of quantum mechanics use a so-called Schrödinger equation, where the size of the wave function gives the probability that the particle will be found in that position and the rate, at which the wave function varies from point to point, gives the speed of the particle. The deterministic view in quantum mechanics presumes that the wave function contains all the information of the particle, both its position and its speed. Thus, following deterministic framework, scientists use the "wave function at one time" in Schrödinger equation to calculate the "wave function at other times." Even though quantum mechanics is not exactly the type of determinism that Laplace proposed, it is based on a mechanistic principle and cannot place itself above the notion of determinism (it is still an attempt toward predicting the wave function accurately) that we commonly find in the scientific research.

Interestingly, even though the theory of quantum mechanics is known to the scientist since the last several decades, it is not well conceptualized by many and thus is not applied in many fields of studies in modern science. As a result, many claims in different fields of science are quite contrary to the claims of quantum mechanics. Many scientists (e.g., biologists) have a deep emotional attachment to determinism. They have thus continually claimed that life can be understood by certain precise chemical pathways and intricate mechanical arrangements. However, despite their emotional claims, they have not managed to provide answers to two simple questions: (1) What is the minimum number of parts that are essential for a living organism to survive? (2) By what mechanism do these parts get assembled together? Thus, scientists have not learnt the lesson of their incapability to know things under the deterministic framework. Scientists must understand that reality is not a slave of their preconceived notions and scientific models. The article entitled "Idols of the Mind vs. True Reality" (http://scienceandscientist.org/Darwin/2015/12/27/idols-of-the-mind-vs-true-reality) by Sripad Bhakti Madhava Puri Maharaja, Ph.D., scientifically elaborates this most important point.

Many have the presumed notion, especially in the field of biology that matter is a well understood concept. However, to avoid unnecessary confusions, it is important for biologists to deeply understand the concept of matter before they could claim things like "life is a chance combination of matter." Quantum mechanics is a progress from observer-independent classical physics to an observer-dependent description of reality, where it has shown us that we are not directly dealing with the science of object but we are dealing with the science of knowledge of the object. Hence, physics has realized that matter does not have an independent existence apart from consciousness. This is a direct challenge to the prevailing dogmatic faith on scientific reductionism: every single process in nature can be broken down into its constituent parts and can be described scientifically.

4 Vedāntic View of Life, Its Origin, and Evolution

The ontological view of the organism as a complex machine presumes life as just a chance occurrence, without any inner purpose. This approach in science leaves no room for the subjective aspect of consciousness in its attempt to know the world as the relationships among forces, atoms, and molecules. On the other hand, the Vedāntic view states that the origin of everything material and nonmaterial is sentient and absolute (unconditioned). Thus, sentient life is primitive and reproductive of itself – *omne vivum* ex vivo – life comes from life. This is the scientifically verified law of experience. Life is essentially cognitive and conscious [11]. And, consciousness, which is fundamental, manifests itself in the gradational forms of all sentient and insentient nature. In contrast to the idea of objective evolution of bodies, as envisioned by Darwin and followers, Vedānta advocates the idea of subjective evolution of consciousness as the developing principle of the world.

In Eastern tradition, also we can find many schools of atheistic philosophy, and most famous atheist in Indian philosophy was Carvaka Muni. The philosophy of Carvaka Muni is paralleled by the many Western philosophers, and it clearly reflects in the materialism that is predominantly practiced in modern science. As we have discussed above, the followers of this philosophy believe that consciousness is the by-product of the chemical combinations of different material substances (atoms and molecules), and thus in this concept with the dissolution of this physical body, neither soul nor consciousness remains. Following this line of thinking, many argue that just as the combination of different chemicals produces something more than the individual chemicals themselves, the mechanical and chemical combination/accumulation of different material elements produce consciousness. Thus some of them believe that with the dissolution of this fleshy (gross) body, nothing remains, and Epicurus first advocated this philosophy in the West. In Eastern tradition, the Buddhists believe that when the physical (gross) body is dissolved, the subtle body (the mental system) goes on to take another birth. Buddhists accept transmigration (reincarnation or *metempsychosis*) from one body to the next. According to Buddhists, although the gross body may vanish, the individual living entity must enter another body following the law of karma. Buddhism proposes that if one practices a particular way of living, then he/she can dissolve the subtle body also and thus nothing remains (*śūnya*). Thus, according to Buddhists, there is no eternal soul. According to Vedāntic literature, Lord Buddha taught this philosophy of *śūnyavāda* to stop the animal killing that was practiced at the beginning of present age – *Kali-yuga* by the unqualified persons on the name of sacrifice. The verse 9 of *Śrī Daśāvatāra-stotra* by Srila Jayadeva Goswami explains:

Nindasi yajña-vidheḥ śruti-jātam
Sadaya-hṛdaya darśita paśu-ghātam
keśava dhṛta-buddha-śarīra jaya jagadīśa hare

Translation Oh Keśava! Oh Lord of the universe! Oh Lord Hari, who have assumed the form of Buddha! All glories to You! Oh Buddha of compassionate heart, you decry the slaughtering of poor animals performed according to the rules of Vedic sacrifice.

Śrīpād Ādi Śankarācārya also gave a similar philosophy with a slight difference. Buddhist school claims that there is no eternal individual soul, and Śrīpād Ādi Śankarācārya also said that there is no permanent individual soul. However, the philosophy of Śrīpād Ādi Śankarācārya accepts the existence or conscious substance, Brahman, and for the followers of this philosophy, the Brahman is the ultimate reality. Buddhists believe that ultimately there is nothing (*śūnyavāda*), and the followers of Śrīpād Ādi Śankarācārya believe that ultimately there is an eternal conscious substance and that is Brahman (*Kevala Advaita* or *Māyāvāda*). This is the major difference between Buddhist school and the school of Śrīpād Ādi Śankarācārya. Thus the followers of *Māyāvāda* believe the existence of consciousness; however, for them, the consciousness of separate existence is false. Under this

Māyāvāda philosophy, there is no scope for the existence of individual soul, and any notion of individual soul is a mere reflection of the ultimate conscious substance. Śrīpād Ādi Śankarācārya gave the example of reflection of the moon in mirror, and if we remove the mirror, then there is no reflection. Therefore, under this philosophy, different individual souls are mere reflection of common source – Brahman. This school claims that in reality, all individual souls are one and the same with Brahman. They also believe that by certain practice, one can dissolve the metal system, and thus the consciousness of individuality vanishes in Brahman.

Famous Nobel Prize-winning Austrian physicist Erwin Schrödinger also developed his thoughts on the nature of consciousness and self, specifically from Śrīpād Ādi Śankarācārya's perspective of Vedānta. Quoting Vedānta in line with *Kevala Advaita* or *Māyāvāda*, Schrödinger was mainly trying to explain that consciousness is only one, singular, identifiable with its universal source (*Brahman*), and he believed that the perceived spatial and temporal plurality of consciousness is merely an appearance or illusion (*māyā*). However, it is a common misconception that is found among the monists (Śrīpād Ādi Śankarācārya's *Kevala Advaita* or *Māyāvāda* philosophy) in Indian Vedāntic tradition. The verse 2.12 from *Śrīmad Bhagavad-gītā* [12, 13] completely refutes the idea of singularity of consciousness, where Bhagavān Śrī Kṛṣṇa says to Arjuna: "*na tv evāhaṁ jātu nāśaṁ na tvaṁ neme janādhipāḥ na caiva na bhaviṣyāmaḥ sarve vayam ataḥ param* – Never was there a time when you, I or all these kings did not exist, just as we exist in the present, so have we existed in the past, so shall we continue to exist in the future." Therefore, according to the Vedāntic view, the plurality of individuals is an eternal fact, and it is confirmed in other Vedic sources (Kaṭha Upaniṣad 2.2.13 says: *nityo nityānāṁ cetanaś cetanānām* – We are eternal, we are many, and Supreme Absolute is also eternal, but He is one) and by authentic teachers like Śrīpād Rāmānuja Ācārya and other Vaiṣṇava Ācāryas. NPR also reported in 2010, "there are 10 times more microbial cells on and in our bodies than there are human cells. That means that we're 90 percent microbial and 10 percent human..." [14]. Apart from our own individuality, we must also accept the individualities of all those microbes on and in our bodies. We cannot deny the individuality of all those microbes, by stating that their individuality is mere illusion (*māyā*). In the healthy body of a multicellular organism, every individual cell, despite having its own individuality, is meant to work for the welfare of the whole body. Similarly, Vedānta advocates that we are living in an "Organic Whole," and every individual unit of this whole is meant to dedicate itself for the satisfaction of the center – the *ādi-puruṣa* or primeval personal absolute. In contrast to Darwinism, symbiogenesis proclaims that life did not take over the globe by competition but by cooperation. In the body of an organism, there are different organs like the heart, kidneys, lungs, and so on, which perform different tasks to serve the function of the body as a whole. One organ does not try to become another. In the similar manner, different living entities and also their environment are related to each other like an organic whole. Evidence in symbiotic exchanges confirms that the sphere of life is like a net, with the different species representing the nodes of that net (network). If changes occur in the network as a whole, then the various nodes (species) change accordingly, to maintain the harmony of the network of life. This viewpoint is completely ignored by many modern evolutionists.

Vedāntic literature explains that wherever life is present, the soul (*ātman*) is there within, and following the "laws of karma," the soul (*ātman*) in the human body may obtain bodies of nonhuman species and vice versa. By advancement, the soul (*ātman*) can obtain the human form, and by degradation, it can also go back to other forms of life. The soul (*ātman*) is endowed with freewill, and by misutilizing freewill, a soul (*ātman*) may do many misdeeds. The acquired reactions from those misdeeds are known as karmic reactions. "Laws of karma" check the freewill of the soul (*ātman*) by providing new bodies and throwing into different suffering conditions. This ancient theory of evolution is based on the subjective evolution of consciousness [15], and the Darwinian objective evolution theory of bodies is a perverted representation of this ancient wisdom. In Darwinism, evolution means transformation of bodies, and in Vedāntic view, evolution means transformation of consciousness. Twenty-first century biology also teaches us that we should not inflict our ideas on nature; let nature reveal herself to us. Life and its evolution cannot be understood by imposing simplistic Darwinian mechanistic reductionism on sentient biological systems. Evidence is forcing biologists to go beyond physics and chemistry to properly comprehend the science of consciousness. Vedānta holds that different forms (species) are original archetypes that accommodate different varieties of consciousness through which the transmigration of the soul (*ātman*) takes place on the basis of the evolution of consciousness. The body is a biological illusion of the consciousness of the soul (*ātman*), and from an amoeba to a human being, all the different varieties of forms are representations of different stages of conditioned consciousness. Following an endless cycle of birth and death ("transmigration of the soul" or *Metempsychosis* in Greek), the soul (*ātman*) keeps on wandering in different grades of conditioned states of consciousness (subjective evolution of consciousness) by obtaining a body suitable to that consciousness until it attains the pure consciousness.

Abiogenesis and the theory of evolution explain that the first life came from the accumulation of inert matter and that biodiversity is a result of random mutation and natural selection. Evolutionary theory and the principles in biology are applied directly to behavior, and they avoid psychological or cognitive level analysis. Both abiogenesis and evolution theory are outcomes of mechanistic or reductionistic thinking and that is why they cannot explain how organisms have cognitive features like thinking, feeling, and willing. These concepts also do not explain how matter developed the two fundamental characteristics that life has (*Naturzweck* and *bildende Kraft*). Therefore, both the origin and evolution of life must be rewritten on the basis of sentience.

5 Conclusion

To make any real progress in our understanding of origin and evolution of life, we have to first understand our real self and thus try to overcome the life that we are living in a notion of false self (identifying material body as our true self). In an ordinary consciousness, we only observe the objects (or matter) that we can experience, and thus we do not try to seriously think about "who is the seer," "who is the listener," "who is the knower," and so on. Therefore, in a material analysis, we simply experience matter and ignore the "self" or the "subject": "listener," "knower," and so on, because we do not experience them. In Vedāntic philosophy, the path of self-realization or spiritual life begins when the subject becomes the object of its own study. As we progress to a higher level in our spiritual journey, we can also experience the self and that is known as self-consciousness or self-realization, which is beyond the material sphere. In such a spiritual plane, both object and subject are spiritual, and the subject-object duality is negated by simultaneous identity and difference between the subject and object. In Vaiṣṇava Vedāntic view of Śrī Caitanya Mahāprabhu, it is known as *acintya bheda-abheda-tattva* – simultaneous difference and non-difference. In material sphere sometimes human being is also conscious of consciousness, and yet he/she is also conscious of matter (nonconscious body or bodily consciousness). Under the guidance of an expert self-realized saint, one can transcend this transient plane of matter (bodily consciousness) and can attain the spiritual plane where everything is made of the same conscious principle. It is known as science of the soul in Vedāntic philosophy. In Western philosophy, Aristotle called it pure form without matter (*noesis noesis*). The cultivation of this science of self is missing in modern objective science, and without including the study of self (scientist) in our scientific studies, we cannot achieve a complete scientific understanding of reality. This fundamental flaw in materialistic science is the reason that we do not have any scientific solution to our understanding of life and its origin.

References

1. Shanta BN. Life and consciousness – the Vedantic view. Commun Integr Biol. 2015;8(5): e1085138. https://doi.org/10.1080/19420889.2015.1085138.
2. Mazzocchi F. The limits of reductionism in biology: what alternatives? E-LOGOS: Electron J Philos. 2011;11:1–19; http://nb.vse.cz/kfil/elogos/science/mazzocchi11.pdf.
3. Martin ACR, et al. Protein folds and functions. Structure. 1999;6:875–84. https://doi.org/10.1016/S0969-2126(98)00089-6.
4. Van Regenmortel MHV. Biosensors and the search for structure–activity correlations. J Mol Recogn. 1999;12:277–8. https://doi.org/10.1002/(SICI)1099-1352(199909/10)12:5<277::AID-JMR466>3.0.CO;2-1.
5. Needleman D, Brugues J. Determining physical principles of subcellular organization. Dev Cell. 2014;29:135–8. https://doi.org/10.1016/j.devcel.2014.04.018.
6. Todd RB. The four causes: Aristotle's exposition and the ancients. J Hist Ideas. 1976;37:319–22. https://doi.org/10.2307/2708827.
7. Berger R. Understanding science: why causes are not enough. Philos Sci. 1998;65:306–32; http://www.jstor.org/stable/188263?seq=1#page_scan_tab_contents.
8. Waldrop MM. Goodbye to the warm little pond? Science. 1990;250:1078–80. https://doi.org/10.1126/science.11536475.
9. Kelley D. Is it time to throw out 'primordial soup' theory? NPR, February 7, 2010. http://www.npr.org/templates/story/story.php?storyId=123447937.

10. Shapiro JA. Evolution: A view from the 21st century. Upper Saddle River: FT Press; 2011.
11. Shanta BN, Muni BV. Why biology is beyond physical sciences? Adv Life Sci. 2016;6 (1):13–30. https://doi.org/10.5923/j.als.20160601.03.
12. Sridhar BR. Śrīmad Bhagavad-gītā: the hidden treasure of the sweet Absolute. Kolerganj: Sri Chaitanya Saraswat Math; 2006.
13. Prabhupada ACBS. Bhagavad-gītā as it is. Los Angeles: Bhaktivedanta Book Trust; 1972.
14. Bacterial bonanza: microbes keep us alive. NPR: http://www.npr.org/templates/story/story.php?storyId=129862107. Accessed 07 July 2015.
15. Sridhar BR. Subjective evolution of consciousness – the play of the sweet Absolute. California: Sri Chaitanya Saraswat Math; 1989.

Chapter 18
Is Nāgārjuna's *Śūnyavāda* Complimentary to Quantum Field Theory?

Shreekala Nair

Abstract During the colonial and post-colonial period, philosophers in India were found making desperate attempts to depict Indian philosophy as something at par with European philosophy, especially with the one evolved out from modernity. Barring the merit that such attempts made Indian philosophers view their own tradition critically, it only helped accelerate the fall of genuine philosophizing in this land; the community kept itself busy trying to match up all that they had with ideas that came from the West, which they took to be the standard or marker of true philosophy. Among these attempts, one was to draw a parallel between Indian idealist theories of consciousness and some of the theories in contemporary physics, especially the Quantum theory. Sundar Sarukkai has argued that modern science, particularly Quantum theory, is a discourse that seems unique in many respects and hence to claim that some elements of it are actually what the ancient Indian thinkers were talking about is to mistake the nature of both Indian philosophy and modern science. Even as I agree with him in principle, there remains a strong feeling that if one is willing to conceive theoretical exercises, be it philosophy or science, as belonging to one single clan, there is a bright possibility to find some of our ancient theories as being complementary to many theories in modern science. Buddhist *Śūnyavāda* propounded by Nāgārjuna seems to be a bright example for this: though the theory leaves a general feeling in us that Nāgārjuna is attempting to establish scepticism, at the heart of it, the theory does much more than we explicitly acknowledge. The present paper argues that Nāgārjuna's thesis should be identified as an attempt to provide theoretical explanation to the question why metaphysical realism fails, more than that it provides an alternative to metaphysical realism in conceiving reality.

Keywords Nāgārjuna · Śūnyavāda · Complimentarity · Majorment · Pratītyasamutpāda · Praṇam

---

S. Nair (✉)
Department of Philosophy, Sree Sankaracharya University of Sanskrit, Kalady, Kerala, India

S. R. Bhatt (ed.), *Quantum Reality and Theory of Śūnya*,
https://doi.org/10.1007/978-981-13-1957-0_18

1 Introduction

During the colonial and post-colonial period, Indian philosophers have made desperate attempts to depict the philosophy of this land as something at par with European philosophy, particularly with modern Western philosophy. Many argued that though such attempts would have made them view their own tradition critically, apart from this single merit, this juxtaposition to an alien philosophical tradition offered little; it only helped accelerate the fall of genuine philosophizing in this land. While the philosophical community kept itself busy trying to match up all that they had with ideas that came from the West, which they took to be the standard or marker of true philosophy, original thinking and continuance of debates on core philosophical issues gave way to comparative studies and cross-matchings with Western ideas. Though there is some truth in this argument, there are instances of meritorious handworks of such studies. One among such attempts of juxtapositions, of course a much later one, was the effort to draw a parallel between Indian idealist theories of consciousness and some of the theories in contemporary physics, especially the Quantum theory. Sundar Sarukkai has famously argued that modern science, particularly Quantum theory, is a discourse that seems unique in many respects, and hence to claim that some elements of it are actually what the ancient Indian thinkers were talking about is to mistake the nature of both Indian philosophy and modern science. Even as I agree with him in principle, there remains a strong feeling that if one is willing to conceive the theoretical exercises, be it of philosophy or science, as belonging to one single clan, there is a bright possibility to find some of our ancient theories as being complementary to many theories in modern science.

The present paper argues that if we conceive theoretical exegesis in human evolution as a single narrative, the Buddhist Śūnyavāda should definitely be considered as an important chapter, thus making a continuous story of human conceptual evolution across space and time. Though the theory leaves a general feeling in us that Nāgārjuna is attempting to establish scepticism, at the heart of it, the theory does much more than we explicitly acknowledge. The paper argues that Nāgārjuna's thesis should be identified as an attempt to provide an epistemic explanation to the question why metaphysical realism fails; further, it provides an alternative to metaphysical realism to conceive reality. With regard to the latter, the paper proposes that Nāgārjuna marches far ahead of contemporary theoretical physics and to that extent his theory may be considered relevant in current discussions on the nature of reality.

Over the last 20 years, there were conscious efforts from various quarters within Indian philosophical community to run dialogues between Buddhism and science. The prime merit of these conferences/seminars has been that both the parties involved were open to the knowledge system of the other, both eager to expand their horizons by partaking the insights of the other, yet critical of the methods employed and presuppositions involved in the *other* that is juxtaposed. These meetings of course, in some way, have promoted cross-cultural as well as interdisciplinary dialogue, especially between science and philosophy. While a few among them went astray, most of these attempts were fruitful since they conducted dialogical discourse.

While it would be a fallacy to reduce science to Buddhism, it would be equally unwise to suppose that the philosophical questions addressed by both traditions were mutually exclusive. We need to strike an Aristotelian golden mean, I suppose: one can safely say that the questions explored by modern physics echo the investigations of Buddhism in unexpected ways. A discerning mind can draw parallels between the two since Buddhism has long been asking questions which science now addresses, questions like, "Can separate, indivisible particles be the "building blocks" of the world?" "Do they really exist, or are they just concepts that help us understand reality?" "Are the laws of physics immutable, and do they have an intrinsic existence, like Platonic ideals?" etc. While not exaggerating superficial similarities, a study of both the differences and points of agreement between science and Buddhism may help us to deepen our understanding of the world.

Up through the nineteenth century, classical science argued that objects had intrinsic existence governed by well-determined laws of cause and effect. But quantum mechanics, which was developed at the beginning of the twentieth century, seriously undermined the idea that the basic ingredients of matter have such a definite existence and also raised doubts about whether the world was governed by strict rules of cause and effect. Eventually quantum came to be identified as the basis of modern physics, potent to explain the nature and behaviour of matter and energy on the atomic and subatomic levels. The theory had a kick start with Max Plank, who proposed that energy was made of individual units or quanta. In 1924 Louis de Broglie came up with a novel assumption that there is no fundamental difference in the makeup and behaviour of energy and matter on the atomic and subatomic levels, either may behave as if made of either particles or waves, a theory later became popular as wave – particle duality [1]. Eventually Heisenberg proved that simultaneous measurement of two complementary values such as the position and momentum of a subatomic particle would be impossible; the more precisely one value is measured, the more flawed will be the measurement of the other value, and the theory became known as the uncertainty principle.

Resolving this paradox, Neils Bohr introduced the concept of complementarity in quantum physics, according to which the wave form and the particle form are not two separate forms that contradict and exclude each other but are mutually complementary forms and that only together they can provide a complete description of physical manifestation [2]. By complementarity Bohr meant that in the quantum world, it is impossible to speak about independent quantum objects because they are in an interactive relationship with each other as well as with the instrument of measurement, in case of measurements. It has been confirmed that measurement procedures while attempting to establish electrons/photons as particles, for instance, destroy the interference of the same being conceived as waves and vice versa. This can be translated to a principle called *superposition* that claims that while we do not know what the state of any object is, it is actually in all possible states simultaneously as long as we don't look to check. This led him to propose that physical reality is fundamentally interactive in nature and complementary quantum objects. Complementarity, however, does not usually imply that classical logic is at fault, rather that the composition of physical properties at the base level does not obey the rules of classical propositional logic.

In classical physics, we speak of a world of things that exists somewhere in the outside world, and we describe their nature. Quantum physics reminds us that physical sciences are not sciences of nature but sciences of statements about nature. Neils Bohr explicates this better by saying, "There is no world of quantum, there is only a quantum mechanical description." Today it is largely accepted that physics deals with the conditions of physical reality and creates models with which it examines reality. While the models may not provide an isomorphic representation of reality, they are not purely adventitious either and normally are not expected to be deceptive. It is envisaged that these models represent the concept/object of the phenomena, and there is some approximation of a structural similarity between a good physical model and the corresponding physical reality. The central core of quantum physics consists of this novel concept of reality that no longer perceives singular independent elements as the fundamental unit of reality but rather concepts. Objects in relation, viz., proton and electron, quark and anti – quark, wave and measuring instrument, two photons, elementary particle and field of force, law of nature and matter etc. They are neither identical nor have individual identities.

A seemingly parallel stream of thinking to that of Quantum theory can be detected in Buddhism as well. The quantum physicists have pointed out that the act of measurement causes the collapse of either the wave or the particle function, depending upon the system of measurement used in the experiment; only upon the measurement does the potential becomes actual. Quite like quantum physics, Mādhyamaka metaphysics too maintains that we cannot speak of the reality of a discrete entity outside the context of its range of inter relation with its environment and other phenomena including language, concepts and other conventions. Taking clues from the commonality in perceiving reality, a few philosophers have argued that both quantum physics and Buddhist metaphysics are attempts to unravel the mysterious illusionary experiences that are generally taken as truth, with this difference that quantum physics is only halfway home while Buddhism seems to have made it all the way. Whatever be the case, the fact remains that both quantum physics and Buddhist metaphysics view that conceiving mind-independent reality is something spurious.

More than basic Buddhist metaphysics, it is Nāgārjuna's *Śūnyavāda*, which receives attention of contemporary scientists, a theory which popularly proclaims that what arises dependently is substanceless. Nāgārjuna's philosophy consists principally of two aspects: the first aspect is an exposition of a concept of reality, according to which, fundamental reality has no firm core and does not consist of independent, substantial components but is of a two body system, say of material and mental bodies, which reciprocally affect each other. The second aspect is an answer to the inner contradictions of four extreme concepts of reality. In *MūlamādhyamikaKārikā*, Nāgārjuna points out that thing does not arise and does not exist in the true sense of the term. That which has arisen dependently on this or that has not arisen substantially (*svabhāvataḥ*), and what has not arisen substantially, how can it literally be called arisen? Similarly, that which originates due to a cause and does not abide without certain conditions but disappears when the conditions are absent, how can it be understood as to exist [3]? Clearly subscribing to the basic Buddhist metaphysis, Nāgārjuna argues that *to exist* implies grasping after eternity, to *not to exist* implies philosophical annihilation, and all that we experience at the phenomenal level lies in between the two, between existence and nonexistence. Invoking the *pratītyasamutpāda* theory, Nāgārjuna unravels the mystery of the interdependence of the phenomenal existence: thus all *bhāvas* are *svabhāvaśūnya*; everything existent is substanceless. Vātsyāyana criticizing Nāgārjuna points out that if everything is devoid of *svabhāva* or essence his own theory is no exception to this, and therefore, the thesis gets cancelled at once. Nāgārjuna in his V*igrahavyāvartinī* rebuts this charge. A clear depiction of Nāgārjuna's counter argument to this charge is found in Matilal's *perception*: "…it is quite possible that every thesis lacks essence or *svabhāva* and this will remain so even if there is nobody (not even Nāgārjuna,) who asserts it as a thesis. To put it in another way, this thesis will get falsified only if anyone asserted it. We can imagine a possible world when all assertions made are empty, but there is nobody to make the crucial assertion that all assertions are empty" [4]. In brief, Nāgārjuna views that though everything lacks *svabh¡va*, this must remain unsaid, for to assert to it is to falsify it.

The realist schools in Indian Philosophy establish their ontology with three key propositions: First, that truth involves a correspondence between the belief and the state of affairs in the external world, second, reality is mind independent and third, there is one theory that perfectly fit the reality. Nāgārjuna rejects all these propositions: he points out that the journey from belief based on *pramā¸* as to the assertion of factual entities, invoking the intuition of correspondence fails, because while constructing the *pramā* theory itself, we carry over the assumptions about reality. A deeper look at the theory of correspondence between the subjective beliefs based on *pramāṇas* and the state of affairs in the eternal world would make it clear that a reflective equilibrium is largely being used. We start with a set of basic beliefs and look for the causal condition that produced such beliefs in us. And then there would be a series of adjustments worked out between the pramāṇas and our beliefs, and the process continues until we arrive at a perfect equilibrium, which we mistake to be a one-to-one correspondence. This way any belief could be justified irrespective of the objective evidential support [5]. In brief, what we arrive at as the objective account of reality is only one possible account of multiple alternatives before us. Since we cannot know which of these schemes, the one we have arrived at or one of the alternatives we might have arrived at under different assumptions, is correct, we do not in fact possess knowledge. In other words, as it is possible to tie up our beliefs with reality employing multiple schemes, the notion of the one right fit between beliefs and world is empty. This seems to echo the antirealist reflections that it is quite not possible to give a conceptual understanding of the way the world is independently of our knowing.

It is also interesting to note that the Quantum theory proposing a dynamic flux underlying the fundamental unity in quantum field theory is quite akin to the Buddhist *Kṣaṇikavāda.* Einstein, for instance, demonstrated the spontaneous and random movements of atoms, called Brownian motion, a theory which proposes that there is nothing constant static state in this universe. All the particles in this universe are in a dynamic motion with relation to each other, and their tendency to execute the random movements is restricted due to the strong gravitational and electromagnetic forces of each other.

Quite like Quantum Mechanics, Nāgārjuna too argues that the ways of knowing (pramāṇas) and the objects of knowledge (*prameyas*) are mutually dependant. Pramāṇas can be established only with the help of certain assumptions about the world and vice versa, which leaves the message that the phenomenal world is closely tied up with the individual measuring apparatus. Nāgārjuna heavily comes down on Naiyāyikas for employing correspondence to establish metaphysical realism, and his resentment is well expressed by Mark Siderits: "...to say of a set of beliefs that they correspond to reality is to pay them an empty compliment. We would do better to say that they help make things go smoothly for us, that they cohere with other beliefs we hold and with our needs, interests and institutions" [6]. Realists may object here saying that even if we grant that proposition *p* belongs to the agent S and not to the world, truth outruns justifications. It is our experience that often we take propositions to be justified and subsequently find them untrue. But note that here we smuggle in an omniscient agency, perhaps God, who is in possession of all truth. In short, Nāgārjuna is found to have arrived at the same terminal point as his modern counterparts that we need to necessarily revise our popular conception of reality, the metaphysical realists' notion of a mind-independent reality, for a reality that can be mirrored in our cognition is a myth. With this Nāgārjuna succeeds in impressing us that our empirical world is nothing but a conceptual fabrication (*prapañca*). In V*igrahavyāvartinī,* he persuades us to recognize that our *pramā*¸ doctrine is nothing but a convenient myth making, the inherent value of which lies only in making day-to-day life work smoothly and render intersubjective communication. A thing is not independent of its conditions; at the same time, it is not identical with them either. A seer is not the same as the seen object, but an object without a seer does not exist, nor a seer without an object [7]. As an antirealist, like Dummett, he argued that we cannot give content to the metaphysical realist's notion of mind-independent reality. Mark Siderits has popularly argued that it is unfair to depict Nāgārjuna as a sceptic; he may best be described as an antirealist and that Nāgārjuna's ancient version of antirealism in Indian metaphysical scenario tacitly closes a common route to metaphysical realism via theory of *pramāṇas.*

The question that confront us is how from the point of view of physics do we reconcile our common sense notions of everyday world of objects and their properties on the one hand and the bizarre world of Quantum mechanics on the other? The Dalai Lama in one of his speeches has observed that scientists often diligently behave as realists inside their laboratories and once outside the laboratory most would say that nothing really exists without the apparatus defining it. Nāgārjuna in fact moves further ahead of the quantum theorists, for he resolves the paradox surfaced by quantum physics, namely, the quantum measurement paradox: since quantum mechanics predicts only possibilities and associated probabilities, what then manifests the unique world of actuality is left unsaid. Nāgārjuna resolves this paradox by bringing in consciousness to the centre stage. To resolve this paradox, Nāgārjuna in his metaphysical position invokes two truth systems – the conventional and the ultimate – relating, respectively, to the everyday world of experience and to things and events in their ultimate model of being on the level of emptiness. On the conventional level, we can speak of a pluralistic world of things and events with distinct identities and causation. This is the realm, where we can also expect the laws of cause and effect and the laws of logic such as the principles of identity, contradiction and law of excluded middle to operate without violation. This world of empirical experience is not an illusion, nor is it unreal; it is real in that we experience it. However, from the perspective of the ultimate truth, things and events do not possess discrete independent realities. Their ultimate ontological status is empty in that nothing possesses any kind of essence or intrinsic being [8].

Let's note here that the fundamental definition of consciousness according to Buddhism is "clarity of cognition": it is difficult to equate it with nothingness, better it could be explained as a field of potential experience with perception as its core function. This definition of consciousness carries an inner tension: the fundamental nature of *awareness–consciousness* is singular, undivided flow of consciousness, while its function, namely, cognition is a process that involves duality. To resolve this, it is claimed that the non-dual awareness–consciousness or *jñana* spontaneously divides itself into dualistic appearances – *vijñana*, the *vi*prefix here added to represent the divide in the unitary consciousness. This paradoxical nature unfolds the riddle of existence; within the quantum consciousness lies the understanding of the process of the experiential, the dualistic, the seeming reality, which in fact is nothing but an endless perceptual process giving rise to various manifestations.

This paradoxical nature of reality revealed in both Buddhist philosophy of emptiness and modern physics represents the limits of human knowledge. The essence of the problem is epistemological in nature: How do we conceptualize and understand reality coherently? Here we may draw the insights of Nāgārjuna articulated in *Vigrahavyāvartinī*, which is in fact a response to Naiyayika's *Prāmāṇyavāda*. In an elaborate argument, he points out that Nyāya attempts to establish the validity of *pramāṇas* is circular in nature; nevertheless, Nāgārjuna points out that, at the conventional level, knowledge is required, and to acquire true knowledge, we may resort to coherence rather than correspondence for our empirical world is nothing but a conceptual fabrication (*Prapañca*), and to speak about the truth of it, we need to resort to the coherence of our ideas or concepts [9]. Note that majority of contemporary Western philosophers too suggest a replacement of coherence in the place of correspondence.

Buddhism outruns science also by pointing out that it is not enough to discover methodically that our own consciousness is intimately bound up with the whole of reality. Our minds must assimilate the implications of this discovery, and our lives must change accordingly. Practicing Buddhists know that when they perceive their own interdependence with the world, they are filled with an irresistible compassion toward every living being – a compassion that radically transforms their existence. As far as the Buddhists are concerned, the method of discovery and transformation are generally gradual processes. It begins with absorption and study, proceeds with intellectual analysis, culminating in a new way of looking at things and of behaving. In this context, to *meditate* means becoming *familiar* with this new perception of the world. Comprehension leads to meditation, which is then expressed in actions. We, thus, pass directly from knowledge to inner accomplishment and finally to active ethics. This part, a journey from knowledge to the transformation of the self, something quite alien to the science, appears to be the need of the hour, and it is envisaged that a dialogue between Buddhism and contemporary science would enable the latter to revise its vision and mission, incorporating such ideals in its projects.

References

1. Nadeau R, Kafatos M. The non-local universe: the new physics and matters of the mind. Oxford: Oxford University Press; 1999.
2. Goswami A. God is not dead: what quantum physics tells us about our origins and how we should live. Mumbai: Jaico Publishing House; 2009.
3. Wolnsky S. Quantum consciousness: the guide to experiencing quantum psychology. Connecticut: Bramble Books; 1993.
4. Bohr N. Atomic theory and the description of nature. Cambridge: Cambridge University Press; 1961.
5. Nair SM. "Is Nāgārjuna an anti – realist or a skeptic?". Brahmavidya: Adyar Libr Bull. 2014;Vol. 78–79.
6. Matilal BK. Perception: an essay on classical indian theories of knowledge. Oxford: Clarendon Press; 1986.
7. Siderites M. Nāgārjuna as anti – realist. In: Perrett RW, editor. Indian philosophy: a collection of readings. New York & London: Garland Publishing, INC; 2000.
8. Einstein A, Podolsky B, Rosen N. Can quantum mechanical description of physical reality be considered complete? Phys Rev Lett. 1935;47:777–80.
9. Goswami A. The idealist interpretation of quantum mechanics. Phys Essays. 1989;2:385–400.

Chapter 19
Quantum Reality and Concept of Śūnya: Need for an Integrative Approach

Surendra Singh Pokharna

Abstract Quantum physics used for microscopic particles of matter like electrons and light shows that they behave both like wave and particle, the two contradictory aspects in classical physics. Also quantum field theory used to explain properties of elementary particles necessitated the introduction of vacuum states with no particle but huge energy in flat space-time domain. But the curved space-time indicates a different scenario. Here one observes a vacuum state in one reference frame but shows a state with particles, when looked from a different reference frame. Quantum field theory to explain superfluidity and superconductivity shows a need to introduce a vacuum state with an order and elementary excitations. David Bohm's concept of explicate and implicate orders in quantum physics to explain hidden variables brought a new dimension to look at the reality. This implicate order looks like another type of vacuum state. Prof. Adrain Klein from Israel has developed a model for sub-quantum physics (domain below the Planck level) and found that this domain does not have any matter; it just has information. This can be again treated as some kind of vacuum state. We find similar concepts in Jainism involving their theory of karma and their concept of two parts of the universe. One finds a similar concept known as "Form and Emptiness" by Nagarjuna of Buddhist tradition. Finally it is suggested that Jain theory of anekāntavāda (multiple truths) should be taken seriously by scientists. Actually since the process of measurement in quantum physics results into wave-function reduction in which one finds only one eigen state out of many, the process itself takes away information about other eigen states; hence it can be treated as vacuum state of some type. Penrose believes that human decision-making is neither deterministic nor random and nor is it computable; hence a Śūnya appears in our formalism at a very fundamental level, when we talk of any decision towards understanding the reality.

S. S. Pokharna (✉)
B-71, Pruthvi Tower, Jodhpur Char Rasta, Ahmedabad, Gujarat, India

Bhagawan Mahavir International Research Centre (BMIRC), Jain Vishwa Bharti, Ladnun, Nagaur, Rajasthan, India

Science and Spirituality Research Institute, Ahmedabad, Gujarat, India

S. R. Bhatt (ed.), *Quantum Reality and Theory of Śūnya*,
https://doi.org/10.1007/978-981-13-1957-0_19

Even if we look at the classical physics, we find that the approach using the conservation laws defined for a closed isolated system implies ignoring interaction at a very basic level. This can be treated as another type of Śūnyatā. While treating the biological, social and human systems as open systems implies one more type of Śūnya, here entropy/order and information are crucial for description of these systems, whereas other parameters like energy are secondary. Even in the process of measurement, one only looks at a few parameters of a system while ignoring other attributes which are treated as Śūnya. When one looks at the concept of logic in this whole scenario, one finds that Gödel's incompleteness theorems imply our inability to understand even the simplest property of mathematical numbers.

Keywords Microscopic particles · Superfluidity · Superconductivity · Vacuum state · Explicate and implicate orders · Sub-quantum physics · Theory of karma · Anekāntavāda

1 Introduction

We present an integrated review paper of Śūnya and Śūnyatā in the most generalized way. We try to look at these two concepts from widely different angles covering quantum physics, classical physics, state of the present environment, Jain philosophy and some recent advancements in quantum physics. The first type of Śūnyatā appears in the general theory of relativity itself when one finds that there can be a solution of the Einstein equation when there is a gravitational field even in the absence of matter, when one expects nothing. Quantum physics used for microscopic particles of matter like electrons and light shows that they behave both like wave and particle, the two contradictory aspects in view of classical physics. One is Śūnya with respect to the other. Also quantum field theory used to explain properties of elementary particles and necessitated the introduction of vacuum states with no particle but huge energy in flat space-time domain. But in the curved space-time, one observes a vacuum state in one reference frame but shows a state with particles, when looked from a different reference frame. Quantum field theory as used to explain superfluidity and superconductivity also shows a need to introduce a vacuum state, but with an order, but without elementary excitations. David Bohm's concept of explicate and implicate orders in quantum physics to explain hidden variables brought a new way to treat the implicate order as another type of vacuum state. Actually since the process of measurement in quantum physics results into wave-function reduction in which one finds only one eigenstate out of many, the process itself takes away information about other eigenstates; hence it can be treated as vacuum state of some type. Penrose believes that human decision-making is neither deterministic nor random and nor it is computable; hence a Śūnya appears in our formalism at a very fundamental level, when we talk of any decision towards understanding the reality.

Even if we look at the classical physics, we find that the approach using the conservation laws defined for closed isolated system implies ignoring interaction at a very basic level. This can be treated as another type of Śūnyatā. Treating the biological, social and human systems as open systems implies one more type of Śūnya because here entropy/order and information are crucial for description of these systems, whereas other parameters like energy are secondary. Even in the process of measurement, one only looks at a few parameters of a system while ignoring other attributes which are treated as Śūnya. When one looks at a concept of logic in this whole scenario, then one finds that Gödel's incompleteness theorems imply our inability to understand even the simplest property of mathematical numbers.

Similar ideas about the limitations of our knowledge are described through syādvāda and anekāntavāda from Jainism. Jainism also talks of an interesting model of soul and matter which is very similar to the model of quantum field theory, implying relative Śūnyatā of either the material world or the spiritual world.

One finds a similar concept known as "Form and Emptiness" by Ācārya Nāgarjuna of Buddhist tradition. An attempt has been made to illustrate that form and emptiness can be related through time, and both can be treated as some kind of parts of oscillations. Some more examples are given of Śūnyatā from widely different fields including parapsychology and others.

If we look at the present situation of the society and the environment, we find that the development of science, technology and their economic exploitations has resulted into a very grim situation of the world such that now only few years are left before the beginning of the irreversible changes in the climatological processes, huge damage to the biological species and decrease in the sperm counts in the males. All these can be traced to ignorance of the scientists about future exploitation of their experiments through technology development and economic advancement by economists, technologists, politicians and other, whose consequences on the environment are of little concern (Śūnya) to all the later parties involved in the game.

Lastly two recent developments are mentioned where again one finds some type of Śūnya and Śūnyatā. Prof. Adrain Klein [1] has developed a model for sub-quantum physics (domain below the Planck level) and found that this domain does not have any matter but has information fluxes. This can be again treated as some kind of vacuum state. Neppe and Close [2] have developed a new formalism for theory of everything (TOE) using calculus of distinction in which a new mathematical formalism is being explored. They claim that their model can explain everything in the world, but the formalism is still finite in the domain of otherwise an infinite reality.

The concept of Śūnya has been described from widely different perspectives in Indian philosophy in disciplines like metaphysics, cosmology, physics, mathematics, religion, yoga, etc. Bhatt SR [3]. Due to its vast importance in various disciplines, it is high time that this concept is explored in the modern perspective in general and modern science and its impact on the society in particular. Such an analysis has been attempted in this paper in a somewhat exploratory way. Such an analysis was possible because of several ideas that came into my mind while preparing for this conference and also while listening to other participants. Although the main focus on quantum physics has been retained, but looking at the seriousness of the problems of environment and the society, an attempt has been also made to look at the concept of Śūnya in widely different fields. This will probably ensure justification of the title of this paper, but the author asks for an apology right in the beginning for taking the readers to a very broad canvas of ideas, which has made this paper a review paper.

Quantum physics has brought a big revolution in our understanding of the microscopic world in general and reality in particular. This was all due to the dual concept of wave and particle nature for microscopic particles like an electron or a photon whose outcome is dependent totally on the process of measurement. When these particles show wave aspect, then particle aspect is Śūnya and vice versa. Some kind of the uncertainty also comes into the picture due to process of measurement of state of a particle which is known as a superposition state. This superposition state of a quantum system can essentially have many eigenstates, and which state will be found in a given experiment depends on some probability and is never certain. In addition, in the process of measurement, we find the system in only one eigenstate with certain probability, whereas the other eigenstates are destroyed in this process of measurement. Hence these other states become Śūnya. If we do not attempt measurements, then we will never know the actual state of the system. Hence it is a state of Śūnya (zero) from one perspective if we do not do any experiment. There is another source of uncertainty, which comes because of the Heisenberg's uncertainty principle. This principle states that unlike classical physics, one cannot simultaneously determine the position and linear momentum of a particle perfectly which are essential to know its complete trajectory in the future (or in the past). If one determines one parameter perfectly, then the other becomes completely uncertain (hence Śūnya-like situation).

However, the concept of Śūnya appears in quantum physics in many more ways. The concept of vacuum is there when quantum field theory (QFT) is applied to understand the particle physics. Here some scalar, vector or tensor fields are assumed, and their quantized solutions are derived under various situations say in a flat space-time domain or the curved space-time domain [4]. In both cases, one gets some vacuum state over which some elementary excitations are observed. Under some conditions, these elementary excitations are then treated as fundamental particles of matter. It has been found that in flat space-time domain, there is only one type of vacuum state irrespective of the frame of reference used to look at it. However, in the case of curved space-time geometry, a vacuum state observed from one frame of reference is a vacuum without any particle. But if this vacuum state is looked from another frame of reference, then it is no longer a vacuum state, and it actually contains some particles. In this sense, it is clear that the concept of vacuum and hence Śūnya is relative and not absolute.

The concept of vacuum comes in the quantum field theory of superfluid helium and superconductivity also (Pokharna [5], Khaltnikova [6] and Andreas [7]). It is found that when the temperature goes down to an absolute zero, both superfluid helium and superconductors start exhibiting unusual properties. These are explained through the quantum field theory such that there is a perfectly ordered state at absolute zero (leaving aside the zero-point fluctuations). As the temperature goes up, this order decreases, and the disorder increases. The state of the superfluid helium or superconductors at this absolute-zero temperature can be also interpreted as something similar to Śūnya or a vacuum state because it is without elementary excitation but with a highest state of orderliness.

The concept of Śūnya also comes into quantum physics through what is known as implicate order, a concept introduced by Prof. David Bohm [8], who believed in a state of reality beyond the visible reality at quantum level. This implicate order is very different from what he defines an "explicate order" which explains the ordinary reality.

The concept of vacuum (Śūnya) appears in the sub-quantum physics [1] also. When one goes below Planck's length that is 10^{-35} cm, then only non-physical information fluxes are there without any matter, and there is a possibility that information can travel both in plus time and minus time direction. In another model developed by Neppe and Close [2] which is nine dimensional in nature, an attempt has been made to develop a theory of everything (TOE) by treating quantization property of nature and developing a system using a new mathematics known as calculus of distinction. The author states that although it is the best model of TOE, it is still finite in the ocean of infinite reality. So Śūnya appears here also.

As per Prof. Penrose [9] human decision-making is neither deterministic nor random but is even non-computable. It all leads to confirmation about the existence of Śūnya at a very fundamental level regarding our total process of decision-making.

Leaving aside these examples from quantum physics, attempt has been made to look for the concept of Śūnya in the basic definitions of scientific methodology used to understand the reality. It is appears that the current problems of the environment and the society like greenhouse effects, large disappearance of biological species and living beings, too much focus on economic development, increase in focus on automation and decrease in job opportunities, decrease in interaction among human beings and increase in dependence on the machines are all related to the basic methodology of science, that is, studying only a few aspects of reality using certain boundary conditions (ignoring other aspects and treating them as Śūnya), definition of conservations laws defined ideally for isolated closed systems (by treating the interaction as Śūnya) and others.

As an alternative, the method of treating biological, social and human systems as open systems that leads to description using information and entropy/order using the general systems theory is better in some respect but treats other parameters like energy as secondary. These ideas take us towards spiritual development (which has certain goal) which is very important but is treated as Śūnya by those who are thinking totally of economic development (which does not have any goal). Similarly the phenomena of extrasensory perception like telepathy and clairvoyance are like Śūnya for the current science because they do not follow the rigid conditions of space-time invariance, but there are strong evidences to show that they do exist and follow some other set of rules, for "hard core science".

To this, one can also add the Gödel's incompleteness theorems, which describe limitations of any methodology including the scientific procedures or any language in describing the truth. Actually in science, we have either unknown or known part of reality, but there can be a reality beyond this which is unknowable. This is also something like Śūnya.

In this context, a theory is proposed to explain the contradictory properties of a pure soul which looks like Śūnya from one perspective and infinite from the other. Thus an entity is being introduced for which causality is neither obeyed nor violated. It is illustrated that this entity can have two contradictory properties like Śūnya and anant together.

A concept similar to Śūnya also comes in Jainism in their theory of soul and matter. It is mentioned that a pure soul completely free from material attachment has infinite knowledge and other properties [10, 11]. Thus a state without any matter is defined, which is Śūnya from one perspective but not from the other. Detail of this theory popularly known as karma siddhānta is described which is similar to quantum field theory. Another example of Śūnya is a term used for a part of space (alokakasha) which is beyond the observable space in which we live (known as lokakasha) defined in Jainism [10, 12]. The former does not have any constituent of the latter like matter, time, living being, medium of rest, medium of motion, etc. Finally it is suggested that the theory of non-absolutism found in Jainism in general and their concept of anekāntavāda and syādvāda in particular should be taken seriously by scientists.

In the end, we suggest that the concept of form and emptiness of Ācārya Nāgarjuna from Buddhism can be linked together through time dimension, and the two do constitute something like oscillations in time domain, and the knowledge of the two can provide better methods of living the life. They also relatively explain Śūnya-like concept. Finally several examples are given to illustrate the concept of form and emptiness from real life.

2 Śūnyatā in Different Branches of Quantum Physics

In this section, we describe Śūnya and Śūnyatā within the domain of the quantum physics and related areas.

2.1 Wave Function of Ignorance and Śūnyatā: The Process of Measurement in Quantum Physics Takes Away the Information from Us

Quantum physics provides a new way of looking at reality. It is quite different from what classical physics describes. In quantum physics, any microscopic particle like an electron or a photon obeys quantum physics, that is, these particles can show either wave aspect or particle aspect. It all depends on the type of the experiment being designed to study these particles (interference or photoelectric effect). When one aspect (wave) is known, then the other (particle) appears to be Śūnya and vice versa. Also as per classical physics, if the initial spatial coordinates and momentum of a particle are known, then its location in the future can be determined using the Newton's laws of motion, both in the presence of a force or its absence. But in quantum physics, location and momentum of a particle are both uncertain and are governed by the Heisenberg's uncertainty principle through the Planck's constant. If one is known accurately, then the other becomes highly uncertain (like Śūnya) and vice versa. Actually, in quantum physics, the state of a particle is determined by the Schrodinger's equation using the probability function, known as wave function which describes only the probability of finding a particle at a given location (and not the actual value). Hence there is an inbuilt uncertainty in our description of the reality or our knowledge about reality. So we do not know about the actual reality. We only know what our process of measurement and our methodology inform us about reality. Since we have only information about quantum reality containing probability, hence we never know the actual reality. It may not be knowable also.

There is another way to describe a quantum system, which can exist in a superposition state, consisting of several eigenstates, such that when an experiment is conducted to understand the system, then in the process of measurement, the superposition state reduces into one of the possible eigenstate, depending on the experimental set-up designed by us. The quantum formulation provides just probability of finding a system in one of the possible eigenstates. In view of this, one scientist has recently termed the wave function as a wave function of the ignorance because while measuring one of the eigenstates (with certain probability), other eigenstates are lost (I have found it in a book *Science and Ultimate Reality* 2008), but not able to locate it again. Now we obtain some information about one of the eigenstates and lose others. We cannot say that other eigenstates fall in the category of Śūnya, although from another perspective, it is actually Śūnya with respect to the known eigenstate.

2.2 *Quantum Field Theory: Single Vacuum State in Flat Space-Time and Many Vacuum States in Curved Space-Time*

Quantum field theory is basically an approach to explain the properties of elementary particles in nature. Here an attempt is made to reconcile the theory of relativity with the quantum physics. Thus quantum electrodynamics was developed to explain the particle aspects of light known as photons. There are two different types of QFT, depending on whether used for special relativity (flat space-time) or general relativity (curved space-time). In all these cases, some types of fields are assumed which can be either scalar, vector or tensors. Lagrangian equations are written down in terms of these fields and their gradients. Attempts are then made to convert them into Hamiltonian forms. The solutions are then found for these fields in terms of some basis vectors and some kind of operators. These operators are known as annihilation and creation operators. Under some conditions, these operators result either into production of the elementary excitations or their annihilation. For a system with "n" elementary excitations, if an annihilation operator is constantly applied to the state of the system, then after nth application, a state will be found which will not have any elementary excitation. One gets what is known as a vacuum state without any particle. This vacuum state is just like a "Śūnya". The elementary particles are then equated with the particles. However, in almost all QFT types, one finds energy being associated with these vacuum states. It could be infinite also.

In the case of the flat space-time, the vacuum state found is unique for all inertial frames (moving with different velocities but obeying Lorentz transformation) and has energy analogous to zero-point energy for simple harmonic oscillator. Here the number of elementary excitations is same but has different momenta (for different inertial frames). However, the vacuum energy can be infinite and can lead to abnormal results. Hence they are avoided through a procedure known as "renormalization", through which some energy is subtracted to get a finite energy for the vacuum state. These subtractions are highly subjective, and one is free to select whatever amount one wants to subtract so as to make an adjustment according to the particle detector being used to study and detect the particles. This vacuum state is just like Śūnya state as there is no matter in it.

However, if the space-time is curved, then depending on the type of frame chosen and the basis vectors selected to describe the field, the vacuum states are different. That is, even the number of particles found in a given vacuum state can be different if observed from one frame (through one set of basis vector) from those found with respect to some other reference frames (another set of basis vectors). For example, from one frame a vacuum state may not have any particle at all, that is, it is an absolute vacuum. But if the same vacuum state is observed from another frame, then it can have many particles. These different types of vacuum represent different types of Śūnya. This is therefore also an example of non-absolute description of reality (see section below from Jainism also).

2.3 Quantum Field Theory and Śūnyatā in Superfluids and Superconductors

The concept of quantum field theory has been also used to explain the properties of condensed matter like superfluids and superconductors. Different properties of systems originate from the type of elementary excitations. The interaction among various excitations causes scattering among them. The relative number of various elementary excitations present in a system is a function of temperature. These excitations determine the disorder present in these systems and also say thermal and acoustic properties of the superfluids and superconductors at different temperatures [5, 6, 13, 14]. The concept of elementary excitations can also explain the phenomena of phase transitions in these systems. Thus phonons, rotons and vortices determine the properties of superfluid helium at different temperatures and determine propagation of sound and its attenuation in liquid helium. If the temperature is absolute zero, then liquid helium is in a perfectly ordered state with no excitations and total superfluid state. Similarly in case of superconductors, there is a superconducting state with very high order at absolute-zero temperature with zero resistance and free flow of current. There are no elementary excitations (known as Cooper pairs in this case) in this minimum energy state. As temperature goes up, the number of Cooper pairs increases and some resistance starts developing. A state without elementary excitations can be treated as a type of vacuum state with all particles (original ones) in one single quantum coherent state but a Śūnya from one perspective.

2.4 *Implicate Order of Quantum Physics and Śūnyatā*

Implicate order of quantum physics, a term introduced by Prof. David Bohm, is also relevant here. In proposing this concept of implicate order, Bohm explicitly challenged a number of tenets that are fundamental to the scientific work. According to David Bohm, in the enfolded [or implicate] order, space and time are no longer the dominant factors determining the relationships of dependence or independence of different elements. Rather, an entirely different sort of basic connection of elements is possible, from which our ordinary notions of space and time, along with those of separately existent material particles, are abstracted as forms derived from the deeper order. These ordinary notions in fact appear in what is called the "explicate" or "unfolded" order, which is a special and distinguished form contained within the general totality of the implicate order.

In Bohm's conception of order, primacy is given to the undivided whole and the implicate order inherent within the whole, rather than in parts of the whole, such as particles, quantum states and continua. For Bohm, the whole encompasses all things, structures, abstractions and processes, including processes that result in (relatively) stable structures as well as those that involve metamorphosis of structures or things. In this view, parts may be the entities normally regarded as physical, such as atoms or subatomic particles, but they may also be abstract entities, such as quantum states. Whatever their nature and character, according to Bohm, these parts are considered in terms of the whole, and in such terms, they constitute relatively autonomous and independent "sub-totalities". The implication of the view is, therefore, that nothing is entirely separate or autonomous. This implicate order can be therefore treated as Śūnya from the point of view of ordinary description of material reality.

2.5 Concept of Śūnyatā in Sub-quantum Physics

We know that in quantum gravity, scientists talk about Planck level of reality conceptualized at 10^{-35} cm and 10^{-43} seconds. However, some scientists are trying to go beyond these levels. Prof. Adrain Klein talks of sub-quantum physics, that is, physics below this spatial level of 10^{-35} cm, and in this domain, laws of physics no longer hold true. Thus laws of quantum physics and theory of relativity are no longer valid, and there is no matter like substance in this domain. Here one talks only of non-physical information fluxes, known as SQ fluxes, and their combinatorials. Here non-physical information can propagate at velocity ranging from zero to infinity (this is in contrast to the velocity of light which is fixed by the special theory of relativity). Information signals can propagate both in positive- and negative-time directions. Hence here we encounter a new concept of vacuum where there is no matter but only SQ information fluxes. The author (Klein) claims that perfect determinism can be restored in this sub-quantum domain. Also Klein claims that all psychological processes are connected with the processes taking place at this sub-quantum level. The brain has some direct access to this sub-quantum domain which might control various processes in the brain through these information fluxes. Due to bidirectional propagation of information in two time directions (both positive and negative), it is argued that these characteristics can explain even phenomena involving extrasensory perception like telepathy, clairvoyance, etc. In view of this model, even astrology which is ignored by scientists and hence treated as a Śūnya from this perspective may turn out to be a science of a different type. This concept of vacuum is quite close to the concept of Śūnya defined in Indian philosophy in general and Jainism in particular, because this state is similar to Śūnya as there is no matter but is not vacuum because there is a huge amount of SQ information fluxes and their combinatorials. So it is analogous to state of a pure soul without any matter.

2.6 Consciousness could be a Non-physical, Non-local Infinite Reality with Infinite Knowledge

The recent formulation of Bell's inequality to understand the famous thought experiment known as EPR paradox (Einstein, Podolsky and Rosen paradox) and its rejection demonstrated by the Alain Aspect [35] experiment has clearly shown that the photon and electrons (and hence other elementary particles) do not exist till they are observed through some scientific experiments [15]. This is a mind-boggling conclusion, and it implies that the Copenhagen interpretation of the quantum physics is correct, and hence Neils Bohr was correct and not Albert Einstein. If this conclusion is correct, then it would imply that the actual reality is nonlocal, that is, there is a quantum entanglement, and what one observes is only the effect of that reality on the instrument or the observer and not the actual reality. Similarly the extension of the double-slit experiment through the "Wheeler's delayed-choice experiment" shows that photons and electrons do not exist until some "appropriate" measurements are made through some experiments. Such ideas have been now experimentally verified using just single particles. Another important analysis carried out Edward Close is that the ultimate observer is always a "conscious observer", a human being in the case of science. Now using the inferences drawn by EPR experimental verifications of violation of Bell inequality and the Wheeler's delayed choice experiments, neuroscience and information sciences and a new mathematics known as "calculus of distinction" he explains that a consciousness has to be non-physical, continuous and infinite and all knowledge must be structured in such a consciousness. This type of non-physical consciousness is therefore again a sunya from the point of view of physical reality.

2.7 *Could Annie Besant see Quarks, Subquarks of Atoms and Dark Matter and Dark Energy Through Clairovyance*

If the above ideas given by Dr. Close are true—that knowledge is structured in the consciousness—then it is very appropriate to mention the work of Occult Chemistry carried out by Shrimati Annie Besant and her co-workers [16, 17] who have used the Anima siddhi of Patanjali Yoga Sutras to see quarks and subquarks like entities as early as 1905 or so, that is, when the name of quark had not come into existence at all. These people also claimed that there are four types of ether present in the nature and described their properties. In a further study of their work in the light of the TDVP model of Neppe and Close, Pokharna et al. have estimated the number of microscopic constituents found by Annie Besant et al. and those coming from TDVP model. They are in agreement within an accuracy of 10% for 21 elements. Not only this, even the magnitude of dark matter and dark energy found from the work of Annie Besant et al. agrees well within 10% of the estimated value in the modern science. This research will provide new evidence that knowledge is actually structured in the consciousness which is non-physical and nonlocal with infinite knowledge, and hence not Śūnya when looked from a physical perspective (details are to be published in Pokharna et al. [18]).

3 A Concept of Śūnya (Vacuum) in the General Theory of Relativity

As we know Einstein was the main scientist behind developing both special theory of relativity and the general theory of relativity. The later was developed to understand the phenomena of gravitational force. By combining space and time, he put forward the idea of space-time geometry. In the special theory of relativity, various inertial frames of references have different values of lengths, times, etc. and are linked through the Lorentz transformation. In case of general theory of relativity, one has non-inertial frames, and tensors play the key role in developing the whole formalism. Through them only, Einstein and others developed the formalism and equations of general theory of relativity (GTR). However, it was surprising to find that in four-dimensional space-time, the Einstein equations can have non-trivial solutions even in regions where there are no sources, i.e. in the regions of space-time, that are void of matter and radiation (even in the absence of dark energy). In the absence of sources, $[T\mu\nu]$ should be $= 0$, thereby Ricci tensor $R\mu\nu = 0$, but this does not imply that Riemann tensor should be 0. It means that the space-time is curved and then the metric tensor $[g_{\mu\nu}]$ is different from the trivial minkowski tensor $[\eta_{\mu\nu}]$. Thus Einstein equations imply gravitational fields even in empty space. This is a type of Śūnya where there is no matter or radiation, but still there are some traces of gravitational field.

4 Different Concepts of Śūnyatā at Very Fundamental Level Within the Modern Science

In this section, we describe the concept of Śūnya tva from widely different perspectives within the modern science.

4.1 Process of Scientific Measurement Under Controlled Conditions and Concept of Śūnyatā

If we go little deeper into the methodology of science, then we find that any phenomenon is called scientific if it can be verified in a laboratory under a given set of controlled conditions and is also reproducible at any point of time and at any other place (space). This condition is called space-time invariance condition in science. Let us take an example of simple pendulum to measure the magnitude of acceleration due to gravity. A long thread and a metallic bob are generally used for this. Bob is tied to the thread and is hanged from a support on a wall, and oscillations of the bob are counted along with the length of the thread. Standard relations are established to estimate the constant of acceleration due to gravity. Now leaving aside some minor differences in the value of acceleration due to gravity, from place to place, this experiment will give the similar values at different places on the Earth. However, if we wish to measure the tensile strength of the thread or its structure, then different set of experiments will be required. Similarly if we wish to measure the molecular structure of the bob or the coefficient of thermal expansion of the bob, then again new set of experiments will have to be designed. In general, one can ask for several properties of these two objects and will have to design several experiments to study other properties of these two objects. So when one is studying one set of properties, then other properties are ignored as if they are Śūnya, which is not the truth. In principle, there can be several properties which one can think of, and so an equal number of Śūnyas will be there, till they are studied. (As we see below, in Jainism, one talks of infinite attributes of any object when studied in relation to other objects and their different aspects.)

4.2 The Relation Between Science, Technology, Economics and Śūnyatā

There is also a need to go into details of scientific methodology and how it is explored and exploited through technology and economics in the name of development. If we denote a scientific experiment as Truth-A, its conversion into technology as Truth-B, the exploitation through economics in the name of the development as Truth-C and its total impact on the environment as Truth-D, then the total truth of the system for this chain (one leading to the other) can be mentioned in the following way:

Scientific experiment -> technology -> economic development-> environment
Truth-A + Truth-B + Truth-C + Truth-D which we call as Total Truth (Truth-T)

then obviously scientific experiment (Truth-A) is a small truth (small subset) compared to the total truth (Truth-T), in this chain of truths. Hence when a scientist is performing an experiment, then the rest of these truths are being treated as Śūnyas (futuristic ones), but the reality is not that because these Śūnyas (futuristic ones) for the scientist are part of another type of reality. In addition, as we move from left to right in the above equation, more and more information is getting added up. Hence total truth will be much more than the one implied by Truth-A of the scientific experiment.

It is the cumulative effect of ignoring these Śūnyas like concepts and situations in the chain of science, technology and economics (being treated as concept of the development) that the world has come to the present state of affairs where the climatological processes will become irreversible to change in a very short time, damage done to the biodiversity and others. It is mostly due to the rigid and tight boundary conditions under which scientific phenomena are defined and experimental observations are taken and studied but whose results are then captured by technologists, economists, industrialists, politicians and others to "exploit them in the interest of the humanity". These other factors are treated as Śūnya by the scientists, whereas they are actually not.

4.3 Close Systems, Conservations Laws, Interactions and Śūnyatā

In addition, we define conservation laws of physics which are the foundation of all scientific measurements. Thus we have conservation laws for energy, linear momentum, angular momentum, etc. Now all these conservation laws are defined for isolated closed systems by approximating the nature. For example, the energy is defined as that variable for a closed isolated system which does not change over time [19]. But in principle we can never have a totally closed isolated system. Similarly linear momentum is defined as that property of a closed isolated system which remains invariant with respect to any spatial displacement, and lastly angular momentum is defined as that parameter for an isolated closed system which remains unchanged due to rotation in space. Hence the mere definitions of conservation laws are such that in their basic definitions, interaction (electromagnetic, gravitational, etc., which can never be zero) is assumed as Śūnya, because it is ignored in the definition of conservation laws themselves. However, the interaction among systems is then studied by considering the nature and magnitude of the interaction.

4.4 Description of Physical Systems by Treating Them as Open Systems

In Indian system, matter is defined through colour, taste, smell, touch and other such characteristics, by treating them as open and macroscopic systems [10]. Here arrangement of matter of different types will describe some type of order say in form of arrangement of colour, touch, smell and taste. This type of description can be termed as partly Śūnya from the point of view of the modern science and vice versa. However, one should admit that both the types of descriptions are true and must be respected.

4.5 Open Systems, Entropy, Order, Information and Śūnyatā

For biological systems which are so strongly interacting with each other, the above type of formalism cannot be applied in a satisfactory way in a real sense. Actually all living beings are essentially irreversible in nature, that is, they grow and decay, and they are also open systems compared to the physical systems which are closed systems. These open systems constantly interact with the environment like human beings, who take information and oxygen from atmosphere and release carbon dioxide and other waste products. Certain order is maintained in them through this process. Human systems also have memory, free will, creativity and tendency to interact strongly with other fellow beings and the environment. Furthermore there are micro controls in the form of thought processes which cannot be easily adjusted in any planned "scientific experiment". These are also goal-directed systems, and so here it is the effect which determines the cause and not the reverse-like cause determining the effect, which is true for the physical systems. Hence biological systems, human systems and social systems cannot be subjected to the process of measurement, and hence they are not exactly describable in the strict terminology of the physical sciences. So it is very difficult to perform experiments on human systems and predicting events about them from a purely scientific perspective. Therefore such systems are studied in a different way by using statistical procedures. In such procedures only some rough trends or patterns can be found. Hence these living systems are better studied using the general systems theory. Here entropy, order and information are very important. Energy comes next to them. Thus for description of such systems, energy-like variable becomes secondary.

5 A Finite Nine-Dimensional Model of the Universe and Śūnyatā of a Different Type: Finite Reality Embedded Into an Infinite Reality

Recently an important model to describe the reality is being proposed by Neppe and Close [2]. They have put forward a very innovative approach in which nine dimensions of reality are visualized. They have probably first time introduced a concept of consciousness in physics in a quantitative way. They define three dimensions of space, three dimensions of time and three dimensions of soul. Again they have introduced a completely new physical concept in which quantum nature of reality is explicitly taken into account. This type of digital model of reality requires a new type of mathematics where one treats reality as consisting of quantized units, instead of treating it as a continuous. This new mathematics known as calculus of distinction (COD) is different from the standard mathematics and the calculus, which treat the reality as continuous system. Here the distinction is given more importance than the other variables. They have used a symbolic system for quantitative analysis. This model has been found to be the ideal model as "Theory of Everything (TOE)" including particle physics and physics along with biology and psychology because consciousness has been mathematically introduced. It has been given 39/39 marks by many scientists around the world compared to about 23 for other TOE models [2].

However, the point of interest for us here is that this model assumes that the reality which we perceive through model is still finite and is embedded in an infinite reality. Hence out of any such successful model which is so exhaustive to explain, everything which can be explained through our limited method of description is still a very small subset of reality. So the rest of the reality can be still treated as vacuum, beyond our best possible attempts, but it actually may *not* be Śūnya.

6 Human Decision-Making Is Neither Deterministic Nor Random and Nor Computable (Śūnyatā at the Most Fundamental Level)

According to Prof. Penrose [9], human decision-making is neither deterministic nor random but is even non-computational. It shows our incapability to say anything about reality. Some kind of Śūnyatā is inbuilt into our intrinsic process of decision-making itself. It may be therefore inferred that concept of Śūnya is essentially inbuilt into our process of doing anything or any action. Actually at the beginning of the twenty-first century, it is being recognized that some experiments related to the human decision-making processes could not be explained by the conventional classical decision theory, but the same could be explained by the models based on quantum mechanics and the mathematical formalism used in quantum physics [20]. This new finding will have far-reaching consequences on the way we look at the world and the methodology of describing anything including the modern science. The concept of Śūnyatā is therefore very profound.

7 Extrasensory Perception, Science and Śūnyatā

Extrasensory perception like telepathy and clairvoyance is now established as true phenomena, but they do not fall into the domain of the modern science, because they are not repeatable in space and time. So for scientists, they fall into the domain of Śūnyatā, but these phenomena require a different type of approach because they involve living systems, for which experiments of science cannot be conducted accurately. Hence for modern science, these phenomena will fall into the domain of Śūnyatā, but it is not so, because they do exist. Some scientists like Adrian Klein [1] and Neppe and Close [2] have developed even theories to explain them.

8 Known, Unknown and Unknowable (Śūnya)

The main question is whether everything can be known about everything or knowledge has any limitations. Science generally divides knowledge into two parts: known and unknown. When an object or phenomenon is studied, unknown becomes known, and eventually, one believes that everything becomes known and nothing remains unknown. However, it has been realized, in view of recent developments in modern logic, that knowledge has severe limitations. To illustrate the point, we cite two examples here. Thus due to the finite velocity of light, we are able to observe only the part of the universe; defined by the distance, light has travelled over the age of the universe. The velocity of light is about 3×10^8 metres per second, and the universe is about 13.4 billion years (1.1×10^{15} sec) old. Light from beyond

~1.3×10^{23} km can never reach us, and we will never be able to observe what lies beyond the visible universe of the radius of 1.3×10^{23} km. If the universe is infinitely big, then much of it lies beyond this limit, and we see only a small part of it. We take the second example of our brain. As we learn about our brain, this information itself evolves the brain, and we will never be able to know the state of the brain completely; we can only know its previous state before we examined it. Such a case has given rise to Cantor's paradox of infinite sets, in which a set cannot contain all the sets (including itself). We consider two other simple cases here: an electron and water. An electron, for example, has mass, charge, spin, magnetic moment, etc., but by observing these properties, we can never find out that electron sometimes behaves like a wave. Similarly water has colour, molecular structure, certain chemical composition and boiling point, freezing point, etc. However, it also has other properties because of which it is so critical for life processes. Simultaneously it has interesting effects on the human body under different circumstances. Hot water makes one feel good in winter, but cold water feels good in summer. It is difficult to understand all properties and effects of water, like how much thirsts it will quench; it will always remain hidden and come into play at some other time. Thus it should be realized that whenever one is describing that it is electron or water, then due to limitation of language in expressing knowledge and the context, it is not possible to describe all its aspects in totality. Hence there is a state of Śūnyatā of a different type.

On the other hand in Jainism, one postulates that the knowledge is of three types: known, unknown and unknowable [21]. Whereas unknown can be converted into known, unknowable can never be known, at least by sensory organs or mind, and can only be experienced by consciousness. The perception of sensory organs (and mind) has limitations, because they are not perfect and are always in a process of learning and improving, but this is only a trivial reason. The main reason is that everything does not manifest all its aspects (modes) at once and manifests its various aspects, depending on context, cause, interaction with other things, time and place, etc. Actually any object therefore will have infinite attributes, out of which only few can be known at any time.

9 Gödel's Incompleteness Theorems, Science and Śūnyatā

The most attractive aspect of scientific knowledge is its mathematical basis. We generally feel that this mathematical representation of various scientific facts makes our knowledge more precise and accurate. However, from the following theorems which have been put forward by the great mathematician Kurtz Gödel, we find that any mathematical representation of any physical reality limits our knowledge of that reality. Not only this but the theorem also implies that none of the languages or representations can express the reality of nature with perfection. Complete knowledge must necessarily have its foundation in an unexpressed, unmanifested field of intelligence. It is being termed as consciousness. The two Gödel's incompleteness theorems are given below:

9.1 Gödel's First Incompleteness Theorem

This theorem says that the truth of a formalism (which describes any phenomenon) cannot be proved. Thus no finite expression of mathematical knowledge can ever provide a basis for comprehensive knowledge even of the most elementary properties of the counting numbers. Thus if one starts with a collection C of symbolic mathematical (or any other) axioms which is specifiable by a finite number of mechanical rules, and if C is consistent, then there will be a true statement about the counting numbers which cannot be proved from the axioms C, using the standard rules of mathematical logic. The proof of this theorem shows that from C, one can construct a sentence S in the simple mathematical language of elementary number theory whose meaning is: This sentence is not provable from C. Once S is constructed, it follows easily that S must be true but not provable from C. Thus on the basis of any finitely specifiable collection of axioms C, one cannot prove all true propositions about the counting numbers.

9.2 Gödel's Second Incompleteness Theorem

A formal language (mathematical or any other) if consistent cannot define its own truth, i.e. the definition of truth for a theory must be of a higher order than the theory itself. We can also say that the consistency of any specifiable collection of axioms can never be established on the basis of mathematical arguments which can be justified by these axioms. Thus to establish the validity of any single mathematical system, one must necessarily utilize a more comprehensive system; to validate the latter system, one has to investigate an even more comprehensive system and so on.

In these examples, Gödel's theorems are applied to some cases in abstract mathematics, but they are now being explored in various other branches of sciences also. However, as discussed above, a concept of consciousness is required to go beyond the current scientific thinking. Hence when we talk of science and reality, then we must keep in the mind about limitations of science in describing the reality. What we describe is a partial truth. Whatever is left out is Śūnya for us, but actually it may not be.

10 Śūnya and Infinity Together: Can a Soul (Consciousness) Be an Entity for Which Causality Is Neither Obeyed nor Violated?

Due to some of these concepts and the properties of soul as mentioned in the Indian philosophy, Pokharna [22] has introduced an abstract entity for which causality is neither obeyed nor violated. The principle of causality points out that for every effect, there is a definite cause behind it (or a large number of causes behind it) and every effect in turn becomes a cause of some other effect. But when we say that causality is violated, then it can have the following three different meanings: (1) when the causality is violated, then this can mean that temporally the role of cause and effect has been reversed so that first we have an effect and then there is a cause. This is the situation believed to be existing in the case of tachyons (particles moving faster than light). They are assumed to be moving backward in time, so it is said that causality is violated by tachyons. (2) Violation of causality can also mean that there may be some effects (or causes) whose causes (or effects) may not be known to us or no physically reasonable causes (or effects) can be searched for given effects (or causes). This is the situation which we have found in quantum mechanics where we have found that there are some uncontrollable fluctuations in the predictability of the state of a system in a given measurement process. These fluctuations are such that we cannot assign any known factors or causes responsible for them. (3) Causality seems to be violated in the case of biological and social systems also where one always defines certain goals and then plans a set of activities such that the goal is achieved. In this case effect is fixed and is defined in the form of a goal, and then causes are organized in such a way that the given goal is achieved. Here the relation between cause and effect does exist, but the temporal relation is reversed in the sense that now cause does not determine the effect (natural time flow) but the effect determines the causes (time is somewhat in the opposite direction to the conventional time).

We therefore make an attempt to study the problem of soul in an abstract way. Our analysis has become an abstract formalism because we are interested in looking for a possibility of perfect determinism within quantum mechanics. We have introduced two abstract entities to deal with this problem. The first is an abstract causal world described by the classical physics in which causality is perfectly obeyed, and we have a state of perfect determinism, completely independent of any measuring apparatus or an observer. We call this entity A.

It may be noted that the assumption of this kind of reality or entity was considered to be irrelevant and physically meaningless by Heisenberg and others, but it does not mean that they have completely ruled out the existence of this reality. This is so because Heisenberg very clearly asserted that the laws of nature which we formulate deal no longer with the nature itself but with our knowledge of the nature. But it may also be mentioned that Einstein had a firm belief in the idea of existence of a reality where the law of causality is perfectly obeyed and there is state of perfect determinism, even after the discovery of quantum mechanics.

We then introduce a second abstract entity for which causality is neither obeyed nor violated and call it B. If we treat this new entity as an abstract observer, then the interrelation between these two entities A and B indicates that this new entity can be interpreted in two contradictory ways. According to one interpretation, the entity B is physically meaningless because it is void of any physical properties, but according to the second interpretation, this new entity B can perceive the whole causal world in one single perception. The past, present and future of the causal world are perceived simultaneously. Thus we have tried to describe the properties of the Laplacian imaginary demon who can perceive the whole classical world in one glance.

This paradoxical situation appears to be very much similar to the one which has been described in different schools of thought of the Indian philosophy whose main conclusion is that the true knowledge of this world is structured in the consciousness and hence can be obtained only by enlightenment, that is, by completely detaching from the material world. It is argued that an enlightened consciousness is capable of perceiving the whole causal world in one perception from one point of view, but from the other point of view, this enlightened consciousness is a null, void of anything [22]. (Prof. B D Josephson has appreciated this idea, the author is very grateful to him.) This work is comparable to the Laplacian demon which was defined to understand the whole Universe due to perfect determinism built in due to classical physics [23].

11 Science, Society, Environment and Śūnyatā: Current Problems of the World Needs a More Fundamental Analysis

We now try to explore the concept of Śūnya from our day-to-day life experiences. We know that science and technology and their use in economic developments and commercialization have revolutionized the whole world in such a way that everything appears to have changed in the last 100–200 years. Developments in the field of space technology, atomic energy, electronics, biotechnology, modern agriculture, telecommunication and manufacturing systems are some of the examples of these changes. These changes also have played a key role in making this world truly global. However, they have also resulted in an increase in population, depletion of natural resources, damage to the environment, increase in terrorism, threats of nuclear wars and so on. Some consequences of science and criteria of economic development can be summarized in what follows:

Population is increasing and resources are shrinking.
Automation/computerization is increasing but jobs are decreasing.
Competition is increasing and love and affection are decreasing.
Risk and uncertainty in life are increasing.
Development is interpreted in the economic sense only. But gap between rich and poor is increasing.
Sperm counts in males have gone down by 50 percent in last 50 years. Means maleness among men is decreasing.
Females are developing male-like characteristics.
Around 25,000 biological species are getting extinct every year. (http://wwf.panda.org/about_our_earth/biodiversity/biodiversity/).

From 1970 to 2010, around 50% of animals have disappeared from the Earth. (http://www.telegraph.co.uk/news/earth/wildlife/11129163/Half-of-worlds-animals-have-disappeared-since-1970.html).

Every year around 150 billion animals are killed for food. The damage done to the environment is unbelievably huge (http://www.adaptt.org/killcounter.html).

Human-human interaction and human-environment interactions have gone down.

Environment is getting polluted and care for animals is decreasing. Climatological changes due to greenhouse gases have become so serious that we are left with very little time before the beginning of irreversible processes in the climate which will have very serious consequences for the future generations. The last year summer was the hottest in last several years. In the summer of May 2016, the average temperature observed at 20 places in the world showed a rise of around 4–5 degree centigrade.

All these consequences have resulted in a large-scale increase in entropy in the world at different levels in different fields [24]. In addition, because of these changes and domination of science and technology in all walks of life, an impression has been created that scientific knowledge is the supreme and anything other, which does not fall into this domain, is not very relevant. However as we have seen in Sect. 9.1, scientific knowledge has severe limitations, and there could be other dimensions of knowledge like spiritual knowledge which is being treated as Śūnya by modern scientists. But the spiritual master treats the worldly knowledge or the materialistic knowledge as insignificant (Śūnya like) compared to the spiritual knowledge.

12 Economic Development, Spiritual Development and Śūnya

In view of these problems and ultimate consequences of science, it is high time for scientists to recognize the concept of consciousness which is treated as the basic source of knowledge and being discussed in great length in Indian system of thought. Actually Indian philosophy talks about spiritual development and spiritual evolution. This is in contrast to the concept of economic development which appears to have become the only concept of development in the modern time. Actually too much focus on this and its proliferation due to science and technology has destroyed the environment and is degrading the life support system irreparably. Economic development is required for smooth running of a society, and everybody wants it. However, with the advancements of science and technology, a huge industrial revolution has taken place in the whole world. Due to this, economic activity started concentrating in a few places and in a few cities. These activities started polluting the natural resources in a highly damaging way. The whole process of polluting the environment can be traced to these increased industrial activities and increase in population due to decrease in the death rates. Due to the congestion of the cities, there results a cut-throat competition among people for survival. The old value system based on simplicity, honesty and sincerity started getting replaced by complexity, consumerism, dishonesty and unwanted domination of certain groups and countries over others. It has also resulted in a large-scale corruption in many places in the world.

The emphasis on economic development has become so much so that all other types of developments have been set aside. Thus women in many countries who used to work towards spiritual evolution, religious activities and family welfare have all started working for economic development only. Also indices based on economic growth are so much dominating the people's mind that the impact of economic development on the environment is totally ignored. Thus when we say that economic growth of a country is 7%, then nobody talks of its spiritual growth or spiritual evolution, whether they are negatively or inversely related.

India is known more for spiritual evolution. There is virtually not much focus on the economic development. Rather it is being suggested to minimize the material things and material growth and focus on the spiritual evolution only. Rishis and Munis who are involved in spiritual evolution consume minimum resources and take maximum care of the environment because they have found a Śūnya in the economic growth, and something great in spiritual evolution, which may be closer to the infinite reality. For modern economists, spiritual growth may be Śūnya or negative and vice versa.

13 Some Examples of Śūnyatā from Jainism

Jainism, traditionally known as Jain Dharma, is an Indian religion that prescribes a path of non-violence towards all living beings and emphasizes spiritual independence and equality between all forms of life. The essence of Jainism is concern for the welfare of every living being in the universe. Practitioners believe that non-violence and self-control are the means by which they can obtain liberation.

Jains believe in the notion that truth and reality are perceived differently from diverse points of view. This doctrine is formally called anekāntavāda. According to it, human beings are limited in their perception and knowledge about the reality. They can thus have only partial information. Philosophical and theological disputes arise only because of the partial knowledge of human beings. The Jain doctrine stresses on the existence of soul. According to Jainism, every living being has a soul and has a separate existence from the body that houses it. This soul undergoes a cycle of reincarnations. The word Jainism is derived from a Sanskrit verb "Jina" which means to conquer. It refers to a battle with the passions and bodily pleasures that the Jaina ascetics undertake. Those who win this battle are termed as Jina (conqueror). The term Jaina is thus used to refer to laymen and ascetics of this tradition alike.

Jainism is one of the oldest religions in the world. Jains traditionally trace their history through a succession of 24 propagators of their faith known as tirthankara with Tirthankar Ādinātha as the first tirthankara and Lord Tirthankar Mahāvīra as the last. For long periods of time, Jainism was the state religion of Indian kingdoms and widely adopted in the Indian subcontinent (For details about Jainism, see Bhandari [25], Pragnya et al. [26] and General Books on Jainism [27].

In Jainism, the world is assumed to be consisting of six distinct substances. They are Akashastikaya (Space), Pudgalastikaya (Matter), Jeevastikaya (Living beings having a soul), Kala (Time), Dharmastikay (Medium of motion) and Adharmastikaya (Medium of rest). The concept of akash, matter, soul and time are quite common to many philosophies. The medium of motion and medium of rest are two special substances described only in Jainism. The medium of motion is interpreted as something which causes motion of matter and soul and living beings. This is sometimes interpreted as equivalent to some kind of force like gravitational force or ether, whereas medium of rest is responsible for position (stability) of matter and is interpreted as inertia or mass. (The meanings of these terms should be reexamined in the light of problems of dark matter and dark energy.) Matter and soul are defined as two completely different substances. Space is defined as a substance which contains all the above five substances. Time is treated as a substance which helps in describing change or is also a cause of the changes in matter, soul, medium of rest and medium of motion [10].

13.1 A Quantum Field Theory-Like Model for Soul-Matter Interaction in Jainism

Pokharna [5, 14] has found an interesting analogy between the karma theory of Jainism and quantum field theory. The karma theory of Jain philosophy deals with the soul-matter interaction. Jainism has defined the soul as the basic constituents of all living beings. Soul has the following properties (Jin Dhammo, Aacharya Shri Nanesh [28] in Hindi and Kachhara [10]). A pure soul has infinite knowledge, infinite intuition, infinite bliss and infinite power (actually 47 characteristics have been defined for a soul, see [10]). Although a pure soul has many characteristics, but the knowledge has been regarded as the chief characteristic of soul. Kundakunda [11] has stated that although from the empirical point of view there is a difference between the soul and knowledge, yet from the transcendental point of view, it is sufficient to say that the soul is knower and nothing else. He further said that there is no difference between the knower and his knowledge. From an empirical point of view, an omniscient (kevali, i.e. a pure soul) perceives and knows the whole of reality, and from the transcendental point of view, he perceives and knows nothing. That is, infinity and Śūnya are the two aspects of this reality. This state of soul with infinite knowledge is like a vacuum state in quantum field theory without matter but infinite energy. As we shall see below, there is a very interesting analogy between their karma theory and the quantum field theory.

As mentioned above, a pure soul can have an infinite knowledge, infinite intuition, infinite bliss and infinite power (four infinities). But from an infinite time in the past, it has been infected by matter. This matter has been held responsible for the disorder present in the worldly souls. This form of matter has been termed as karma. The word karma in Jainism does not mean work, but it means a type of very fine category of matter. The whole universe is full of that kind of matter which can become karma. Due to the presence of different types or karmas in different quantities, different characteristics are manifested by worldly souls, i.e. by different living beings. But actually these karmas obstruct the true powers of a pure soul.

At every moment, an empirical soul is attracting this matter towards him by his actions through mind and body. This matter which has now become karma remains latent (in memory) in the empirical soul for some time which is determined by the passions at the time of arrival of new matter. Passions in turn are determined by karmas which are already present with the soul and some other conditions around the person. Karmas have the following four characteristics: (A) nature, (B) number, (C) lifetime and (D) intensity.

13.1.1 Nature of karmas

Jainism claims that all properties of living beings can be explained in terms of 8 different types of karmas and their 158 categories. Each one of these is held responsible for different types of disorder present in the empirical self and impedes the manifestation of true nature of soul. The following are the eight categories of karmas and their subcategories are given in the paranthesies.

Knowledge-obscuring karmas (5)
Intuition-obscuring karmas (9)
Feeling-producing karmas (2)
Age-determining karmas (4)
Belief- and conduct-producing karmas (28)
Body-determining karmas (103)
Status-determining karmas (2)
Power-hindering karmas (5)

Different properties of different living beings are explained using the above karmas (for details, see [10]). Thus knowledge-obscuring karmas obscure the real knowledge of a living being which is assumed to be structured in the consciousness. The situation is similar to the ideas of elementary excitations developed in the condensed matter theories for matter at a very low temperature. Thus phonons, rotons and vortices are meant for different properties of superfluid helium at different temperatures. They reduce the off-diagonal long-range order (ODLRO) present in the superfluid, which is the highest at the absolute-zero temperature. Similarly in case of superconductors, there is a superconducting state with very high order (zero electric resistance), and there are elementary excitations over it in the form of Cooper pairs, which actually reduce the ODLRO present in the superconductor at higher temperature. These elementary excitations also explain various properties of these liquids at low temperature like velocity and attenuation of sound and thermal properties, etc.

13.1.2 Number of karmas

The number of karmas changes from one animate to another, and within a given animate, they vary from time to time, depending on one's mental state and state of evolution. Their relative quantities are given by the following rule: The age-determining species receive the smallest part; a greater portion goes to the body-determining and status-determining ones, both of which obtain an equal portion. More than that goes to the knowledge-obscuring, intuition-obscuring and power-hindering species, each of which gets an equal portion. Still a larger part than this goes to the belief- and conduct-obscuring species, and the greatest of all goes to the feeling-producing species. It is obvious that these karmas are not very similar to the matter which constitute the bodies of living beings but are different and could represent a new state of matter.

This difference in the number will then determine different properties of different animates. In case of concept of elementary excitations, it is again the number of these excitations which determine the disorder in the state of these systems, say at different temperatures like thermal and acoustic properties of these superfluids and superconductors.

13.1.3 Lifetime of karmas

Similarly the lifetime of the incoming karmas depends upon their interaction with karmas already present which actually determine one's passions. This lifetime is then determined by karma-karma interaction. It is similar to the interaction among various excitations causing scattering among them and influencing the properties of these superfluids and superconductors. Thus in a particular state, an excitation stays only for some definite time.

13.1.4 Intensity of karmas

Pure matter is neutral. The various effects are manifested because of its association with the soul. The effect which these karmas show depends upon their rasa (juice) which is determined by the passions of the empirical soul. This intensity is analogous to the energies of different excitations which show different properties. Thus phonons and rotons determine propagation of sound in liquid helium. The properties of sound will depend on the contribution from different parts of the spectrums of these excitations.

The concept of elementary excitations can also explain the phenomena of phase transitions. Thus different phases of helium have been explained in terms of elementary excitation picture. Similar situation occurs here in this theory also. There are 14 stages known as guṇasthānas which have been recognized in this theory through which one passes before acquiring the pure soul, starting from a state of highest sinfulness. These have been nicely explained in terms of various karmas and their mutual interaction.

Table 19.1 below gives a comparison between the two models.

At this juncture, it would be quite appropriate to mention the subject of hard problem of consciousness discussed by Chalmer [29] to bridge the relation between subjects of experience (SOE) with the physical activity. As per Anand Rangarajan [30], "the concept of hard problem of consciousness is stated as follows "To elaborate using naturalistic language, why are some physical process and events associated with an interior- a qualitative space which is felt from inside". To answer this question, Anand Rangarajan mentions the work of "Strawson [31], who has made a yeoman effort in resurrecting the more important concept of a 'thin subject'. A thin subject is an SOE which is spatially and temporally bounded, arises and decays and has a phenomenal content associated with it during its (typically) short existence". He has therefore invoked a concept of quantum field theory and introduced an elementary excitation and termed it as Selfon generated from a quantum field. It will be interesting to compare the karma theory with these concepts, because karma is treated as a physical entity but is linked with psychological processes in the brain. It looks like karma like a particle of Jainism is conceptualized in this hypthesis with shunyas. Hence karma theory may provide new insights to understand the problem of consciousness in general and the issue of "hard problem" of Chalmer in particular with QFT like concept forbrain sciences.

13.2 *Alokakasha: An Interesting Example of "Śūnya" from Jainism*

As we have seen above, Jainism talks about six types of substances which have been described earlier. They are soul, matter, medium of motion, medium of rest, space and time. Out of these six, space which is actually called as akash (space) consists of two parts, viz. lokakasha and alokakasha. Lokakasha is assumed to be equivalent to our present universe which has space, matter, medium of motion, medium of rest, living beings and time. However, they also define another part of akash which is termed as alokakasha. This is defined to be empty space which does not contain any

Table 19.1 Analogy between quantum field theory and soul-karma theory of Jainism

SN	Parameters of quantum field theory	Parameters of Jain theory of soul and karmas
1	Highly ordered vacuum state	Perfectly ordered soul with four infinities and without matter
2	Elementary excitations	Material karmas
3	Phonon, rotons, Cooper pairs determining the properties of normal fluid	Eight types of karmas are defined which obstructs the true properties of a soul like knowledge, intuition, power and bliss obscuring karmas
4	Phase transitions	Guṇasthānas

of these above substances. It is infinite in size and consists of infinite pradeshas (discrete individual quanta like entities). As this does not have any of the contents of ordinary lokakasha, so this can be taken as an example of "Śūnya". It would mean that if one believes in the big bang theory, then the big bang (finite) took place in this alokakasha which is infinite. In addition, expansion of the Universe can be also explained with the infinite alokaaksha.

13.3 Definition of Matter in Jainism May Help in Better Understanding of the Difference Between Soul and Matter (Śūnya-Like State Compared to Matter)

These days, a lot of discussion is going on whether soul or consciousness is obtainable from matter as some kind of emergent property or it is totally different from matter [32]. However, the definition of matter in Jainism is very interesting and may be useful in the context of distinguishing soul from matter. As per Jainism, the term used to describe matter is known as *pudgal*, a Sanskrit word. The word *pudgal* consists of two parts: *pud* means to combine, and *gal* means to dissociate. If we look at chemical reactions or the process of making mixtures and compounds or even in radioactivity or in particle physics, then this definition holds quite well, because in all these cases, the process of making and breaking is going on at all levels. But this property is not present in case of other substances of nature like soul. Thus if we use the definition of matter as given in Jainism, then it can probably better help in understanding the difference between matter and soul. Actually, one is Śūnya with respect to the other.

13.4 Concept of Infinite Attributes of a System in Jainism and Śūnyatā

In Jainism, it is mentioned that all types of matter and worldly souls have infinite attributes, depending on the conditions in which they are studied. At one time, we shall find only one or few attributes. This means that under different set of conditions at other times, other attributes will be shown by the same system. Hence a conclusion drawn with the limited observations cannot be generalized to other conditions. This again shows some type of Śūnya for a given set of properties observed by a system under given set of conditions, because one believes that the system has only those properties which were observed under a given set of conditions, but this is not true, as it will show other attributes under different set of conditions.

13.5 Implicate Order of Quantum Physics and Keval Jnana of Jainism

This implicate order of quantum mechanics as interpreted by David Bohm seems to be quite close to the Jain concept of Keval Jnana (absolute knowledge) as described in Jainism. It is mentioned in scriptures known as *Agamas* [10] that when all worldly knowledge disappears, then in this state, only absolute knowledge is left. In this state, a pure soul perceives all aspects of all substances in the universe. Total past, total present and total future of everything knowable in the universe for all material entities, all living beings, medium of motion, medium of rest and their all infinite properties are perceived in one single glance. Once a person acquires this knowledge, all obstruction to the knowledge disappears. However, implicate order is Śūnya while looked up on from the point of view of explicate order; similar is probably the situation when keval jnana is looked up on from the point of view of worldly knowledge.

13.6 Concepts of Non-absolutism in Jainism and an Excellent Example of Śūnya

Actually Jainism does not believe in an absolute concept of reality at all. It talks of non-absolute nature of fundamental reality. As per this concept, a substance can have more than one attribute in general and can have infinite attributes depending on the capability of knower and his perspective. These attributes are exhibited under different situations and different set of conditions. All of them are equally meaningful. At a given time, it may show just one attribute, but it can have other attributes also which might be hidden when one attribute is being observed. Anekāntavāda means to recognize this multiplicity of attributes. Hence when we look at one attribute, then a doubt will develop whether this attribute is a truth or not or whether we can perceive the total truth at all. Other attributes are treated as Śūnya till they are observed in some set of conditions. As an example, let us quote the quantum aspect of reality. In view of the multiple attributes of particles discovered in quantum physics like wave and particle attributes observed at different times under different set of experiments, the eminent scientist Professor Daulat Singh Kothari (former chairman of University Grant Commission and founder of DRDO) explored *anekāntavāda* and *syādvāda* from Jainism to explore this dual nature of quantum particles [33]. It is given below:

1. May be elctron behaves like a particle.
2. May be elctron behaves like a wave.
3. May be electron behaves both like a particle and a wave.
4. May be electron is neither a particle nor a wave.
5. May be electron is not a wave and is indescribable.

6. May be electron is not a particle and indescribable.
7. May be electron is neither a particle nor a wave and is indescribable.

This concept known as *saptabhangi* (sevenfold judgement) introduces the concept of indescribability (*avyakta*) which states that some of these seven states are indeterminate. This concept of *avyakta* is scientifically somewhat intriguing as it means an indeterminate, indescribable or unmanifested state. Something like Śūnya being inbuilt into the reality.

If we go little further, then we find that Jainism talks of seven different ways to describe the reality. They are known as *nayas*. (Those seven ways are completely different from the sevenfold judgement given above to explain wave-particle duality.) They exhaust all possibilities to understand the reality. They come into picture because one can understand only one aspect of reality at one time. The seven *nayas* are mentioned in [10].

The combination of *nayavad*, *anekantavada* and *syādvāda* can provide new avenues of thought towards concept of our "understanding" of Śūnya and quantum reality. These concepts must be appreciated in the light of Gödel's incompleteness theorems also.

14 Form and Emptiness (Śūnyatā) as Periodic Oscillations in Time

Up till now, we have tried to search and understand the concept of Śūnya in widely different fields from different angles. Actually a concept similar to this known as "form and emptiness (Śūnya)" is extensively discussed by the great Buddhist saint Nāgarjuna (Jan [34]). The form is equated with observable material world and emptiness with some kind of Śūnya. Here it would be quite appropriate to explain these terms in a simplified way so as to have practical applications in day-to-day life. Thus when a person eats food, then before he eats food, his mind is busy in visualizing the food which he may be served and while eating also, he will be looking at the food, its quantity, shape, size, taste, etc. If a person is hungry, then his visualization will be very strong, and while eating also his focus will be there totally on the food items. However, after he completes eating the food, all interest in food will disappear, and his attention may be focused on some other subjects. So from the point of view of food, he will have a feeling of emptiness. Again after few hours, the food will take some form in his mind, and the cycle will be repeated several times in a week or so. This example may explain the concept of form and emptiness in a simplified way. Several other examples can be given like examination or any challenge or any activity in our day-to-day life. Thus before examination begins, he will prepare extensively for the examination. Here the form will depend on the type of examination involved and the items or objects and processes involved in the examination. But once the examination is over, all items, objects and processes involved in the examination or their images will disappear in his mind. Hence it appears that time plays an important role in this description of form and emptiness. Also in the case of food, one should respect food and use it properly. The same holds for all other situations. Thus a process-based thinking is required to have a somewhat better view of reality by taking form and emptiness together along with their periodic repetitions. Also there will be a lesson to respect both the situations with equal importance.

15 Some Applications and Examples of Form and Emptiness (Śūnyatā)

In the following section, we provide some more examples from day-to-day life to understand the importance of Śūnyatā in general and the concept of form and emptiness in particular.

15.1 Man-Machine Interplay and the Game Between Finite and Infinite

Some latest electronic gadgets like computers, machines, mobiles, iPhones, the Internet and social networking sites are keeping people so much busy that their interaction with actual human beings and the environment has significantly decreased. It is indicating that machines are now dominating the men and their minds. Temporarily machines and these instruments are of much importance for the users, and the rest of the world is Śūnya, till they return back to the real world. Here it is important to recognize that as per Indian philosophy, human consciousness has infinite potential including infinite knowledge. But it appears that temporarily people are not recognizing this infinite source of knowledge and its potential and are getting confined to the finite domain of the machines. This difference should be recognized at the earliest and appropriate action should be taken to have a balance in one's life. The domination of finite over the infinite is due to lack of proper knowledge of oneself and one's consciousness.

15.2 Google Network and an Ant

Another example of interest would be to compare a Google network with a small ant. Google Web Search Engine is a huge network involving millions of servers and is a satellite-linked system containing and having a huge amount of information. However, it does not have a life compared to a small ant. But an ant cannot process so much information as Google can. So we have a concept of Śūnya with respect to each other. A balanced mind will look at both of them with equal importance.

15.3 Criteria of Richness

The modern criteria of richness also need to be reexamined as there is a lot of tilt towards generation of non-biological structures at the cost of renewable resources. For example, we have 500 fortune companies in the world who are generating maximum revenue every year and are known as the richest companies in the world. However, a majority of them are involved in producing non-biological structures, which do not have life and are dealing with non-renewable resources and matter. While manufacturing their products or providing services they are doing maximum damage to the society and the environment because of their process of automations through machines and computers, there is net unemployment in the world and also huge natural resources are either consumed or are degraded because of their activities. But they are not taken into account while making balance sheets. Again types of states or systems are ignored as if they are Śūnyas, which are actually not. In the olden days, criteria of richness were measured by land and animals possessed by certain individual or a group. It was much closer to the nature. The difference between form and emptiness was less.

15.4 Order and Disorder

Nature has a large amount of ordered systems built into it, which are crucial for the life-supporting system. Thus we find that systems with ordered systems with circular processes are extremely important for the process of life. We have climatological cycles, astronomical cycles, agricultural cycles, soil cycles, water cycles, nutrient cycles, biorhythms, blood circulatory systems, endochronological system and several others. While conducting any scientific experiment, nobody talks about the long-term effect of their experiments on some of these cycles. It is so because this order is ignored while conducting any scientific experiment. Different types of disorders mentioned in Sect. 9.2 are ultimately a consequence of this unidirectional process of doing scientific experiments.

15.5 Discovery of Higgs Bosons and Śūnya

Let us take the example of the study of Higgs bosons in the Large Hadron Collider (LHC). While conducting this experiment, nobody estimates or even try to know how much damage was done to the environment due to this LHC experiment. It may include pollution of water, soil and air. One finds that while performing the LHC experiment, the focus is totally on Higgs boson, and other factors are not relevant, and no scientist even talks about these factors. Hence for them, these factors can be treated as Śūnya. But we know that these factors are not Śūnya.

15.6 "Survival of Fittest" Versus "Live and Let Live"

We have the Darwin's principle of evolution which talks of "survival of the fittest", whereas in the field of spiritual development, one talks of "Live and Let live". The two appear to be contradictory to each other, and each one appears to be Śūnya for the other, but actually a very meaningful concept is there in both the cases.

Several such examples can be given from widely different fields.

16 Implications of This Study and Conclusion

In this paper an attempt has been made to explore the concept of "Śūnya", Śūnyatā or the so-called vacuum from widely different perspectives covering developments in quantum physics and several areas which are of current interests including environment, fundamental elements of science based on closed systems or open systems approaches and others including some interesting ideas from Jainism. It appears that nature or reality is essentially multidimensional or hyperdimensional and also various systems or disciplines are interconnected among them at different levels through various mechanisms. The number of dimensions can be even infinite as per Jainism (anant paryaya). Process of scientific studies, or scientific measurements or human visualization or understanding, is highly restricted due to limitations put by us due to intrinsic definitions of methodologies in "our understanding of a system" along with the limitations of the sense organs and the intrinsic uncertainties at various levels. Because of this one observes only a few dimensions at a time or during certain time interval and "understand" the world as per these dimensions. In the process, a large number of dimensions are left aside or ignored. and one treats them as Śūnya or Śūnyatā.

As per the Indian philosophy, it is our ignorance and ahankar (ego) which comes in the way and keeps us away from the actual reality. In order to solve the current problems of the society and the environment, new fundamental changes in our perspectives are required including meaning of information, knowledge, our perception of reality and so on. Even the directions of development need critical examination, and one should think of how to optimally combine the concept of economic development with spiritual development. In this era of satellite communication, marketing and advertisements using nonliving machines, it appears that finite perception of the reality is dominating over the infinite reality. However, the present technologies like social networking, Workspace, Facebook, etc. should be used to spread multidisciplinary concepts of reality present in many schools of thoughts. Otherwise, we are not only making the world very complex for our future generations, but many other knowledge systems will be lost due to domination of the so-called scientific truth.

Lastly the concept of order may turn out to be more important than any other parameter in any type of study including the scientific studies.

References

1. Adrin Klein. Consciousness and causality: a sub-quantum time symmetric approach. In: Presented at the Fourth International Conference of Science and scientists, organized by Sri Chaitanya Saraswat Institute and the Bangalore University and held at Bangalore University November 5–6, 2016. See pre conference Proceedings, pp. 72–85.
2. Neppe VM, Close ER. How some conundrums of reality can be solved by applying a finite 9-D spinning model. IQ Nexus J. 2015;7(2):7–94.
3. Bhatt SR. Concept note for the International Conference on Quantum Reality and Theory of Śūnya, organized by ICCR, December 09–10, 2016 in Delhi (Abstracts of this conference).
4. Carrol SM. An introduction to general relativity: space time and geometry. San Franscisco: Addison Wesley; 2004.
5. Pokharna Surendra Singh. Study of the effect of the phonon and roton interactions on the transport and thermodynamic properties of liquid helium, A Ph.D. thesis approved by the University of Udaipur, Rajasthan, India. 1977.
6. Khaltnikova IM. Theory of superfluidity. New York: Benjamin; 1965.
7. Andreas S. Introduction to superfluidity, field-theoretical approach and applications. Heidelberg: Springer; 2014.
8. Bohm D. Wholeness and the implicate order. London: Rontledge & Kegan Paul; 1980.
9. Roger P. 2017. http://nautil.us/issue/47/consciousness/roger-penrose-on-why-consciousness-does-not-compute
10. Kachhara NL. Scientific explorations of Jain doctrine, part 1 and 2. Delhi: Motilal Banarasidas Publishers Pvt. Ltd; 2014.
11. Mehta ML. Jain psychology: an introduction. Varanasi: ParashvanathVidyapeeth; 2002.
12. Mehta ML. Jain philosophy: an introduction. Bangalore: Bhartiya Vidya Bhawan; 1998.
13. Hugenhold NM, Pines D. Ground state energy and excitation spectrum of a system of interacting bosons. In: Pines D, editor. The many body problem. New York: Benjamin; 1962. p. 332.
14. Pokharna Surendra Singh. Quantum field theory like model of consciousness and matter in Jainism. In: Proceedings of International conference on Quantum Mechanics and Quantum Consciousness, being organized by Amal Jyothi college of Engineering, Kottayam, Kerala, India from January 6–9, 2015, pp. 93–101.
15. Close ER. Transcendental physics. San Jose: i.Universe.com; 1997.
16. Besant A, Besant A, Leadbeater CW, Jinarajadasa C. Occult chemistry: clairvoyance observations on the chemical elements. Cosimo Classics; 2007, original published in 1908.
17. Stephen P. Extrasensory perception of subatomic particles I. historical evidence. J Sci Explor. 1995; 9(4): 489–525.
18. Surendra Singh Pokharna, Samani Chaitanya Prajna, Manohar Lal Kalra, Rajmal Jain, Jagat Singh Pokharna. Some Quantitative Evidences indicating that knowledge is structured in the consciousness: A recent Theory of Everything supports this possibility. ISJS trans. January-March 2019. New Delhi.
19. Landau LD, Lifshitz. Quantum mechanics (non-relativistic theory). Oxford: Pergamon Press; 1964, pp. 28 for definition of energy and pp. 42 for definition of momentum in QM.

20. Agarwal PM, Sharda R. OR-forum, quantum mechanics and human decision making. Oper Res. 2013; 61, 1–16.
21. Narendra B, Pokharna Surendra Singh. Syadavada and Anekantavada in the modern scientific context, Compendium on Science and Mathematics in Jainism, collection of white papers, Bhagwan Mahavir International Research Center, for Scientific Research and Innovative Studies in Social Sciences, Ladnun (Rajasthan), India. 2017.
22. Pokharna SS. A new investigation into the problem of perfect determinism in modern science. Indian Philos Q. 1985;XII(1):67–83.
23. Laplacian Demon in https://en.wikipedia.org/wiki/Laplace%27s_demon.
24. Singh PS. Knowledge, order and evolution proceedings. Gyan Sagar Foundation, presented in Mumbai conference January 7–8, 2012.
25. Narendra B. Jainism: the eternal and universal path to enlightenment (a scientific synthesis). Jaipur: PrakritBharti Academy; 2015.
26. Prajna C, Kachhara NL, Bhandari N. Scientific perspectives of Jainism. In: Proceedings of a two day conference organized by Bhagwan Mahavir International Center for Scientific Research and Social Innovative Studies, Jain Vishwa Bharti, Ladnun, Rajasthan, India. 2016.
27. Jainism (General Books on Jainism) see the link www.jainlibrary.org/english.php.
28. Dhammo J. Aacharya Shri Nanesh (Nana Lalji Maharasahab). Bikaner: Akhil Bhartiya Sadhumargi Jain Shravak Sangh; 2008. p. 202–9. (In Hindi)
29. Chalmer DJ. The conscious mind: in the search of a fundamental theory. In: Philosophy of mind. revised ed. Oxford: Oxford University Press; 1997.
30. Rangarajana A. Can a quantum field theory ontology help resolve the problem of consciousness. In: Pre-proceedings of the International conference on Quantum Reality and Theory of Śūnya, organized by Indian Council for Cultural Relations (ICCR) in New Delhi on December 9–10, 2016.
31. Strawson G. Consciousness and its place in nature: does physicalism entail. Thorverton: Imprint Academic; 2011.
32. Niskama SB. Life and consciousness–the Vedāntic view. Commun Integr Biol. 2015;8(5): e1085138.1–11 (electronic version).
33. Kothari DS. In: French AP, Kennedy PJ, editors. The complementarity principle and eastern philosophy, Neils Bohr centenary volume. Cambridge: Harvard University Press; 1985. p. 325–31.
34. Westerhoff J. Nāgarjuna Madhymaka, a philosophical introduction. Oxford: Oxford University Press; 2009.
35. Aspect A, Grangier P, Roger G. Experimental realization of Einstein-Podolsky-Rosen-Bohm Gedanken experiment - a new violation of bell inequalities. Phys Rev Lett. 1982;49(2):91–4.

Chapter 20
The Concept of Śūnya with Reference to the Pañcasakhās, the Mahimā, and the Modern Quantum Reality

Bijayananda Kar

Abstract His paper deals with the concept of *Śūnyapuruṣa* propounded by Mahima Swami, Bhima Bhoi, and Pañcasakhas. This concept has been compared and contrasted with the concept of quantum reality. It is also contented that comparison with Buddhist conception of reality also may not be apt.

Keywords Śūnyapuruṣa · Pañcasakhās · Jagannātha · Mahimā · Mādyamika · Quantum reality · *Bhakti*

Advaita Vedānta and Buddhist leaning at the background of Śrī Jagannāth and Mahimā Movement its Budhistic and Advaita Vedānta impact and its social reformative steps. The modern concept of Quantum Reality its role, as based on New Physics and its bearing on metaphysics.

1 Introduction

The concept of *śūnya* is not alien to classical Indian philosophical sources. In common parlance, it is used as void/empty. That general use is also found to have been referred to in certain philosophical tradition like Śaṅkara Advaita where '*Śūnyata*' is ridiculed as, a philosophical stand. Even though Śaṅkara himself interpreted his Advaita as, indeterminate, like śūnya (*savisesa rahit atvāt śūnyavat*).[1] It is Nāgārjuna, the noted Màhāyana Buddhist, advocating *śūnyavāda*, holds *śūnyatā* as the ultimate *tattva* (essence) and that which is directly realized and as calm, where

[1]Vide his commentary on the *Visnusahasranama* (Qtd. Also in T. Patnaik, *Śūnya Purusa*, New Delhi: D.K. Print world, 2005, p. 103)

B. Kar (✉)
Department of Philosophy, Utkal University, Bhubaneswar, Odisha, India

S. R. Bhatt (ed.), *Quantum Reality and Theory of Śūnya*,
https://doi.org/10.1007/978-981-13-1957-0_20

all forms of plurality get merged.[2] All the same, the concept of *śūnya* does not remain as applicable to the talk of the highest reality, i.e., *Īśvara*. In other words, the allied concept of *Śūnya Puruṣa* is not noticed in some later *Upaniṣads* and *Smṛtis* like *Mahopaniṣad, Śrībhāgavatam*. These hold that *Parama Brahman* is *śūnyavat* but not *śūnya* as such.

It is again notable that for Nāgārjuna, śūnya is beyond fourfold categorization. In that way, existence, nonexistence, neither a combination of both nor a negation of both can be predicated of *śūnya*. But, at the same context, it cannot be gainsaid that there still remains some similarity between the Advaitin's concept of *Parama Puruṣa*/Brahman and the Buddhist's concept of *śūnya*. Both are elucidated in terms of non-categorical discrimination, transcending all qualities, indescribable and inconceivable. Even then, the Śaṅkarite position is said to be in favor of the highest reality (*Paramasat Brahman*), while the Mādhyamika position is interpreted in some quarters as totally nihilistic and not in favor of committing to any form of absolute or highest Reality. In modern time, it is interpreted that negation of all categorization alone is the mark of conceptual analysis without committing to any metaphysical stand [1].

It is held that around the fifteenth century, the Oḍiyā rendering of *Mahābhārata* was composed by a Sudra sage: Sāralā Das in which there was not only the narration of the great Indian epic but also there was frequent inclusion of Oḍiyā localized anecdotes in it for having a popular appeal to the then masses. And, as a result, it is even now considered as a valuable literary piece in the region. It presented locally the most favored deity: Jagannāth as both Buddha and Kṛṣṇa.[3] Sāralā Dās presented Jagannāth as *nirākāra* (shapeless), *Alekha* (unwritable), *Niraṇjana* (pure and untainted). He further said Jagannāth as śūnya purruṣa. Thus Kṛṣṇa, in Jagannātha-cult, was turned into *Śūnya Puruṣa*-Buddha Jagannāth. This turning, in later phase, was taken up with great effort, by the subsequent writers of the *Pañcasakhā* group. It became not only popular in the region but it has brought out new identification of Ọdishan Vaiṣṇavism with a touch of both traditional Vedāntic *Bhakti* movement and Buddhist emphasis on *nirguṇa* (devoid of quality) *tattva*. It gave rise to the new coinage of the expression: *Jñāna-miśrā bhakti* i.e., devotion being blended with knowledge: This sort of amalgamation is something strange in contrast to Bengal or Gauḍiya Vaiṣṇavism which is singled out as *prema bhakti* (devotional love).

Not only Sarala Das nor the subsequent *pañcasakhās* but also numbers of later Oḍiyā writers are cling to have devotion to all pervasive cosmic Lord without having any belief/faith in personal god. Such is the rendering of *Jñāna-miśrā* bhakti of the Jagannātha-cult. It has the touch of both Buddhism of a specific form of later period and also bhakti tradition of the old established pattern. It is claimed that the santhas of Buddhist Tāntrism, specially Vajrayāna and Sahajayāna where there was the advocacy of *bhakti sadhanā* through typical yogic practices of hidden type, not of

[2]"*aparapratyayam santam prapancairapara pancitam, nirvikalpam anarthamr etat tattvasya laksanam*" *Mādhyamika Kārikā*, XVIII. 9

[3]T. Patnaik, op.cit., pp. 90–91 and 110–133

any personal god or goddess but some form of Advaitic harmonious whole. Such practices gained currency in some parts of Oḍishā but remarkably were somewhat obscure. Such obscure and esoteric practices did not find favor of the *Pañcasakhās* and others. The influences of Buddhist non-dualism as well as non-categorization at the knowledge sector was approved with enthusiasm for unifying the same with impersonal devotion to cosmic Lord. The present-day scholars on Jagannāth mode of daily rituals points out that in the usual *pujā* of Jagannāth both individual and cosmic consciousness in different parts of the human body, viz., heart and head.[4] Such is the brief note on the essential feature of the Jagannāth-cult, in its different facets with special reference to the *Pañcasakhās*.

But, despite all these positive renderings, by way of attempting to detail out the metaphysical basis of Jagannāth theological formulation, it seems there persist some difficulty and obscurity in estimating the concept of *Śūnya-Puruṣa* in the framework of Jagannāth.

First, in what sense, the negative connotation of any idea or anything what so ever (in form of non-predication, non-categorization, indescribability) is also treated as *Puruṣa* (in whatever sense, i.e., individual *Puruṣa* or cosmic *Puruṣa* or absolute *Puruṣa*)? That is clearly of the mark of some description of some unique form of categorization like infinite/Highest Reality; and, again that is to point at the Entity/ Being for the purpose of devotion, aiming at self-surrender or praying for His grace or aiming at salvation. Is such explanation cognitively comprehensible or is it a mark of obscure emotional outburst?

Secondly, there is no show of any reasonable basis as to how the *nirguṇa tattva* is compatible with *bhakti* toward *Brahmeśuara* (i.e., not the same as Brahman, the absolute Reality, as advocated by Advaita). The concept *Īśvara* is the product of *māyā* and, in that way, it is recognized as of only *vyavahāra-paraka* (i.e., practical convenience). In case of the theist Vedāntins, the concept of *Brahmeśvara* is introduced only in redefining Brahman tacitly to suit their interpretation. This shows that different Vedāntins advance their respective rendering to suit their purpose of verbal manipulation, least caring for the objective mode of description which is clearly implicit in the expression itself.

Further, originally Buddha has never proclaimed himself as God (*Īśvara*), nor as the incarnation of Him (*avatāra*). In later Buddhist sources, in some quarters, Buddha is adored as Lord (*Bhagavān* or the great man, guiding the followers and supporters) but not as creator-God. So, depicting as an incarnate of the supreme Lord (*Īśvara*) is inadmissible in the Buddhist frame-work. Most probably, after the departure of the Buddhist followers (in large scale) to outside India, the traditional Vedic spokesmen and their followers have tried (with some amount of success) to incorporate the Hindu dharmic views and beliefs in Buddhism, to point out that the two systems are not radically different, but there is the inner presence of mutual harmony and adjustment. Whether such rereading and reappraisal is coherent/

[4]G.C. Tripathi (Qtd. By T. Patnaik, op. cit., p. 143)

consistent to the original standpoint, formulated by the two traditions does appear to be debatable.

The attempt made by some thinkers to present *Pañcasakhās'* point of view as basically based on Tāntric Buddhism of Sahajiā cult is found to be questionable. True, the Sahajiās emphasize on yoga as the background *Sahaja sadhanā*. That has led some to surmise the Pañcasakhās' approval of general Tāntric practices in their set up which have their roots in Tāntric Buddhism of a particular variety. But the esoteric practices, being based on some form of sex appeal, supported by *Sahajayāna* yogic procedure is not found to have been advocated by the Pañcasakhās' spokesmen.

Again, the attempt made by some to claim that Buddha is the incarnate of Jagannāth and the anecdote that Buddha's tooth is preserved in the body of Jagannāth-idol is a matter of illicit assumption having least evidential validity as a matter of fact. Such assumption is rationally not cogent and convincing.

Further, the attempt to unite Buddha with Jagannāth by such forced assumption and make-belief only reveal to justify, somehow or other, the greatness of Jagannāth as supreme and the best among the rest. This indirectly brings cheap localized sentiment to the fore-front and paves the way for unhealthy regional so-called supremacy. The same point can also be raised with regard to the *Pañcasakhās'* claim that Kṛṣṇa is the incarnate of Jagannāth of Puri. Such renderings, really speaking, are neither theologically aspiring nor philosophically reasonable. It is therefore evident that to characterize Pañcasakhā movement as boosting up both Buddha and Viṣṇu on par with Jagannāth appears to be somewhat far-fetched.

The concept of *Śūnya-Puruṣa* may be a novel attempt by the then Ọdishān thinkers for harmonization and unification of Buddha and Hindu dharma; but it is religio-philosophically not that convincing. As a matter of fact, such claim advanced by some Hindu theistic group does not find parallel echo among the Buddhist adherents.

The concept of *Śūnya-Puruṣa* with the background *Śūnya*-bound philosophical absolutism is again advocated by one reactionary as well as reformative religious movement in Ọdishā (during the eighteenth-nineteenth century) called as Mahimā dharma darshana. Mahimā Svāmī, the founder of the movement was there during the last part of the eighteenth century. Later, it was Bhīma Bhoi, who was a Kondha by birth, propagated this religio-cum-social movement through his Oḍiyā lyrics (*bhajana*) which are, even now, of great appeal to the local people, mostly of the down-trodden. He championed the cause of *śūnya* as well as *śūnya-dehi*, *arūpa puruṣa* in his own writing.[5] This movement, it is said, is originally having the Buddhistic flavor by holding the firm belief that Buddha would appear for the spread of knowledge proper.[6]

---

[5]Vide Bhīma Bhoi's bhajana "*ruparekha nahi he śūnya dehi, sunye ude hoi*" (The śūnya-bodied, having no form, has manifested in Śūnya)

[6]N.N. Vasu, *The Modern Buddhism and its Flowers in Orissā* (Qtd. In T. Patnaik, op.cit., p. 78)

The Mahimāits, consequently, have been viewed by some scholars as upholding the main ideals of Buddhism.[7] Their advocacy of *alekha-śūnya* etc. are highlighted in this connection. But, it seems that Bhīma Bhoi (as indicated before) who was virtually the first to compose Mahimā-dharma ideas by way of lyrical expressions has clearly opted for personification of *Śūnya* as *Śūnya-dehī*. So, in this sense, the idea of impersonal theistic Lord (*Śūnya Puruṣa*) cannot be ruled out from Bhīma's thought.

And, perhaps, this is the ground because of which, the later Mahimāits in the twentieth century have moved for unifying Mahimā dharma under the generally approved Vedāntic fold, i.e., the Advaita concept with some notable modification. Accepting *Puruṣa* as *Śūnya*, they continue to be rejecting any form of idol-worship in form of an image. Biśwanātha Bābā is the foremost leader of this new modification. He has interpreted and named the Mahimā movement as Viśuddhādvaita darsána/Alekha parama Brahma Darshanam. According to him, Śaṅkara's *māyā* does not fit to the true spirit of Advaita. For Brahman, even being *nirākāra*, *alekha*, etc., is not different from *Īśvara*, the alt pervasive (cosmic) *Puruṣa*, to whom prayer in the non-idol (*śūnya*) form is admitted by them. That means in this new trend, Mahimāit's thoughts and ideas continue in the fold of essential feature of Vedānta and Buddhism.

However, one thing is to be noted here; the Mahimāites are in favor of *Śūnya Puruṣa*, both under the group of Advaita and Buddha *Śūnya-vāda*, but not in favor of Vaiṣṇavism of any variety that accepts idol-worship, devotional group/singing, praising and celebrating similar public rites (*kirtan*). The Mahimāits have their own places (like their religious establishment, i.e., *maṭha*) where they advocate and reiterate their view-points, but not any such public propaganda or demonstration (i.e., *nagara kirtana*). In this sense, it seems, perhaps, better for Mahimā stand's originality as distinct from bothg Buddhism and Vaiṣṇavism. Again the theological *śūnya*-bound absolutism is emphasized without any idol-worship, personal god/goddess and those being supposed to be born again and again.

Another novel and befitting idea to the present secular fabric, the Mahimā thought-pattern is in favor of equality of all castes without having any gradation of higher and lower caste and holding a pure non-discriminative temper all through. This, however, has a notable social reformative step.

So far, the Mahimā point of view is essentially touched upon. But, there is also scope for its critical appraisal. The Mahimāits have not stated any reason-based grounds for their spiritual-cum-metaphysical stand. Unlike Śaṅkara, Rāmānuja, and others, the Mahimāits have not commented upon the *Brahma-sūtras*. So, Biśwanātha Bābā's nomenclature of his Mahimā-stand as *Viśudhādvaita Darsána* (without being a commentary) lacks classical sanction of Advaita Vedānta. So also, Bhīma Bhoi's interesting lyrics with the splendid touch of human fellow-feeling is a great boosting for humanist outlook. But, the secular spirit underneath humanism becomes somewhat beclouded with the clear tone of shapeless, all pervasive theologism

[7] T. Patnaik, op. cit., p. 225

(*Alekha, Arūpa Īśvara*). For, the humanist outlook rests upon a non-theological, secular stand.

To advocate against caste-discrimination is a sound reformative step, in this Oḍishān religious thought of India. But, it is not brought out clearly as to why such idea of caste-rigidity is to be necessarily tied down to a particular theological set up. That is virtually a social reformative movement which is entertained in an open-setting anywhere, among any group or locality/region without being necessary linked up with a theological setting. Secular outlook is not grounded under any religious setting what so ever. Rather theologism of any sort is to obstruct the free flow of secular human spirit in its openness.

Now, as per the arranged plan, a look into the modern scientific concept of Quantum reality is taken up at the context of *śūnya*. The modern physicists, it is held "do not have a single picture of the world". To them "there are eight ideas of quantum reality". Those are:

(a) There is no deep reality.
(b) Reality is created by observation.
(c) Reality is an undivided wholeness.
(d) Reality consists of a steadily increasing number of parallel universe.
(e) The world obeys a non-human kind of reasoning.
(f) The world is made of ordinary objects.
(g) Consciousness creates reality.
(h) The world is twofold, consisting of potential and actuals.

The principal features of quantum theory contradict the "cause and effect" relationship by assuming that random, spontaneous events can occur, as specified by the Heisenberg Uncertainty Principle.[8]

Such basic note about quantum reality clarifies at least one point, i.e., it is a new physics, different from traditional physics. It is named as quantum physics that deals with sub-atomic particles. It is a specialized science, different from Newtonian physics. It explores and explains as to how everything in the world comes to existence. It is also held "that it determines every aspect of reality, including *but not limited to all aspects of your life*."[9]

In this connection, a new popular science book (1985) has come out by noted physicist Nick Herbert, i.e., *Quantum Reality* – Beyond the new physics: An excursion into metaphysics (From Wikipedia: The Free Encyclopedia). It tries to the ontology of quantum objects, their attributes, and their interactions. In introducing quantum objects, Herbert explains "how quantum properties inhere in a proxy wave function."[10]

---

[8]News Letter and Articles Video of Books Papers Presentations speakers links. (Source: http://www.commonsense-science.org/quantumreality.html)

[9]Source: http://www.abundance-and-happiness.com/quantumphysics.html

[10]https://en.wikipedia/org/wiki/QuantumReality

Another interesting writing, namely "Common Grounds between Buddhism, Quantum physics and the Baha' i Faith" has come out by Jack Coleman.[11] Here the writer tries to show notable "Common grounds among Buddhist, Quantum physics and Baha' i Faith concepts". According to him, "Voidness (*Śūnyatā*) is used to characterize the Absolute, the Ultimate Reality. *He is not creator*. Behind this changing world and its illusions, there is the Buddhist Void or Absolute Reality (God). The "void" or "nothingness" of quantum physics is distinctly different from than the "Void" of Buddhism".[12]

From such brief but clear citations, it is evident that quantum reality is a topic, introduced as beyond the new physics. Therein mentioning quantum objects, it is said as to how "quantum properties inhere in a proxy wave function".[13] Quantum, a term introduced and used in physics, means "a very small quantity of electromagnetic energy".[14] As a theory "it is based on the idea that energy exists in units that cannot be divided".[15] As such, it deals with force and movement in pieces of matter, smaller than atoms. That is why; it is said as dealing with some atomic particles.

With acceptance of all these explanations, it seems to be fairly clear that quantum reality is acknowledged as something positive and definite. In this sense, prima-facie, it is found to be against all forms of negation or negativity. If this is the case, then to locate the similarity between quantum reality and the Buddhist concept of total void or *Śūnyatā* (as taken for granted in certain quarters) becomes implausible. And, in this manner, one can find justification in the view that while physics/New physics deals with positive phenomenon/phenomena, Buddhism, as a religion (in the broader sense, i.e., beyond the original meaning of it, as established mainly in Christianity and some other religious tradition) is different from the quantum reality that deals with sub-atomic particles like electron and proton/neutron (a very small piece of matter).

Again (as already indicated before), in the study of quantum in new physics "there is no deep reality"; whereas those Buddhist scholars and others have the advocacy of *śūnya* as not simply void/empty but as directing toward absolute Reality, i.e., *Puruṣa* in the superlative sense, equivalent to the concept of God that appears to be theo-centric, hinting at deeper sense of reality. This shows that the new physicists' exposition of quantum reality as twofold of both potential and actual (as said before), the Buddhists' advocacy of *śūnya* as not void but as absolutely positive real existence (in the equivalent sense of theistic God/*Īśvara*) are fully different in meaning.

New physics as indicated in "Quantum Reality" is, of course, distinct from classical/traditional physics. The talk of reality is there inclusive of both observation

[11]Source: https://bahai-library.com/coleman_common_grounds_buddhism

[12]op. cit, pp. 1-2-3 (italics mine)

[13]op. cit (Sl.No.11), p. 1

[14]*Oxford Advanced Learner's Dictionary*, OUP, 2005, p. 1233

[15]ibid

and experiment. Whatever it may be, it continues to have the open-textured study and testing, whereas the concepts of *Śūnya-Puruṣa*, as advocated by tradition-bound Advaita Vedantins, inclusive of both *Pañcasakhā* and Mahimā followers, are in their own way, have theological speculation/belief-assumption, and have the print of metaphysical stipulation that does not carry scientific backing. It is in predominant form of religiously coated and faith oriented stand. As such, it does not yield to scientific methodological justification.

Further, the admittance of both potential and actual in the world of quantum reality cannot be taken as similar to the Buddhist talk of *śūnyatā* which is clearly stated as beyond all attributions/categorization or qualifications. Even in that sense, to bring in there the sense of undivided wholeness surely appears to be improper and unwarranted.

It is again notable here that Herbert (as stated before) has sub-titled his popular work: *Quantum Reality* as an "excursion into metaphysics". The use of the expression: metaphysics, in this context, needs certain clarification. It goes without saying that metaphysics means, as usual, beyond physics. That is, it is not physics as traditionally understood. It is also not new physics in the sense it deals with quantum reality as beyond the acceptance of matter as is normally understood as espousing a metaphysical theory of crude materialism. Herbert, due to this, subtitled his book as 'An Excursion into Metaphysics'. Probably to confirm the point held by Quantum physics that it "determines every aspect of reality including, but not limited to all aspects of life" (as stated before).[16] It is needless to point out that physics never claims to deal with each and every aspect of life, realizing fully well that there does exist in life-situation many other issues and problems that need separate mode of enquiry and investigation. That is the reason, in course of living, man, in general, is exposed to different avenues for search of further knowledge. Probably, in this way, the remark made by Herbert becomes meaningful and cogent.

In course of both *Pañcasakhās* and the Mahimāits, unlike the remark held by the modern physicists including the advocates of Quantum Reality, seem to be entirely different particularly from the concept of *Śūnya-Puruṣa*. Therein, the talk on such metaphysical reality is not clearly based on free and open-textured course of reason, but rather more on imaginative stipulation, not being grounded on physical or even new physical form of investigation in terms of various types of observational and experimental mechanism. Further, the result of scientific investigation never aspires to have absolute finality. The sense of revision, modification and even falsification ever remains in scientific enquiry and that is why such probability is ever absent in such metaphysical positions. The point of view advanced by Pañcasakhās and Mahimāits lacks argumentative rigour of both opposing and supportive (*pratipakṣa/svapakṣa*) argumentative processes. Whatever conclusions are adduced by them are more of religio-theological and sometimes, of course, socially reformative-oriented but not so much on free-floating reason-based move.

[16]op. cit (Sl. No.10)

It is, as such, concluded here that whatever is found so far and is traced out in the old spiritual sources, concerning the talk of *Śūnya-Puruṣa* it seems not enough to locate a substantial form of similarity between that and modern physicist conception of Quantum Reality.

Reference

1. Nayak GC. Mādhyamika Śūnyatā. New Delhi: ICPR; 2001.

Chapter 21
Quantum Reality and *Śūnya*

Jayant Burde

Abstract In this paper points of commonality between quantum reality and nature of Śūnya have been discussed. Principles of Max Planck's constant along with particle-wave duality have been analysed. The problematic issues of quantum physics have been approached through a ready solution in Buddhist theory of catuṣkoṭi. Uncertainty principle propounded by Heisenberg has been discussed in the context of theory of Śūnya. Śūnya has been explained as infinity and fullness. Parallels have also been drawn from Upaniṣads 4.

Keywords Quantum theory · Max Planck · Particle-wave duality · Uncertainty principle · Subatomic particles · Quantum mechanics · Catuṣkoṭi · Śūnya · Brahman

1 Introduction

The words "quantum theory" and *śūnya* (void, zero) evoke awe in the minds of lay persons who feel that both these concepts are shrouded in mystery. In this short paper, I shall discuss only briefly quantum reality and the nature of *śūnya* in Indian philosophy and try to find out if a common thread runs through these concepts which apparently occupy two disparate regions of knowledge.

2 Quantum Theory

Two revolutionary theories in physics were born in the twentieth century: relativity and quantum theories. The former banished absolute space and time to be replaced by space-time. Based on Michelson-Morley experiment, the velocity of electromagnetic waves, of which visible light occupies a small range of frequencies, was found

J. Burde (✉)
Bombay University, Mumbai, Maharashtra, India

S. R. Bhatt (ed.), *Quantum Reality and Theory of Śūnya*,
https://doi.org/10.1007/978-981-13-1957-0_21

to be a universal constant. Under certain conditions mass could be converted into energy and vice versa. In the presence of gravitational field, the "flat" space-time becomes "curved".

Quantum theory was weirder and opened up a bizarre microworld at the sub-atomic level. Einstein made a major contribution to both these theories.

There were two theories about electromagnetic radiation. Newton assumed that light consists of fast-moving particles. Huygens maintained that light consists of waves. Both these theories were ad hoc, designed to explain certain properties of light. The correct form of wave equation was found by Maxwell who incidentally vindicated Michelson-Morley's result of constant velocity c of light.

In 1900 Max Planck discovered a strange phenomenon: electromagnetic radiation is absorbed or emitted in discrete packets of energy called quanta. His famous equation is:

$$E = \mathrm{h}\nu$$

where E is the energy, ν is the frequency, and h is "Planck's constant". If light is in the waveform, the discrete absorption or emission cannot be explained. The great dilemma of "is light a wave or a stream of particles" engaged the best brains in physics.

In 1923 de Broglie predicted that even other entities would show the particle-wave duality. This was subsequently proved. Entities like electrons did behave sometimes as wave and sometimes as particles.

3 Particle-Wave Duality

Interference is the well-known property shown by waves. If two waves of the same frequency, amplitude and velocity, meet, the points where two crests (or troughs) fall on each other, the waves reinforce each other. Where the crust of one wave meets the trough of another, there is cancellation. At other points there is partial reinforcement or cancellation (Fig. 21.1).

This effect can be shown by allowing light from a monochromatic source S to pass through two slits E and F equidistant from S. When only one slit is opened, the light illuminates the screen DN (perpendicular to the plane of the paper) almost uniformly. When the intensity of S is reduced, the screen shows spots displaying the corpuscular nature of light. The other slit is now opened. What we see is an interference pattern of wavy nature with alternate bands of brightness and darkness, thus proving that light behaves like a wave (Fig. 21.2).

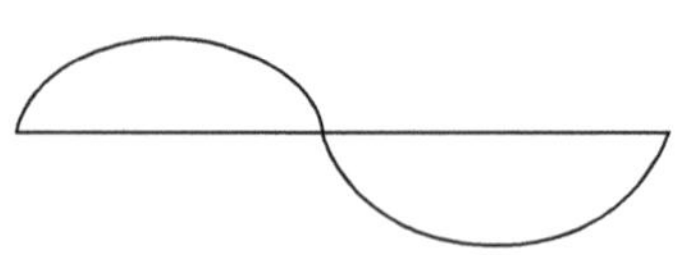

Fig. 21.1 A wave form

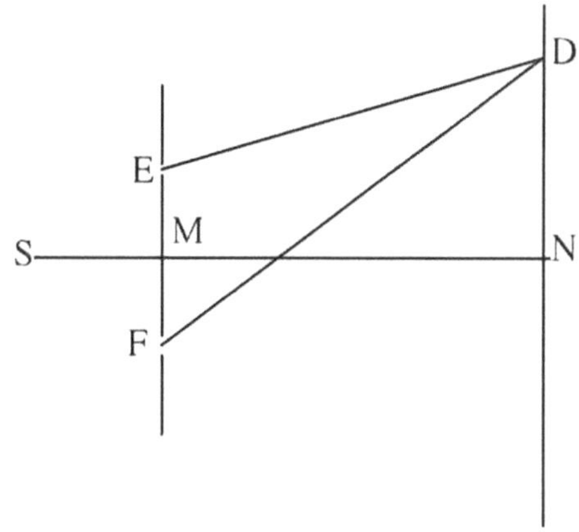

Fig. 21.2 Interference

If the intensity of source S is reduced sufficiently, it can be ensured that not more than one photon at a time passes the vicinity of the slits.[1] According to Penrose, the phenomenon of interference arises even from one single photon. Most physicists agree that a photon (single light particle) cannot split, and hence the question of split parts giving rise to interference does not arise. Experimenters have tried to use a photon detector near a slit to find out which way the photon moved. The moment you keep a detector, the interference pattern disappears. Ostensibly it looks as if the lack of knowledge as to which slit the photon passed through is a precondition for interference to occur.[2]

When a scientist meets a hurdle, he tries to suggest the easiest way out. The experimental dilemma was whether light, a stream of photons (light particles), or a wave is. Some physicists found a ready solution in *catuṣkoṭi*, a doctrine found in Mādhyamika school of Buddhism founded by Nāgārjuna (ca 150–250 CE).

In our ordinary logic, a categorical statement has two and only two alternatives, e.g.

A is rich
A is not rich

In *catuṣkoṭi* there are four alternatives:

A is rich
A is not rich
A is rich and not rich
A is neither rich nor not rich.

It is possible that as in *catuṣkoṭi* we need a four-limb logic which would admit four possibilities:

[1]See Penrose [1].
[2]Ibid. p. 246

x is a particle
x is a not-particle
x is both a particle and a not-particle
x is neither a particle nor a not-particle

This argument is fallacious unless we assume that not-particle equals wave. There is no reason to assume this identity. We may argue that light consists of "wavicles" or "partiwaves" which have dual properties of particles and a wave. We need not deviate from the two-option system of alternatives which are mutually exclusive and also exhaustive. It is necessary to remember that we cannot see photons or subatomic particles; we can only observe their interactions. A photon may divide itself or not. It is irrelevant when we create a third category of wavicles which behave exactly as the light behaves.

We thus assume that electromagnetic radiation consists of wavicles or partiwaves. We do not know their ontological nature, but their epistemology in some aspects is quite clear; they behave exactly as electromagnetic radiation does.

4 Uncertainty Principle

The quantum world is linked with an uncertainty principle first propounded by Heisenberg. This appears as an inequality, viz.

$$\Delta x \; \Delta p \geq h/2\pi$$

where Δx and Δp are accuracies in position and momentum of a particle.

To quote Penrose:

> According to this principle it is not possible to measure (i.e. to magnify to classical level) both the position and momentum of a particle accurately at the same time.[3]
>
> What this means is that the more accurately we try to measure one parameter (say, position), the more inaccuracy appears in the other (momentum). What's more, this is not attributable to an experimental error, but an inherent property of quantum reality.[4]
>
> Before we explore the quantum world further, we should introduce the concept of *śūnya*.

[3]Ibid. p. 248.

[4]This conclusion leads scientists to mysticism. A more rational and scientific view would be to take into account the possibility of epistemic ignorance which in due course can be cleared when they acquire more knowledge of the quantum systems.

5 *ŚŪNYA*

All these words (zero, *śūnya*, nihil, nonbeing, nonexistence, void, emptiness, nothing) convey a common core "nothing". We are aware that these words are not synonyms and the nuances among them can be gleaned only from the context.[5]

For us the variants "emptiness" or "void" are most relevant. Note that while these words connote "nothing", they have a linguistic freedom to expand to infinity spatially (and temporally). This allows philosophers to bridge the gap between zero and infinity, e.g. "infinite void". Vedantic monists treat "sky" (Ākāśa) as equivalent to Brahman, the ultimate reality.

In Sanskrit the system of word numerals (*bhūtasankhyā*) is interesting. Some word numbers for zero are *ananta* (infinite), *kha* (sky) and *gagana* (sky) which suggest a link between *śūnya* and infinity.

6 Is Void Really Empty?

While the concept of infinite void exists in most schools of Indian philosophy, there seems to be some ambiguity about its nature.

In monism Brahman "the ultimate reality" is all-pervading, eternal and, what's more, pure consciousness. It pervades the whole universe outwardly as well as inwardly.

A well-known invocation in Upaniṣads may be quoted.

Om the infinite fullness is that Brahman The same infinite fullness arises from this finite universe. When this same infinite fullness is taken away from the infinite fullness what remains is infinite fullness.

Mathematically this is true of zero as well as infinity:

$$\infty - \infty = \infty; 0 - 0 = 0$$

This juxtaposition is important for us because the word used in the above invocation for fullness is *pūrṇa* which in word number system means zero or *śūnya*. This leads to an oxymoron "void which is full, i.e. not empty".

7 *ŚŪNYATĀ*

The conception of *śūnyatā* found in Nāgārjuna's philosophy seems to have evolved from *śūnya* (void). It can be traced to the inevitable association of negative and positive.

[5]Similarities and differences relating to these words have been extensively discussed in Burde [2].

$$\text{Remember } x - x = 0$$

Śūnyavāda does not imply denial of reality. It considers the phenomenal world unreal. Consider the following arguments:

1. Following the doctrine of conditional origination, anything that is real must not have origination or must not be conditioned. But every reality we know is dependent on some condition. Hence anything we call real is unreal.
2. Reality cannot be called unreal because it does exist, and anything that is unreal cannot come into existence.

These contradictory conclusions should lead us to the following:

3. Reality is both real and nonreal.
4. Reality is neither real nor nonreal. These are nothing but the last two limbs of Nāgārjuna's *catuṣkoṭi* cited earlier.

Śūnyatā has also been defined in the following different ways:

1. It is characterized by what we perceive; or the phenomenal world may be called *śūnyatā.*
2. *Śūnyatā* also means absolute truth which is unutterable or indescribable. One can feel this state when one attains *nirvāṇa.*

8 Subatomic Particles

There was a time when physicists thought that an atom consists of a nucleus which in turn held protons and sometimes protons and neutrons. Electrons, negatively charged particles, revolved around the nucleus. These particles were called elementary particles. However, after several years of laboratory experiments, they have found there are at least 200 elementary particles.

The world of these subatomic particles appears mysterious as the following facts will show:

1. Most particles have a very short life, less than a millionth of a second.
2. Most particles appear in pairs. For example, an electron is negatively charged and has a partner positron of the same mass and has equal positive charge. When a particle and its antiparticle coalesce, the masses vanish and energy is released.
3. When we try to divide a particle x with high energy (E), we have the following:

$$x + E = x + x$$

which means the energy is converted into the same type of particle x. This has made Capra claim "The subatomic particles are destructible and indestructible at the same time" [3].

In high energy physics, relativistic equations have to be used because of high speeds. Today the physicist looks upon the force between two particles as an exchange of particles. These particles are called virtual particles. It is guessed that even the nucleons (protons and neutrons) may emit virtual particles. A few reactions will tell you how strange the subatomic world is

1. $n \rightarrow p + e + \nu(a)$

This means a neutron decays into a proton (+), an electron (−) and an antineutrino $\nu(a)$ which is massless.

2. $0 \rightarrow p + \pi + p(a) \rightarrow 0$

Three particles are created from vacuum, a proton, a pion and an antiproton, but they disappear immediately. This example shows that a vacuum hides the particles which may vanish as soon as they are visible.

3. $n \rightarrow n + \pi \rightarrow n$

A neutron emits a pion which immediately merges with the parent.

If you try to get the "feel" of this quantum world, you can only say that it is a mad, mad world, full of uncertainty. At one moment you find that the space is empty; the next moment you see many actors who dance only for a short time and disappear. You cannot say when energy will transform into particles and vice versa. Discovery of virtual particles suggests that even the boundary between the field of force and substance has become blurred. You cannot keep track of an entity, and it looks as if the concept of identity and individuation has disappeared.

9 Quantum Mathematics

To deal with paradoxes and contradictions, physicists have created a complex mathematics where we can speak of only probabilities. Each quantum particle is associated with a wave function ψ. The strange thing is that when the particle is in the quantum state, it seems to be hiding from the physicist who can measure its parameters (position, momentum, etc.) only when the system is enlarged to the classical level. What's more, there are two different procedures: U (deterministic behaviour) and R (involving uncertainties and probability). The problem physicists face is they do not know whether to use U or R at a critical juncture.

Quantum theory has led to many paradoxes and antinomies. In what is called EPR "thought experiment", suppose a particle having zero spin ("spin" is a parameter) such as a photon splits into two particles: positron p and electron e each of spin 1/2 (Fig. 21.3).

Since the initial spin is zero, their spin axes are in the opposite direction e $\uparrow$ and p $\downarrow$. Suppose after some time we find e comes to the down state $\downarrow$; then the law of

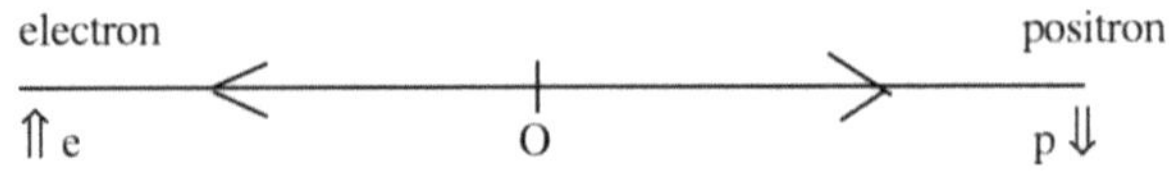

Fig. 21.3 EPR "thought experiment"

conservation of spin implies that the spin axis of p should automatically go to the up state, i.e. p↑. This implies that there is a permanent bond between e and p which may be miles apart. What's more the effect should be instantaneous, though we know that the maximum speed in the universe is c. Of course, in practice the experiment cannot be carried out.

There are so many contradictions, discrepancies and uncertainty gaps in the quantum theory that the only conclusion we can draw is that scientists are still groping in the dark and trying to find a satisfactory theory which would join the microworld of quanta and our macro-world seamlessly.

10 Quantum Reality and *Śūnya*

It is time to recapitulate our understanding of quantum reality and the theory of *śūnya*, compare them impartially and determine if there is any special bond between these two areas.

It is not difficult to see that it is mysticism that seems to pervade both quantum theory and the concept – especially philosophical concept of *śūnya.* The most important feature of mysticism is that an object or a phenomenon is inexpressible, indescribable or ineffable. Instead of saying "x is ineffable", one can also use contradictions and opposites simultaneously, e.g. x exists and does not exist. In the following statements relating to the quantum world and void or *śūnya*, mysticism abounds:

1. *Chāndogya Upaniṣad* mentions that Brahman is Ākāśa outside the body as well as inside the body (5-12-9).
2. According to *Avadhūta Gītā*, there exists neither total void nor voidlessness (1-76), and also "there is no space or absence of space" (5-5). It also says "There is nothing which pervades or is pervaded. There is no abode nor is there lack of abode. How can I speak of void and nonvoid?" (4-9).
3. One could also say since $+0 = -0 = 0$, *śūnya* is both positive and negative, or *śūnya* is neither positive nor negative.
4. *Ánanya* and *ananta* are two sides of the same coin. Both indicate the identity of zero with the infinite.
5. The behaviour of subatomic particles is inexpressible. Particle x may become particle y all of a sudden and then merge with the void giving out energy.
6. A photon is a particle and also a wave. A photon is neither a particle nor a wave.
7. Brahman is indescribable. It is "not this not this" (*netineti*)
8. There was neither non-existence nor existence then (RiṚgveda 10-129)

9. Haṭhayoga refers to Brahman as *Śunyāśūnya* which means the one beyond void as well as nonvoid.
10. The vacuum in quantum physics behaves like a creative void.
11. Another characteristic of mysticism is that the universe should be perceived as a whole. A similar view prevails among quantum physicists. A quantum system can only be described as a whole. Most particles have momentary existence. This reminds one of the doctrine of momentariness (*kṣaṇikavāda*) in Buddhism where the ultimate reality consists of a chain of insubstantial momentary beings.
12. Mysticism often creates multiplicity of realities, usually one mundane and the other the ultimate reality which is usually hidden. It is interesting to see how Nāgārjuna tries to reconcile the two truths in *Mādhyamikaśastra*:

> Despite the fact that from the point of view of conventional truth (*vyavahāra*) the realms *saṃsāra*and *nirvāṇa* ought to be clearly distinguished, from the viewpoint of transcendental truth (*paramārtha*) no such distinction exists. [4][6]

13. We see a similar situation *in the quantum world* which is inaccessible when the particle is in quantum state but is amenable to measurement when the system reaches the classical level.
14. Most mystics who experience mystical states say that these experiences are indescribable and transient. An observation is in order. The case of idealistic philosophies such as Buddhism or Advaita Vedanta or even in Tantra mysticism enters via belief or transcendental matters. Hence the interpreters have no option but to resort to sophistical reasoning or pseudoreconcilation of contradictions. However, it is necessary that scientists should try to solve an intractable problem with patience instead of taking the easy way to a mystical path. In fact they should concentrate on demystification where mystical elements *appear* to be present.

References

1. Penrose R. *The emperor's new mind*. New York: Penguin; 1991. p. 245.
2. Burde J. *Ananya and nothingness*. Delhi: MLBD; 2009.
3. Capra F. The Tao of physics. London: Flamingo; 1993. p. 89.
4. Ichimura S. *Buddhist critical spirituality: Prajñā and Śūnyatā*. Delhi: MLBD; 2001. p. 360.

[6]Quoted in Ichimura [4].

Chapter 22
Ontological Reality: Quantum Theory and Emptiness in Buddhist Philosophy

Geshe Dorji Damdul

Abstract The ultimate reality connotes untainted nature of phenomena by the stains of subjective influence. Since the time of Shakyamuni Buddha and Arya Nāgārjuna, the quest for the ultimate ontology led to the discovery of a plethora of layers of reality detached from the influence of mere belief. The many facets of quantum physics, such as quantum entanglement, quantum vacuum, uncertainty, randomness, observer dependency, and wave-particle duality, all point to the great complementarity of the two traditions – quantum mechanics and Buddhist philosophy. Buddhist philosophy, while probing into the deeper reality of the world, realizes the play of conventionality of dependent origination in the flux fabric of illusion while precisely not defying the efficacy of functional world. Arya Nāgārjuna said in *Mūlamādhyamakakārikā* that only when one sees the emptiness in the light of its true insight of dependent origination can one see the nuance of emptiness to be in stark contrast with nihilism. This legacy of wisdom has greatly profound implication in elevating an individual from the worst of the pains and stress of life. One with exposure to the two traditions can have a great advantage of gaining deeper insight into both the traditions. It is not a good idea to rashly leap to the conclusion that the two thoughts merge identically. Further inquiry is required to explore areas where the two overlap and where the two diverge, keeping in mind that quantum physics is still in its phase of development.

Keywords Quantum theory · Emptiness · Dual truths in Buddhism · Macro- and microworld · Dependent origination of relation · Quantum entanglement · Big Bang theory

G. D. Damdul (✉)
Tibet House, New Delhi, India

S. R. Bhatt (ed.), *Quantum Reality and Theory of Śūnya*,
https://doi.org/10.1007/978-981-13-1957-0_22

1 Introduction

At the core of the manifold crises the world is confronted with today – from terrorism, gender discrimination, high and rising inequality, starvation, and so on – lies ignorance, according to the Buddha. This sentiment resonates with the statement made by Albert Einstein, where he indicates that the illusion created through a misperception of reality is responsible for all problems:

> A human being is a part of the whole called by us universe, a part limited in time and space. He experiences himself, his thoughts and feeling as something separated from the rest, a kind of optical delusion of his consciousness. This delusion is a kind of prison for us, restricting us to our personal desires and to affection for a few persons nearest to us. Our task must be to free ourselves from this prison by widening our circle of compassion to embrace all living creatures and the whole of nature in its beauty.

The approach to remedy ignorance using one's rationality is therefore indicated by two giants of humanity – the Buddha and Albert Einstein.

The basic inspiration in drawing parallels between quantum physics and the Buddhist philosophy of emptiness is grounded on the two premises. First, both quantum physics and Buddhist philosophy explore ontological reality. Second, both resort to rationality and objectivity as the means to find ontological reality. The Buddha said:

> Wise people and the monks,
> Just as the goldsmith tests the purity of the gold
> Through cutting, rubbing and burning the gold,
> Likewise, you should examine my words and put them into practice
> Not simply because you respect me.

2 Profundity of Ontological Reality from the Viewpoint of Buddhism and Quantum Physics

Niels Bohr is known to have famously said, "If quantum physics has not profoundly shocked you, you haven't understood it yet." The first experience of the concept of emptiness as presented in Buddhist philosophy is described as very subtle, with an initial tinge of fear, which later will be replaced by profound joy and bliss. How subtle the concept of ultimate truth is, in Buddhism, is rightly reflected from what the Buddha proclaimed when Indra and Brahma descended to the earth to make a request to the Buddha to start giving teachings after the Buddha remained silent for 49 days post Enlightenment under the Bodhi tree in Bodhgaya.

> Profound, peaceful, freed of elaborations, clear light, non-composite,
> Such a nectar-like reality is what I found.
> Finding no one who can fathom this path,
> In silence I will retire in the woods.

This similarity in the experience of the truth, as presented by both quantum physics and Buddhism, points to the fact of its subtlety in both views. Whether or not the two proceed in parallel requires an extremely nuanced analysis, which this composition partly endeavors to do so.

3 Wave-Particle Duality in Quantum Physics and Dual Truths in Buddhist Philosophy

In quantum physics, depending on the kind of the design of the experiment (such as double-slit experiment), the same electron is seen to behave differently as a wave or as a particle. This duality is totally unacceptable according to the classical Newtonian physics. Likewise, the Buddhist philosophy of emptiness has demonstrated that every object displays two natures depending on which analysis is resorted to, conventional or ultimate. Technically, this is known as the concept of the two truths. The Heart Sūtra clearly elucidates it thus:

> Form is empty;
> Emptiness is form;
> Emptiness is not other than form.
> Form is also not other than emptiness.

In his text *Entry into the Middle Way*, Acārya Candrakīrti (seventh century CE) explains this clearly:

> All entities obtained by the seeing of perfect truth and falsehood,
> Each of these entities bear two realities;
> The object that is seen by perfect vision is 'true suchness,'
> While that which is seen by deceptive vision is taught to be the 'conventional truth.'

4 Physics and Buddhist Philosophy on the Macro- and the Microworld

4.1 Macroworld

For many decades, physicists tried to look for the unified field theory. The latter half of Einstein's life was spent mostly in this quest. Although variant explanations are now emerging from physicists, it can still be seen that quantum theory applies so well to the microworld, as does Einstein's theory of relativity to the macroworld. For instance, the wave-particle duality cannot be seen operating in the macroworld. This points to the reality that the two theories cannot possibly be squarely merged together – squared or merged two different things.

As indicated earlier, the Buddha proclaimed that the two truths can never be seen as compatible, with respect to one object and with reference to a single analysis. With conventional analysis as the reference meant for the macroworld, at the most the functional world of the macro can be seen and not the micro – the ultimate truth. On the contrary, with the ultimate analysis as the reference, one will find the ultimate truth, in which case the macroworld dissolves and turns devoid of meaning and definition.

5 Dependent Origination of Relation (Subjectivity) and Quantum Entanglement (Nonlocal Spooky Action at a Distance)

Relational nature is the fabric of existence of all phenomena in Buddhist thought. This precise concept is known as dependent origination – Pratītyasamutpāda. It has three levels of subtlety, each progressively more so: causal dependent origination, dependent origination of the whole's dependence on parts, and dependent origination of dependence on the mere mental designation.

Quantum theory's phenomenon of quantum entanglement, which Albert Einstein described as spooky action at a distance, is what Buddhist philosophy terms relational existence. Two electrons ejected in opposite directions, which may not materially be connected in any way, can still be related and connected on the basis of the third level of dependent origination from Buddhist philosophical point of view.

6 Substratum of the Universe: Quantum Vacuum in Physics and Expanse of Emptiness in Buddhist Philosophy

Singularity, the trigger point of the Big Bang, is what physicists see in the nature of quantum vacuum. Quantum vacuum is not to be understood as nothingness or total absence of everything, from the point of view of physicists. Instead, as the ultimate cause of the universe, it is conceived to be full of possibilities or potential.

Likewise, the ultimate expanse of emptiness in Buddhist thought should not be seen as nothingness but being of the nature of the middle way, which is freed of the two extremes – nihilism and absolutism – as clearly indicated by Ārya Nāgārjuna (first century CE) in his text *The Fundamental Wisdom of the Middle Way (Mūlamādhyamakakārikā)*:

> That which is dependently originated
> Is posited as empty of objective existence.
> That being dependently designated,
> This is the middle way.

The ultimate sphere of emptiness is the profound fabric that manifests in the form of all facets of the functionality of the universe, as clearly explained by Lama Tsongkhapa in his commentary on Srī Guhyasamāja Root Tantra. Nuanced experience, in its subtlest form, of the truth of emptiness requires one to see emptiness and dependent origination as the two sides of the same coin.

7 Ontological Reality

The mere absence of objectivity of the object is the objectivity of the object. This, according to the highest Buddhist philosophical thought, is referred to as the final ontological reality of phenomena. This requires extensive learning and reflection to gain insight into this view.

From the viewpoint of quantum physics, the absence of the description of the nature of the particles independent of the observer is the ontological reality. In other words, the nature of the particles existing only in dependence on the observer is the final subtle reality of objects and is the ontological reality.

8 Implications of the Two Theories

If applied to the psychology of the humans, the knowledge of the discrepancy between the appearance and the reality – as indicated by Buddhist philosophy of emptiness and quantum physics – has the capacity to revolutionize the world of emotions from negative to positive.

The behaviors of humans determine the plight of the world. The world of today, with unimaginable crises, is a clear indication of the negativity of human behavior. Behavior is determined by emotions, which in turn are directed by the information provided by our senses. The accuracy of the information is dictated by the accuracy of the perception of the world by the senses. When the perception is mistaken, as indicated by the Buddhist philosophy of emptiness and quantum physics, the final result in the form of the plight of the contemporary world – which is unhealthy and in deep crises – is seen.

In summary, the world crises witnessed today are the outcome of a perception, which is entirely faulty. By rectifying the perception, as discovered by the Buddhist philosophy of emptiness and dependent origination, as well as quantum physics, there is a definite hope to see the world of tomorrow as one of greater harmony and peace.

Chapter 23
Logical Interface Between Quantum Physics and Śūnyatā with Special Reference to Theravāda and Mādhyamika Philosophy

Rana Purushottam Kumar Singh

Abstract The world is searching for a new way and means to reinterpret science and philosophy in the twenty-first century. The search for commonality and distinction between science and philosophy is a burning topic among scholars in the world today.

Quantum physics suggests that physical quantities like energy, momentum, and so on can have only certain discrete or discontinuous values. It says that one can never see something behaving as a wave and a particle at the same time. This fact is known as complementarity. According to quantum physics, one cannot observe wave and particle in a single measurement. For example, in some types of experiments, an electron acts like a wave, while in others it acts like a particle. It will never act like a wave and particle at the same time. The difficulty for common sense comes in trying to reconcile the wave behavior at one time with the particle behavior at another.

The main aspect of quantum theory that can be compared with Mādhyamika Philosophy can be illustrated in this way: in quantum theory, the observer does not play a purely passive role. Whether an electron behaves as a wave or a particle depends on the type of experiment being done, and it is the observer who decides what sort of experiment to conduct. This is called "participatory universe." Mādhyamika Philosophy has its own version of the "participatory universe." In line with the general principle of dependent origination, subject and object knower and known and observer and observed exist only in relation to each other. Neither has an independent "objective existence."

Keywords Quantum physics · Śūnyatā · Theravāda · Mādhyamika · Paramārtha · Participatory universe

R. P. K. Singh (✉)
Department of Pali, Nava Nalanda Mahavihara (Deemed University) Ministry Of Culture, Government of India, Nalanda, Bihar, India

S. R. Bhatt (ed.), *Quantum Reality and Theory of Śūnya*,
https://doi.org/10.1007/978-981-13-1957-0_23

1 Introduction

The present century is veritably a century of tireless quests for new ways and means to reinterpret science and philosophy. The search for commonality and distinction between science and philosophy is a burning topic among scholars in the world today.

Quantum physics suggests that physical quantities such as energy, momentum, and so on can have only certain discrete or discontinuous values. It says that one can never see something behaving as a wave and a particle at the same time. This fact is known as complementarity. According to quantum physics, one cannot observe wave and particle in a single measurement. For example, in some types of experiments, an electron acts like a wave, while in others, it acts like a particle. It will never act like a wave and particle at the same time. The difficulty for common sense comes in trying to reconcile the wave behavior at one time with the particle behavior at another.

The essence of quantum theory that it merits comparison with Mādhyamika Philosophy can be understood through an illustration, where the observer does not play a purely passive role. Whether an electron behaves as a wave or a particle depends on the type of experiment being done, it is the observer who decides what sort of experiment to conduct. This is called "participatory universe."

Mādhyamika Philosophy has its own version of the "participatory universe." In line with the general principle of dependent origination, subject and object, knower and known, and observer and observed exist only in relation to each other. Neither has an independent "objective existence." They are all empty of any self-contained, intrinsic nature.

2 Historical Development of the Theories of Knowledge in Buddhism and in Classical Physics

The intellectual history of both physics and Buddhism went through remarkably similar steps that can be enumerated as follows:

1. Transition from the paradigm of substance to the paradigm of causality
2. Replacement of productive causality by law-like succession
3. Transition from causal and law-like succession to co-emergence
4. Criticism of ontological construal of co-emergence and claim that co-emergence itself is relative to the cognitive act that points to it
5. Silent return to the practices of life or agnostic return to the practices of experimental science

To begin with, the first step from substance to causality was taken very early in Buddhism. It involved a strong attack against the belief that substances and selves are permanent beings, and every phenomenon is permanent in nature.

Here we have to discuss the concept of *Śūnyatā* which is a key concept in Buddhism. Ārya Nāgārjuna developed the concept on the basis of the concept of *Tilakkhaṇa*. The present paper would discuss the genesis of this concept. It was used in Pāli literary texts for the first time. According to Pāli-English Dictionary, the meaning of the term *suññatā* or *suñña* is empty, uninhabited, devoid of reality, unsubstantial, phenomenal, and useless also. Again, it has been used to denote something that is empty of permanent substance.

Pāli literature provides concepts of Buddhist schools in nascent form on the basis of which all schools of philosophy subsequently developed. In the Dīghnikāya, understanding *suññatā* or the stage of *suññatā* is a very high experience. In the Dīghnikāya, it is said, Hoti dho so *Bhikkhave, samaaccayena ayaṁ loko vivaṭṭati, vivaṭṭamāne loke suññaṁ Brahmavimānaṁ pātubhavati.*[1,2]

Here, at another place, it has been used in the sense of emptiness of relations. In Dīghnikāya Tīkā, it is said that *Pakatiyāti sabhāvena tassa suññaṁti iminā sambandho.*[3] The Buddha enjoins his followers to always see the world as *suñña*, i.e., a place where nothing like a substance exists. Here, in Sutta nipāta, the experience of *suññatā* is a direct experience of a marked realization that is never stained by any slightest dualistic perception and where the continuity of all mental activity has ceased to exist. The Buddha has suggested that one should always see the world as *suñña* and keep oneself away from the concept of permanent soul. Then only, a man can attain the state of *nibbāna*. Here, the style of exposition is *sappariyāya* which is why the Buddha has used figures of speech and illustrations. It is said like *Suññato lokaṁ avekkhasu, Mogharāja sadā sato, attānudiṭṭhaṁ uhancca, evaṁ maccutaro siyā evaṁ lokaṁ a vekkhantaṁ Maccurājāna passati.*[4] In course of explaining human personality in Mahānidāna Sutta, the Buddha has used the concept of *suññatā*. According to the Buddha, all the 18 elements (*dhātus*) are *suñña* by nature, and each and every moment of it gets changed. The eye itself is *suñña,* and in the same way, the visible objects are also void. In the same way, the eye consciousness originated out of the contact between eye and the visible object, resulting in a process that is also void. In brief, the functioning of human personality is *suñña* or void. Here, a very important point arises from the statement of the Buddha. The concept of a permanent soul is a false view, and the function of the aggregate of personality is *suñña*. It is said *Suññaṁ attena vā attaniyena vā, tasmā suñño loko ti vuccati cakkhu kho Ānanda, suññaṁ, cakkhu viññāṇa suññaṁ, cakkhusamphasso suñño....*[5]

In *Visuddhimagga*, the concept of *suññatā* is taken forward further suggesting that both mind and matter (*nāma* and *rūpa*) are *suñña* in nature. The mind mistakenly apprehends the natural phenomena and manifests as an unbroken string of thoughts

[1]Pāli-English Dictionary, p. 717
[2]Dīgha Nikāya Vol I P.40
[3]Dīghnikāya Tika p. 137
[4]Sutta nipāta p. 125
[5]Mahāniddesa P.329

that have continued one after the other from time without beginning. The mind accumulates many habitual tendencies or *kārmic* imprints. Due to these habitual tendencies, the diverse pleasurable and painful appearances of saṁsāra manifest, although they are nonexistent. As a result, we accumulate further habitual tendencies for more appearance to continue to arise; for as long as we cannot recognize the nature of phenomena as they are at which time the mind characterized by deluded dualistic apprehension, reabsorbs back into the basic space of phenomena, the deluded appearances of the world will never cease to exist. Moreover, the continuity of birth and death from the time when they began until the present will not end.

Everything that we perceive in the present life, from our birth till the time when we die, all our inner mental phenomena, arises continuously in the same way. When the required causes have again gathered, the appearances of the next life, in absence of any obstruction, cannot but arise. In this way, the root of all the various deluded appearances of past and future lives, of happiness and suffering, exists in the mind.

The mind itself also arises from cause and conditions. The cause lies in the previous instance of mind, which resembles it and is of the same kind. The conditions are the gathering of sense objects and sense faculties. That is why the Buddha said that the main method to reach liberation is to purify the mind and to understand nature of mind and matter (*nāma* and *rūpa*) as *suňňa*.

> "Nāmaňca rūpaňca idhatthi saccato,
> Na hettha satto manujo ca vijjati,
> Suňňaṁ idaṁ yantamivābhisaṅkhataṁ.
> Dukkhassa puňjo tiṇṇakatthasādiso" ti.[6]

In the commentary of *Dhammasaṅgaṇi*, the term *suňňatā* has been used in the sense of a tool of meditation. "Uppannattā maggo maggo ārammaṇato suňňatanāmaṁ labhati." Here, understanding the real nature of world as void or suňňa is wisdom. In *Abhidhamma*, when a mendicant goes in the realm of *arūpabhūmi*, he concentrates on the four stages like *ākāsaňcāyatananaṁ*, *Viňňāṇānaňcāyataṁ*, *nevasaňňānasaňňāyataṁ*, and *akiňňcāňňāyataṁ*. Then, he understands the real nature of the world, i.e., the nature of *suňňatā*: "Ākiňcaňňāyataṁ.... Santato manasikaritvā sasseve Viňňāṇaňcāyatanārammaṇa bhūtassa ākāasňcāyatana Viňňānassa abhāvo suňňatā vivattikāro manasikāratabbo."[7]

It can thus be opined that like the Mūlamādhyamika Kārikā, the term *suňňa* or *suňňatā* has also been used in Pāli literature to mean multiple things. Moreover, since the development of the concept in Mūlamādhyamika Kārikā, many Pāli commentators have also given ample space to this concept to elucidate the *Buddhavacana*.

Now, the next aspect of the paper would be logical analysis of the term *Śūnyatā* in Mādhyamika Philosophy.

[6]Vissuddhimagga Vol II p. 228
[7]Ibid Vol I p. 323

Etymologically, the word *śūnya* has been derived from the root śvi, which means "to swell," "to expand." The Buddha is said to be seated in the *Śūnya tattva* in the *śūnya* principle. The word *śūnya* seems to have been used in the ontological sense in such context.

Some scholars believe that the word *śūnya* has not only an ontological signification but a soteriological suggestion as well. However, the word has obviously been used also in an ontological sense with an axiological overtone and soteriological background.

In the ontological sense, *Śūnya* is the void which is also fullness. Because it is nothing in particular, it has the possibility of everything. It has been identified with *Nirvāṇa* with the absolute, with *paramārtha*-sat (the Supreme Reality). Here, Nāgārjuna has said

Apara pratyayam śāntam prapañcairaprapañcitam.
Nirvikalpamanānārthametattattvasya Lakṣaṇam[8]

1. It is *aparapratyayam*- it is the experience which cannot be imparted to anyone by another. It has to be realized by everyone for himself/herself.
2. It is *śāntam*- it is quiescent unaffected by the empirical mind.
3. It is *prapañcairaprapañcitam* i.e., inexprepressible by the verbalizing mind.
4. It is *nirvikalpam* i.e., transcendent to discursive thought.
5. It is *anānārtham* i.e. non-differentiated.

Here, the word *Śūnya* and *Śūnyatā* can be understood in connection with *svabhāva* which means own being or self-nature of a thing. Here Nāgārjuna says:

svabhāvaḥ kṛtako nama bhaviṣyati punaḥ katham
Akṛtrimaḥsvabhāvo hi nirapekṣaḥ paratra ca.[9]

That is really "svabhāva" which is not brought about by anything else. This is unproduced (*akṛtrimaḥ*) implying that which is not dependent on nor relative to anything other than itself and unconditioned.

Here, *śūnya* has to be understood from two points of view. From the point of view of phenomena or empirical reality, it means *svabhāva-Śūnya*, i.e., devoid of *svabhāva* or independent, substantial reality of its own. From the point of view of the absolute, it means *prapañca Śūnya*, i.e., devoid of *prapañca* or thought construct, verbalization, and plurality.

To sum up, the impact of *Śūnyatā* may be understood in six ways. Three of these are given together in this verse:

yaḥ pratītyasamutpādaḥ śūnyatām tām pracakṣamahe
sā prajñapatirupādāya pratipat saiva madhyamā[10]

[8]M.K. XVIII, 9
[9]M.K. XV, 2
[10]M.K XXIV, 18

1. In reference to *vyavahāra* or empirical reality, *Śūnyatā* means naiḥsvābhāvya, i.e., devoidness of self-being and of unconditioned nature. In other words, it connotes conditioned co-production or *pratītyasamutpāda*.
2. This idea is conveyed in another way by the term *upādāyaprajňapti* or derived name. It means that the presence of a name does not mean the reality of named.
3. 3.*Śūnyatā* exposes the folly of accepting any absolute beginning or total cessation, thus connoting madhyamā pratipat, i.e., taking things as they are.
4. In reference to *paramārtha* or ultimate reality, *Śūnyatā* connotes the nonconceptual nature of the absolute.
5. In reference to the aspirant, *Śūnyatā* implies its attitude of *anupalambha* or the skillfulness of non-clinging to the relative as absolute or to the absolute as something specific.
6. The Mahāprajñā Pāramitā Śāstra brings out another implication of the *Śūnyatā* principle, viz., Dharmaiṣṇā, the irrepressible longing for the real.

Now two aspects related with *Śūnyatā* are discussed.

(a) *Śūnyatā* has a soteriological significance.

Śūnyatā is not merely an intellectual concept. Its realization is means of emancipation. When rightly grasped, it leads to the negative of the multiplicity of the *dharmas* and to detachment from the "passing show" of the tempting things of life. Meditation on *Śūnyatā* leads to *prajñā* which brings about the emancipation of the aspirant from spiritual darkness. Nāgārjuna puts quintessence of his teaching about *Śūnyatā* like this:

karmakleśa kṣayānmokṣa karmakleśā vikalpatah
Te prapañcāt prapañcastu śūnyatāyām nirudhyate.[11]

Emancipation is obtained by the dissolution of selfish deeds and passions. All selfish deeds and passions are caused by imaginative constructs which value worthless things as full of worth. The *vikalpas* or imaginative constructs are born as *prapañca*; the verbalizing, imaginary activity of the mind ceases when *Śūnyatā* or hollowness of things are realized.

(b) *Śūnyatā* is not a theory.

Here, it is clear that the pioneer of Mādhyamika uses the dialectic as a criticism of all *dṛsti* as criticism of all *dṛṣṭi* (theories) without theory of his own. By the use of his dialectic, he reaches the conclusion that all *dharmas* are *śūnya* or *nissvabhāva*, i.e., devoid of any independent substantial reality.

It may be thought that *Śūnyatā* itself is a theory. But its interpretation as the Mādhamikas position would be a misleading. In real sense, *Śūnyatā* is not a theory. It is at once the awareness of the importance of reason to realize truth and the urge to rise to a level higher than reason in order to realize it.

[11]M.K. XVIII, 5

Atrabrūmaḥ sūnyatāyām na tvam vetsi prayojanam
Śūnyatāyām śūnyatārtham ca tata evam vihanyase.[12]

You do not know the purpose of *Śūnyatā*. *Śūnyatā* is not used as a theory just for the sake of *Śūnyatā*. Here, important aspects related to *Śūnyatā* have been discussed. There are many common grounds and different platforms to discuss both the quantum theory and *Śūnyatā*.

[12]M.K. XXIV, 7.

Chapter 24
How the Pañcakośa Model of Experience Fits the Understanding of Śūnya and Helps Explain Quantum Reality?

Rashmi M. Shetkar, Alex Hankey, and H. R. Nagendra

Abstract The Vedic system explains the structure of human subjectivity through the idea that human experience is based on various properties and levels of the mind with separate abilities and roles to play in the human makeup, namely, *Manas*, *Buddhi*, *Ahamkāra*, and the Citta that constitutes the underlying driving force. These are regarded as linked to various independent vehicles for conscious experience that exist apart from the gross physical body and which are open to the cognition and action of yogis whose abilities have developed through prolonged practice of deep meditation. Subtle bodies go under the general classification of *Sukṣma* and *Kāraṇa Śariras*. The *Manomaya* and Vijñānamaya Kośa are generally translated as "subtle" (Sukṣma), while the Anandmay kosha is considered "causal" (Kāraṇa).

The paper will discuss that these bodies function on a kind of information different from that of gross senses – physical, digital/entropy information – and point to a different kind of reality prevailing at subtle levels of experience. In particular, the information types can be classified through the approach of Śikṣa, which states that there are four levels of verbal expression, *physical* through sound, *mental*, *ideational*, and *transcendental*. These correspond to digital information; experience information; integrated, higher-order experience information; and totality information. Only quantum reality and its extensions can model these, and only yoga, in the fullest sense of the word as the eight-limbed practice of Aṣṭāṅgayoga as propounded by Maharishi Patanjali, or its equivalents such as Islam's Chist, can provide subjective confirmation of the existence of these states.

Keywords Pañcakośa · Veda · Vedānta · Yoga · Vedāṅga · Citta · Ājñā Cakra

R. M. Shetkar (✉) · A. Hankey · H. R. Nagendra
SVYASA, Bangalore, Karnataka, India

S. R. Bhatt (ed.), *Quantum Reality and Theory of Śūnya*,
https://doi.org/10.1007/978-981-13-1957-0_24

1 Introduction

The Vedic system explains the structure of human subjectivity through the idea that human experience is based on various properties and levels of the mind with separate abilities and roles to play in the human makeup, namely, *Manas*, *Buddhi*, *Ahamkāra*, and the *Citta* that constitutes the underlying driving force. These are regarded as linked to four independent vehicles for conscious experience, or *kośa*, that exist apart from the gross physical body and are open to the cognition of yogis with abilities developed through prolonged practice of meditation. Subtle bodies are generally classified as "subtle" (Sukṣma Sarīra) and "causal" (Kāraṇa Śarīra). The Manomaya Kośa and Vijñānamaya Kośa are subtle, while the Ānandamaya Kośa is considered causal, because it motivates action through the other kośas. These bodies function on a kind of information different from that of gross senses – physical, digital/entropy information. This points to different kinds of reality prevailing at subtle levels of experience. In particular, the information types can be classified through the approach of Śikṣā, which states that there are four levels of verbal expression, *physical* through sound, *mental*, *ideational*, and *transcendental*. These correspond to digital information; experience information; integrated, higher-order experience information; and totality information. Only quantum reality and its extensions can model these, and only yoga, in the fullest sense of the word as the eight-limbed practice of *Aṣṭāṅgayoga* laid out by Mahariśhi Patanjali, can provide subjective confirmation of the existence of these states. Yoga provides tools and techniques to do so, particularly its final three limbs, the *antaraṅgas*, *Dhāraṇā*, *Dhyāna*, and *Samādhi* or pure consciousness. Nāgarjuna characterized the latter from the perspective of the Manomaya Kośa and Vijñanamaya Kośa as a state of Śūnya, emptiness. However, Ādi Śankarācārya, in accordance with the Upaniṣhads, emphasized that these are states of Pūrṇa, fullness or completeness, consistent with the perspective of the Prāṇamaya Kośa, in which the unrestricted fullness of the flow of Prāṇa is enabled.

The ancient Vedic civilization developed a unique approach to understand the human mind and how to optimize its use so as to fulfill all desirable goals in a person's life. Some have suggested that with these abilities they were able to create a state of "Heaven on Earth" for a large fraction of the population for a large fraction of the time. The key knowledge that the civilization possessed was derived from practices of meditation [1] that they perfected, giving them access to deeper levels of their mind and the ability to actualize its full potential.

The Vedic understanding of mind [2] was based on a completely different kind of investigation from those used in contemporary western psychology [3]. Whereas the latter is largely derived from studies of psychopathology [4], and how the mind can go wrong when subject to overloads of stress [5], pain [6], and suffering [7], the eastern approach [8] emphasized the means to leave suffering and its effects behind. A person then learns to live from deeper levels of the mind, completely transcending levels of suffering, to arrive at states of pure *Ānanda*, bliss [9].

2 Results

The Vedic system [10, 11] therefore begins to explain the structure of human experience [12] and its inherent subjectivity [9] through the idea that the mind has different levels [2] that can be explored by directing the attention inwardly in meditation [13]. The different levels present various different properties, which are detailed in specific sections of the Vedic literature [10, 14]. Each has particular abilities and roles to play in the human makeup. They include *Manas*, the level that contains information in the form of ideas generated by sensory perception, together with their emotional associations based on experience; *Buddhi*, the level of the discriminative faculty, decides on intentions and associated courses of action and has the power to control ideas; the *Ahamkāra*, or little ego, is a point-like structure supported by a person's underlying sense of personal existence, i.e., one's sense of individuality with its authorship of action; and finally the *Citta*, or storehouse of past memories and "impressions" reaching back into the deep past, provides the motivation for a person's activities and choices of further action. Citta constitutes the driving force behind all a person's actions. Together these four concepts are grouped into an entity called the Antaḥkaraṇa, a term treated in specific sections of the Vedic literature [15, 16].It has no equivalent in western languages because of their ignorance of internal levels of mind due to lack of techniques to explore them systematically. Western psychology sticks to more limited concepts of the body and mind and does not go beyond them. Neuroscience has taken steps to understand causes behind various mental diseases [17] linked to a person's heart and mind and to their sentience [18]. Kośas provide a missing link between these eastern and western approaches [19].

Traditional Indian psychology [20] based on Vedānta is holistic and adaptable to any lifestyle. It provides the means to manage the stress of life [5] and improve the overall quality of life by pursuing a path to fulfillment [21]. In the Vedic system, and more generally in the East, in Jainism and Buddhism, the three concepts of *Manas*, *Buddhi*, and *Ahamkāra* are linked to various independent vehicles for conscious experience, or kośas [19]. These structures are recognized to exist apart from the physical body and are experienced by advanced practitioners of meditation techniques such as those associated with yoga [22, 23] or the various schools of Buddhism [24]. With appropriate training, yogis can use the techniques to speed reduction of unwanted material in the Citta, which causes ongoing stress. Such practices accelerate their progress on the path to spiritual liberation – freedom from all such leftovers from previous memories and experiences.

The most powerful system to remove problematic and distracting activity of the Citta, the Citta-Vṛttis, is yoga [25], particularly its forms of meditation. The Yoga Sūtras states, *Yoga Citta-Vṛttis Nirodha*: the purpose of Yoga is to purify the Citta of its problem-causing activity [26] originating in unresolved memories and experiences. Modern neuroscience offers possible explanations. For example, focus on the medial prefrontal cortex (mPFC) activates and regulates the attention-forming executive network [27] in the frontal lobes and prevents intrusion of distracted states of functioning caused by the Citta-Vṛttis, which may generate activation or deactivation of the default mode network (DMN) [28]. Such meditative focus replaces DMN activity with more goal-directed processes that can maintain and stabilize inner peace and silence [29]. Only then is the executive network set to function efficiently and prevent the Citta from interfering with its ongoing activity. This promotes efficient functioning of the inner aspects of self, dissected by Damasio [30] as the (1) autobiographical self, (2) protoself, and (3) core self. These constitute the self-referential network [31] and can be mapped to the Pañcakośas [19].

The executive network can then process functioning without interference, mainly in a goal-directed manner like a CEO, i.e., chief executive officer; the Yoga Sūtras state: *Tadā dṛṣṭuḥsvarūpe avasthānam* [32], the personality self is established in its inner self, *Svarupa* [33]. The executive network and the self-referential network can then integrate and work together in coherence and silence, bringing harmony to a person's life and continuing experience of *Śānti*, inner peace [29]. By such means of correct focus of attention, the ability of yogis to enhance the functioning of all four levels of the Antaḥkaraṇa is enlivened and enhanced [16]. This essential focus maintains unrestricted goal-directed activity. It generates a subtle energy that integrates the functioning of the Pañcakośas [34], which come into coherence and harmony naturally [35].

The subtle energy, Prāṇa, is a massive power to activate latent abilities and is the creative energy, known as Śakti, that shines through all great people's lives in their highest accomplishments. [36] One of its major centers is the Ājñā Cakra [23] in the forehead[1]. Focus in the frontal lobes infuses this energy into the system, which is a converging center of the brain, strengthening the intentional network and directly connecting it to the Antaḥkaraṇa and its innate higher abilities [37]. These are described in Yoga Sūtras Pāda III, based on mastery of earlier advanced stages and results of regular meditation practice. Such exercises activate a person's latent abilities by triggering various states of consciousness, based on pure self-referral awareness forming the foundation of the familiar states of consciousness [38], waking, dreaming, and deep sleeping. They do so by fine tuning the three As, attention, awareness, and arousal of meditation. This idea is acknowledged in the eastern sciences in the Māṅdukya [38] and Mundaka [12] Upaniṣhads. Cognitive neuroscience suggests its proof by listing neural correlates of waking, dreaming, and sleeping [39]. The Vedic civilization thus understood how higher abilities can be developed by extended practice of meditation and associated exercises.

To recapitulate, the Vedic system classifies subtle energies under the concept of several, distinct subtle bodies under the general classification of subtle and causal bodies (*Sūkṣma* and *Kāraṇa śarīras*). [40] The Taittirīya Upaniṣhad lists five specific bodies covering the highest "self" or Ātman, the Ānandamaya Kośa, Vijñanamaya Kośa, Manomaya Kośa, Prāṇamaya Kośa, and the Annamaya Kośa, the last being the gross physical body made of food (*anna*). Of these the first, Ānandamaya Kośa,

[1]Swami Madhavananda, Swaroopyog, Pune, personal training & private communication on Science of Meditation, in Natha Sampradaya.

is known as the *Kāraṇa śarīra*, or causal body [40]. Paramahamsa Yogananda states that it is point-like, so it also carries the level of little ego, or *Ahamkāra*, responsible for maintaining the sense of having a separate identity [41]. When actions are impure or done for selfish motives, the Ānandamaya Kośa is clouded or even blocked, e.g., by *Citta-Vṛttis*, and is unable to let the full light of consciousness shine through.

The second and third subtle bodies, the Vijñanamaya Kośa and Manomaya Kośa, generally translated as "subtle" (*Sūkṣma*) bodies, are the residence of the *Buddhi* and *Manas*, respectively. Also, according to the Tattvabodha of Ādi Śankarācārya [11], the Vijñanamaya Kośa has two components, the Buddhi and the Jñānedriyas, the organs of perception and action, both. The Manomaya Kośa also has two components, in this case, *Manas*, or mind, and five Jñānedriyas, giving the body the ability to perceive. The final subtle body, the Prāṇamaya Kośa, or body of subtle energy, consists of the five Prāṇas and five subordinates, Upaprāṇas, together with the five karmendriyas, or organs of action.

3 Discussion

The Vedic sciences see the manifestation of the whole creation as arising from basic imbalances between three fundamental qualities or *guṇas*, *Sattva*, *Rajas*, and *Tamas*, which constitute fundamental spiritual characteristics on various levels of creation. [42–44] More spiritual levels are characterized by *Sattva*; greedier, selfish qualities are due to excess of *Rajas*, while the most unyielding, dark, and heavy are expression of *Tamas*. *Sattva* characterizes the subtle bodies, *Rajas* is the gross, physical body with its various needs and habits, and *Tamas* is caused by lethargy, lack of spiritual knowledge, and lack of ability to discriminate eternal values that liberate, from temporal values that bind the soul back to incarnation and suffering.

An important aspect of the functioning of the kośas is that they do not function on the same kind of information as the gross senses. They function on information of a different kind, the information used in cognition and experience [45, 46]. This is not physical, digital information, acknowledged as similar to information in thermodynamic entropy or disorder. Analysis shows it to be new and different, "experience information." [46] At subtle levels of experience, a different kind of reality operates, far more interconnected than the commonly experienced gross world of *sthūla* forms and phenomena. The world of Śūnya can only be scientifically discussed in these terms.

The various information types used in human conscious experience can be classified through the Vedic science, Śikṣā, the first of the six *Vedāngas*, the limbs of the Veda. [44] Śikṣā tells us that human cognition encompasses four levels of verbal expression: [1] *physical* through sound, *Vaikhari*; [2] *mental speech*, *Madhyamā*; [3] *ideational*, *Paśyanti*; and [4] *transcendental*, Para [44]. These four levels can explain many aspects of human communication that simple neuroscience hypotheses of digital information content. They cover four distinct types of information: digital information; experience information; integrated, higher-order experience information; and totality information [46].

Only quantum reality and its extensions can model these various information types. Experience information in particular requires an extension of quantum field theory to quantum fluctuation fields; the higher-order forms are then easily constructed [47]. Yoga, in the fullest sense of the eight-limbed practice of Aṣṭāṅgayoga [22] laid out by Patanjali in his Yoga Sūtras, provides subjective confirmation of the existence of these states. Only Yoga provides tools and techniques to do so. Certain derivative meditation systems such as those practiced by Tibetan Buddhism, or Islam's mystical Chist practices known to the Sufis, are also known to give access to the deeper levels with the required degree of clarity. [48] The general failure of European science and philosophy in this regard constitutes a severe condemnation of the methods at its disposal.

In the case of yoga, its final three limbs, the Antaraṅgas, provide systematic understanding of how to access deep levels of the mind and transcend them to reach the final state of inner silence. The Antaraṅgas comprise *Dhāraṇā*, focus of the attention on the mental stimulus being used, *Dhyāna*, allowing the stimulus to spontaneously refine so that the attention slips spontaneously to deeper levels of the mind, culminating in its reaching one of the states of *Samādhi* where the state of pure consciousness slowly becomes accessible. [49]

As we know, Nāgārjuna characterized the latter as a state of Śūnya, emptiness [50]. In contrast, Ādi Śankarācārya [9] adopted the Turīya perspective of Manomaya and Vijñanamaya Kośas in accordance with various Upaniṣhads, such as the Māṅdukya [38] and Taittirīya. [51] He emphasized that these are states of Pūrṇa, fullness or completeness [52], which is the way they are experienced through the Prāṇamaya Kośa, because unrestricted fullness of the flow of Prāṇa is enabled.

Once it is accepted that information gained at deeper levels of awareness is valid, the question arises as to how to explain the phenomena that become available at these subtle, *Sūkṣma*, levels of experience. Only by extending quantum physics to include quantum fluctuation fields do explanations for such phenomena become possible. Adopting these new forms of quantum theory extends modern science to include this reality in its fields of theory and understanding. For this reason, the study of states of Śūnya is as important for modern science as for Buddhism and other forms of traditional wisdom like the Vedic sciences.

4 Conclusion

The Vedic sciences explain the detailed structure of subjectivity in terms of various kośas that surround the Ātman or self and can be witnessed by the self [53], through its self-referential cognitive abilities [54]. The various classifications of the Upaniṣhads, Samkhya, and Vedānta can be interrelated by explaining the effects of various stages of Yoga practice [55]. When connected with modern neuropsychology concepts of the default mode network, the executive network, the self-referential network, and the attentional network, strong hypotheses can be formulated concerning the stages of development of unrestricted attention and higher abilities through Yoga practices [56, 20]. Then the processes of cognition and corresponding brain mechanics for the fundamental process in Yoga can be explained: the elimination of problems in the Citta and subsequent shining forth of the self, leading to self-realization and enlightenment [57]. These processes lead to various stages of realization described in different schools of Vedānta linking the apparently different eastern [57–59] and western sciences of self [60] and self-realization.

The future journey of this research will depend on what can next be shown to be scientific and proven as such. That will naturally build on our understanding of the biophysics of meditation [61] and its relationship to self-organized criticality [62, 63]. Subjective accounts of meditation from the first-person perspective [64] and similar reasoning show how meditation benefits mental health [65] and emotional intelligence [37].

References

1. Patanjali M. Yoga Sūtras (Shearer A. Trans. Effortless being). The Yoga Sūtras of Patanjali. London: Crown Publishing; 2010.
2. Isvarakrsna, (Thomas H. Tran.) SankhyaKarika. Philadelphia: Wentworth; 2016.
3. Jung CG. The collected works of C.G. Jung: psychology and religion. New York: Pantheon Books, U. Minnesota; 1989.
4. Weiss E. Sigmund Freud as a consultant – recollections of a pioneer in psychoanalysis. New York: Transaction Publishers; 1991.
5. Selye H. The stress of life. New York: MacGraw-Hill; 1984.
6. Hansen GR, Strelzer J. The psychology of pain. Emerg Med Clin N Am. 2005;23:339–48.
7. Lomranz J, Mostofsky DI. The psychology of pain and suffering. New York: Springer; 1997.
8. Nagarkar M. Gitagyanyoga – The Yoga of Knowledge. Adhyaya 2, Sat BuddhiYog (Sankhyayog). Pune: SwaroopyogPratishthan; 2012.
9. Ādi Śankarācārya (Chinmayananda S. trans) Atmabodha. Mumbai: ChinmayaPrakashan; 2009.
10. Radhakrishnan S. Principal Upaniṣhads. New Delhi: Harper Collins; 1994.
11. Ādi Śankarācārya (Dayananda Saraswati S. Tr & Comment) Tattvabodhah. Chapter 8, p. 112. Chennai: ArshaVidya; 2013.
12. Mundaka Upaniṣhad in Radhakrishnan S. Principal Upaniṣhads. New Delhi: Harper Collins; 1994.
13. Patanjali M. Yoga Sūtras II.54 (Shearer A. Trans. The Yoga Sūtras of Patanjali). London: Crown Publishing; 2010.
14. Isvarakrsna HT. SankhyaKarika. Philadelphia: Wentworth Press; 2016.
15. Nagarkar M. ParamAmrutChintanYog process for supreme knowledge of Swarupa. Pune: SwaroopyogPratishthan; 2016.
16. Ādi Śankarācārya (Chinmayananda S. trans) Atmabodha. Mumbai: ChinmayaPrakashan; 2009.
17. Wasserman T, Wasserman LD. Depathologising psychopathology: the neuroscience of mental illness and its treatment. New York: Springer; 2016.
18. Damasio A, Damasio H, Tranel D. Persistence of feelings and sentience after bilateral damage of the insula. Cereb Cortex. 2012; Apr 3:bhs077
19. Taittriya Upaniṣhad I.1-3 in Radhakrishnan S. Principal Upaniṣhads. New Delhi: Harper Collins; 1994.

20. Nagarkar M. Gitagyanyoga – The Yoga of Knowledge. Adhyaya 17, Sat BuddhiYog (Sankhyayog). Pune: SwaroopyogPratishthan; 2012.
21. Yogi MM. Science of being and art of living: transcendental meditation. New York: Plume; 2001.
22. Patanjali M. Yoga Sūtras Ch.II.29 (Shearer A. Trans. The Yoga Sūtras of Patanjali). London: Crown Publishing; 2010.
23. Nagarkar M. Science of meditation, Dhyanayog. Pune: SwaroopyogPratishthan; 2007.
24. Books LLC (ed). Branches of Buddhism: Hinayana, Pragmatic Buddhism, Schools of Buddhism, Buddhist Modernism, Southern, Eastern and Northern Buddhism. General Books LLC; 2010
25. Patanjali M. Yoga SūtrasI.2 (Shearer A. Trans. The Yoga Sūtras of Patanjali). London: Crown Publishing; 2010.
26. Cohen RA. The neuropsychology of attention. New York: Springer; 2016.
27. Vivekananda S. Chapter 2 in Raja Yoga – the yoga of knowledge. AdvaitaAshrama, H.P.; 2012.
28. Raichle ME, MacLeod AM, Snyder AZ, Powers WJ, Gusnard DA, Shulman GL. Inaugural article: a default mode of brain function. Proc Natl Acad Sci. 2001;98(2):676–8.
29. Yogi MM. Science of being and art of living: transcendental meditation. New York: Plume; 2001.
30. Damasio A. Self comes to mind: constructing the conscious brain. New York: Barnes and Noble; 2010.
31. Northoff G, Heinzel A, de Greck M, Bermpohl F, Dobrowolny H, Panksepp J. Self referential processing in our brain – a meta-analysis of imaging studies on the self. Neuroimage. 2006;31:440–57. https://doi.org/10.1016/j.neuroimage.2005.12.002.
32. Patanjali M. Yoga SūtrasI.3 (Shearer A. Trans. The Yoga Sūtras of Patanjali). London: Crown Publishing; 2010.
33. Madhavananda S. SanskarPradipika (Slokas from Gyaneshwari) 1.1 SwaroopyogaPratishthan, Pune; 2000.
34. Nargarkar M. The divine touch, selected Abhanagas of SantTukaram, Chap 4, pp 106–108. Pune: SwaroopyogPratishthan; 2014.
35. Travis F, Arenander A. Cross-sectional and longitudinal study of effects of transcendental meditation practice on interhemispheric frontal asymmetry and frontal coherence. Int J Neurosci. 2006;116(12):1519–38.
36. Abdul Kalam APJ, Tiwari A. Wings of fire – an autobiography. Hyderabad. Highest accomplishments: Universities Press; 1999.
37. Shetkar R, Hankey A. Optimizing emotional intelligence in management education: a role for vedic sciences, Nitte Management Review (NMR); 2014.
38. Mandukya Upaniṣhad, Ch.1, Verse 3-7, in Radhakrishnan S. Principal Upaniṣhads. New Delhi: Harper Collins; 1994.
39. Fell J, Elger CE, Kurthen M. Do neural correlates of consciousness cause conscious states? Med Hypotheses. 2004;63(2):367–9.
40. Ādi Śankarācārya (Dayananda Saraswati S. Tr & Comment) Tattvabodhah. Chapter 8, p. 112. Chennai: ArshaVidya; 2013.
41. Yogananda P. Autobiography of a Yogi. Kolkata: YogodaSatsangaSoc India; 2009.
42. Venkatesananda S. The concise Yoga Vasishtha. Albany: SUNY Press; 1984.
43. Nagarkar M (2012c) Gitagyanyoga – The Yoga of Knowledge. Adhyaya 13, Sat BuddhiYog (Sankhyayog). SwaroopyogPratishthan, Pune.
44. Olivelle P. Upanisads, Taittiriya Upaniṣhad Cha.1, Verse 1, Oxford world's classics. Oxford : Oxford University Press; 2008. ISBN:978-0-19-954025-9.
45. Husserl E. The phenomenology of internal time consciousness. Bloomington: Indiana University Press; 1964.
46. Hankey A. Complexity biology-based information structures can explain subjectivity, objective reduction of wave packets, and non-computability. Cosmos Hist J Nat Soc Philos. 2014;10 (1):237–50.

47. Hankey A. A complexity basis for phenomenology: how information states at criticality offer a new approach to understanding experience of self, being and time. Prog Biophys Mol Biol. 2015;119(3):288–302.
48. Patanjali M. Yoga SūtrasCh.I.47 (Shearer A. Trans. The Yoga Sūtras of Patanjali). London: Crown Publishing; 2010.
49. Patanjali M. Yoga SūtrasCh.III (Shearer A. Trans. The Yoga Sūtras of Patanjali). London: Crown Publishing; 2010.
50. Potter KH. Nāgārjuna and Madhyamika Buddhism (from Presuppositions of India's Philosophies). Conn: Westport; 1976.
51. Taittriya Upaniṣhad I.1 in Radhakrishnan S. Principal Upaniṣhads. New Delhi: Harper Collins; 1994.
52. Brihadaranyaka Upaniṣhad (5.1.1), in Radhakrishnan S. Principal Upaniṣhads. New Delhi: Harper Collins; 1994.
53. Mundaka Upaniṣhad (3.1) in Radhakrishnan S. Principal Upaniṣhads. New Delhi: Harper Collins; 1994.
54. Carsetti A. Functional models of cognition, self-organizing dynamics and semantic structure in cognitive systems. New York: Springer; 2010.
55. Patanjali M. Yoga SūtrasIII.4 – 6 (Shearer A. Trans. The Yoga Sūtras of Patanjali). London: Crown Publishing; 2010.
56. Patanjali M. Yoga SūtrasCh.III (Shearer A. Trans. The Yoga Sūtras of Patanjali). London: Crown Publishing; 2010.
57. Venkatramiah M. Talks with Sri Ramana Maharishi. 13th ed. Tiruvanamalai: Sri Ramanasramam; 2013. p. 24.
58. Sri Sankara'sVivekachudamani, (Trans. Acharya Pranipata Chaitanya), Chinmaya Mission, India.
59. Sri Jnanadeva'sGita BhavarthaDipika – Jnaneshwari (Trans. M.R.Yardi), BharatiyaVidyaBhavan; 1991.
60. Damasio A. Self comes to mind: constructing the conscious brain. New York: Barnes and Noble; 2010.
61. Shetkar R. Hankey A, Nagendra HR. Biophysics of meditation. Submitted to the Journal of Consciousness Studies; 2018.
62. Bak P, Tang C, Wiesenfeld K. Self-organized criticality: an explanation of the 1/f noise. Phys Rev Lett. 1987;59:381–4. https://doi.org/10.1103/PhysRevLett.59.381.
63. Shetkar R, Hankey A, Nagendra HR. Reason for health benefits of deep meditation: self organized criticality restores regulation to optimal. Eur J Pharm Med R (EJPMR). 2016; 3 (5):435–41.
64. Shetkar R, Hankey A Nagendra HR. First person accounts of Yoga meditation yield clues to the nature of information in experience. Cosmos Hist. 2017;13(1):240–52.
65. Hankey A, Shetkar R. Self-transcending meditation is good for mental health: why this should be the case. Int Rev Psychiatry. 2016;28:236–40.

Chapter 25
How Physics Would Look Like *If Based* on the Metaphysical Guidelines of *Śūnyavada*?

Debajyoti Gangopadhyay

Abstract This paper discusses some overlaps between the foundational issues in physics and philosophy from eastern and western perspective. This provides a basis for science philosophy dialogue. There are numerous approaches attempted by many thinkers. They may be treated as different programs of epistemic extensions. In this paper an alternative answer to issues of epistemic extensions is also provided.

Keywords *Śūnyavada* · Epistemic extension · Realist and anti-realist · Nāgarjuna · Freedom · *Catuṣkoṭi*

1 Introduction

The central issue the Buddha addressed prior to all other can be fairly pinpointed as *Dukkha* (*human sufferings*). During his time almost all the major system of Beliefs originated in India happened to have addressed *Dukkha* as their central concern to take "philosophical" care. The historic Buddha is said to have taught the concept of *Śūnya* in this connection as it is recorded in *Prajñāpāramitā Sūtras*. However, in question of any metaphysical issue, the Buddha is known to have employed a framework of argument admitting four possibilities (the celebrated *Catuṣkoṭi* or Greek *tetralemma*). Clearly this framework is more accommodative than what our common sense approves.

Ācārya Nāgārjuna further substantiated the ontological status of *Śūnya* as *Catuṣkoṭivinirmukta*. Later Buddhist scholars continued to expand their logico-epistemic standpoints keeping the Nāgārjuna's standpoint of *Śūnya* as pivotal.

Śūnyavada thus, right from the very beginning, had differed for common sense ontology as well as standard Aristotelian logic which underlies our classical physical scientific worldview. However, quantum reality seems to clearly outstep the

D. Gangopadhyay (✉)
Department of Physics, Annada College, Hazaribagh, Jharkhand, India
e-mail: humanphysics2014@gmail.com

S. R. Bhatt (ed.), *Quantum Reality and Theory of Śūnya*,
https://doi.org/10.1007/978-981-13-1957-0_25

framework of common sense ontology. With this implication in view, we will address a counterfactual question here – What a physics would look like *if based* on the possible metaphysical as well as logical implications of *Śūnyavada* right from the beginning?

Śūnyatā is the central concept in the *Mādhyamika* approach to reality. *Mādhyamikas* apparently expounds a view contrary to our commonsense realist belief that there *is reality* ontologically out there. Mādhyamika argues that everything is merely designated by our conceptual construction and verbal language.

Quantum mechanics, on the other hand, is a modern outgrowth of the seventeenth century scientific revolution relying at least implicitly on a hypothesis of real world out there. However, right from the beginning of quantum revolution during the third decade of the last century, difficulties to interpret the formalism in *realist terms* became evident. Consequently, following the early developments of quantum formalism, founding fathers were involved in debate about the possible ontological implications of the theory.

Schrodinger is known to have changed his mind regarding the interpretation of quantum mechanics at least three times during his struggle with this theory. Schrodinger's earlier conclusion about the *psy*-function as intermediate-level concept (*hilfbegriff*) was eventually modified with Born's probabilistic interpretation of *psy*-function, and he is known to have decided to teach quantum mechanics, from1928, according to the mainstream version with which the *sticker* "Copenhagen" was attached later. This decision is considered by many as his 'third' interpretation of quantum mechanics. However, during the late 1940s and early 1950s, Schrodinger advanced something more radical in the sense of having the potential promise to get far away from the realist ontology of standard logic and set theory faithful to Leibniz's *principle of identity of indiscernibles*.

In this article, we will try to argue that Schrodinger's last standpoint, which he did not articulate in formal details, is closest to the *Mādhyamika* standpoint, and on the basis of this overlap, we can address a counterfactual question about the *possibility of a physics* more faithful to the ontological guidelines of *Mādhyamika Śūnyavada*. Stated in other words, whether the ontological guidelines of Mādhyamika can be formally translated to meet the requirements of quantum reality?

2 Foundational Debate in Quantum Mechanics: Do We Have Sufficient Grounds So Far to Make a Definite Choice? Strange Coexistence of Phenomenal Success and Baffling Failure!

Ever since the formal developments of quantum mechanics took a reasonable shape at around the late second decade of the last century, founding fathers of the theory started debating about the question of *what the quantum formalism was really about*, that is to say, what suggestions about the underlying reality *(furniture of the world – ontology)* can be figured out from the theory/formalism.

It was quite surprising to note that, though the new theory had been turned out to be the most successful theory in the history of physics by that time, the theory was simply *unable* to make adequate sense of the most primary expectation we could usually ask any theory to be for – expectation about making adequate sense of "*One*" *of a particular kind* in a way to make further sense of any *countable plurality (taking the "individuals" together)* on this basis. Indeed, standard formalism can't be consistently interpreted as instantiating "one" of its particular kind in the sense of, for example, Lowe –"...an object that is differentiated from the others of its kinds in such a fashion that it and they are apt to constitute a countable plurality with each member of such plurality counting for just one, a unit of its kind" [E J Lowe, 1994].

Within the intended scope of this article, we will not get into any further technical details of how quantum formalism can't be consistently interpreted as saving the notion of "one" of a particular kind. There is plenty of literature available even online which addresses this issue [1–5]. Of course, there is a wide divergence of opinions, but the difficulty to preserve the *spatiotemporal meaning* of what constitutes "one" is one of the primarily acknowledged ontological crises in foundational debate of quantum mechanics.

In fact, making sense of "*one*" *of its particular kind* is an ontological priority of any physical theory. But standard version of quantum theory fails to make proper intuitive sense of this obvious ontological priority (of individual/one) within the scope of its own framework, as no consistent space-time picture had yet been emerged to make adequate sense of *individuating* something (as one!) by clearly distinguishing it from similar one of same kind. And exactly in this sense, quantum mechanical entities are termed *indiscernible* leading to an apparent failure of *Leibniz's principle of identity of indiscernible*. Put other way around, it can as well be said that *the* "sense" quantum formalism seems to suggest it is utterly difficult to *reconcile* with the domain of our ordinary experience/perception which seems always to make a distinct sense of *spatiotemporal IS* as *one*. The so-called quantum measurement problem is well known to be all about *the different aspects of the difficulties of reconciliation.*

2.1 Two Choices and Two Different Historio-Epistemic Attitudes

In view of this baffling failure, one is most likely to ask for a *choice between two options -*

Either

as a *Realist,* for whom the existing formalism itself is incomplete or inadequate in some sense, which can, in principle, be taken care of (by supplementing or modifying...) to reinstall an *adequate meaning of "One"* within the scope of its framework. Quite understandably, this option can have several variants[1]

or,

as an anti-realist/phenomenalist, for whom the *ontological expectations themselves are deeply flawed*, at least in a sense that we can't ask for an unambiguous criteria of addressing "One" of a particular kind on the basis of the experimental findings pertaining to the domain of quantum[2]

The first alternative can be easily recognized as a far more preferable option for the practicing physicists for most obvious reason that the *one of a particular kind*, alternatively called *thing/object/entity*, is taken almost for granted as something to be primarily ensured within the scope of any physical theory. This preference can be traced right back to the beginning of the physical theoretical tradition that can be dated back to the seventeenth century dynamics. Possibility of addressing a *spatio-temporally unambiguous entity with determinate identity* is taken as automatically prioritized or ensured within the framework of any physical theory. Indeed, during nearly the last 80 years or so, majority of the physicists, in one way or another, not even bothering much about the *ontological credential of this priority*, remained sympathetic with this option. Einstein's attitude even to renounce quantum mechanics, in view of this failure, can be made more sense as a reflection of his strong implicit commitment or personal conviction to the *ideological jurisdiction of this very choice itself.*

However, the *second option* connects itself directly with the whole *tradition of skepticism* developed mostly by the philosophers in East as well as West. Hume famously concluded that, *belief* in an external world is *rationally* unjustifiable. In fact, Hume and George Berkley as the two famous extreme empiricist voices of *phenomenalism* held that a *physical object is a kind of construction* out of our experiences. Hume in his *Treatise of Human Nature* (*1739*) famously emphasized that we *never see* objects, but only *qualities*, and that it is nothing but our imagination if we consider the observed qualities as properties of an object. The same problem was taken by Kant in his *Critique of Pure Reason* (*1787*). He, in contrast to Hume, systematized the doubt and emphasized that, as the objects of our knowledge are recurrent, and not arbitrary ("permanent possibility of sensation" as described later by John S. Mill), there should be a *well-defined recipe* or *conceptual prescription* about how *knowledge of objects* (of distinct individual type) is *possible* with sense experience as starting point. We need not get here into any further details

[1]One of the earliest defenders of Realism among the modern physicists was Einstein himself. He was even famously known to have almost disowned quantum mechanics [keeping] in view of the apparent *failure of Realism.*

[2]Though the term Reality is not always clearly defined, a wide band of Physicists - particularly those within the ideological jurisdiction of Bohr, seems to be comfortable with the phenomenalistic position. Reality for them is *not objective* in Einstein's strong sense of being there independent of observer, but can only be specified *inseparably* in context of the experimental set up (Experimentally verifiable question addressed)!

of these all philosophical ramifications; but it should be carefully noted that Kant's intension was not to provide any ground of straightforward suspicion against "our" notion of *individual of thing type! On* the contrary, Kant provided a framework of making sense of the basis of metaphysical scaffoldings operative behind normal human thinking. For Kant it was a question of providing epistemic *legitimacy* of the notion rather than endorsing a token of reality for it.

But the Kantian framework of *assessing metaphysics* was aptly employed as *framework of doubt* by some philosophically oriented physicists like Ernst Mach, Gustav Kirchhoff, and H. Clifford particularly during the late nineteenth century in West (so quantum mechanics is not an essential prerequisite to develop this skepticism). This framework provided room for systematic doubt for the first time about the *origin of the thing concept* as well as its different ontological imports hosted, as a consequence, in any physical theory.

Whitehead's commentary, more or less during this time, can also be taken as a remarkable anecdote to the prevailing skepticism *about ontological constructs* among a section of philosophically oriented physicists. He exclaimed that the *belief in an independently existing* entity is the central delusion of 2000 years of Western thought. This was definitely the prelude to what he professed later (during 1928 or so) as *process* philosophy. But that is a different story.

2.2 A Remarkable Cultural Asymmetry Referred to the "Central Delusion"!

There seems to be a remarkable asymmetry in terms of attitudes toward *thing* concept in East compared to Western philosophical/theoretical tradition. It took a surprisingly long time in West to develop even a systematic framework of doubt about whether the concept of thing is really something compelled by the nature herself or is something our mind *needs to attribute to nature* to make sense of our sense experiences! And we must remember that the *framework of doubt* in West was developed and subscribed further mainly by the philosophers, rather than the physicists.

It is only during the late nineteenth century when the classical Newtonian epistemology was critically questioned within a neo-Kantian framework by the physicists like Mach, Kirchhoff, Clifford, and a few followers of them. In fact, our innate psycholinguistic *commitment to the thing concept* as well as its collateral (common sense) ontological paraphernalia was accepted almost *intact* in the formal language of classical mechanics. As mentioned before, Mach and his followers were motivated more or less by Kant's diagnosis for unwarranted metaphysics. But the critical epistemology, within Kantian framework, did not attract any remarkable attention of the majority of physics community then, and the subsequent lead to make *more precise formal sense of this epistemology* was taken over most notably by Russell, Whitehead, as well as the Vienna-based analytical philosophers during the first four decades of the last century – exactly the time when the strange-looking quantum formalism was taking shape with an underlying commitment to thing concept discredited or suspected yet, in one way or another, by the philosophers.[3]

Quantum mechanics in different phases of its formal developments during the very same period, throughout the first three decades of the last century, can definitely be read to have dramatized or amplified this *question of choice we mentioned* by providing compelling grounds of skepticism about the so-called realist ontological suggestions we *normally* expect any theory to provide.

But the cultural asymmetry between Eastern and Western theoretical traditions is quite evident as the full-fledged skepticism about *thing concept* (mostly by the philosophers) in its different looks is seen to be a matter of *not more than 250 years or so* in West, while it was not so in Eastern theoretical traditions particularly with the Buddhist philosophical origin which can be dated back to the fifth century BC! Buddhists are well known to have doubted long back the very basis of *common sense ontology* itself!

2.3 Doubt Mechanism Operative in Buddhism

In Buddhism the thematic origin of the same problem can surely be traced back to Nāgārjuna's (150 A.D) *doubt* (based originally on the Buddha's teaching) about the possibility of acquiring knowledge. Nāgārjuna famously pointed out the self-contradictory character of all commonsense means of acquiring knowledge. Subsequently Dignāga (after a long time at about 450 A.D) provided systematic foundation of what can be termed as logic and epistemology in modern sense within the framework of this doubt. Subsequent to *Dignāga*, a significant part of *philosophical dialog in India* between the Buddhists and non-Buddhists evolved around the different proposals to take care of this issue (more specifically, valid norms of cognition – *pramāṇa-lakṣaṇa*) which continued at least up to the twelfth century. The complex ramification during this time is found in Ratnakīrti – perhaps the last medieval Buddhist philosopher of note, who tackled the problem of *empty term* in his *Kṣaṇabhaṅgasiddhi Vyatirekātmikā* at about 1000–1050 A.D. This concern was shared also by the Chinese and Tibetan Buddhist philosophers.

Within the scope of our article, let us take a bird's-eye view of what *Nāgārjuna* expounded in *Mūlmādhyamakakārikā*.

Any ontological picture, according to Nāgārjuna, is likely to be shown as vulnerable from the point of view of any of the four possible kinds of logical justifications or options (the famous *Catuṣkoṭi*), meaning "four corners." The four corners are identified as four options that one might take in connection with any question – *yes*, *no*, *both*, and *neither*. A *Mādhyamika Buddhist* can't possibly

[3]One of the most comprehensive discussions about the fate of Analytical empiricism can be found in Ref. [6].

indulge any ontological construction within the true spirit of Nāgārjuna. A *Nāgārjunean* would possibly invite him-/herself in an endless debate of deconstructing any ontological picture he/she is enticed to get stuck. In fact this is quite challenging, as human being can't normally wish away mental constructions as a part of his normal cognitive process. But *Nāgārjuna* advocated contradiction to lurk behind all the four logically possible means of acquiring knowledge, and that is why true reality (as this must not involve *contradiction* in any way) is beyond this – *Catuṣkoṭivinirmukta Śūnya*. The advocated framework of showing contradiction is obviously *Pratītyasamutpāda* – the theory of *dependent co-origination* preached by the Buddha himself.[4,5]

So, referring to the doubt mechanism in *Madhyamaka*, we can say that our ordinary psycholinguistic commitment to *individual* seems to have played a historically different *doctrinal* role in the development of Buddhist worldview. Unlike the Western theoretical tradition for a long time, in which the concept of individual was accepted *as given* in the sense of something *forced on us by nature* (as an unambiguous part of what there IS/ontology), a Buddhist (generically speaking) can't hold it as something more than a *construct* – provisionally or conventionally true only within a relational context (technically *Pratītyasamutpāda*).

3 How Physics Would Look Like...

> We have been compelled to dismiss the idea that, a particle is an individual entity which retain its 'sameness' forever. Quite the contrary, we are now obliged to assert that the ultimate constituents of matter have no 'sameness' at all. (Schrodinger, Dublin Lecture, 1950)

How the "Buddhist challenge of commonsense ontology" can serve as an alternative metaphysical backdrop for physics, we are actually provided with some metaphysical/ideological guidelines to portray a worldview in which *change* is prior to *permanence* or, in other words, a worldview in which concept of *individuation* and *identity* need not be considered as *ontologically prior*? But is it really rewarding or of practical worth to think of a physical picture dispensing altogether with the notion of re-identifiable (permanent) individual and identity *right from the*

[4]There are so many standard expositions of these issues ... One of those of my choice is – [7].

[5]Undoubtedly *Dukkha* was addressed as one of the prime issues of perennial concern by almost all early Indian Thinkers based in Bihar. It is important to note that a wave of renunciation and asceticism of diverse origins was brewing over the society of Bihar for a few centuries before the emergence of the historic Buddha. The earliest 'philosophical' concerns regarding *Dukkha* by the later thinkers were surely the outcome of this social situation. *Sankhya*– considered as the earliest doctrinal core preached by *Kapil* (his origin is as usually controversial), is well known to have addressed *Dukkha* as their prime concern. It is not surprising to see that *Kapil* and his immediate followers like *Asuri,Panchshikha*all belonged to the same place of North Bihar not far-away from *Kapilavastu* – the celebrated birthplace of the historic Buddha.

beginning from its metaphysical baggage? Full answer can't be given here within the limited scope of this article. But we will try to give a brief outline of the present state of the arts of this question addressed from logical point of view.

In fact, this is exactly what is addressed by a group of modern logicians.

However, their guiding motivation is not Buddhism but some ideological guidelines outlined by Schrodinger in quantum mechanical context in a series of public lectures given in *Institute for Advanced Studies* in Dublin in 1950 (subsequently published as *Science and Humanism*, 1952).

Based on Schrodinger's appeal in the Dublin lectures, Newton da Costa explored possible semantics for what he termed as Schrodinger's logic. This logic derives its chief motivation from Schrodinger's (almost unheard by physics community audience) appeal in that lecture that the concept of *identity* and *sameness* doesn't make any sense for the elementary particles. Basically da Costa put forward a suggestion that the "right" semantics for his first-order Schrodinger logic (SL) should be done in a different set theory, termed quasi-set theory. This suggestion had been further developed by Krause. He also developed the semantics for higher-order Schrodinger logic (see da Costa and Krause, 1994, 1997). To the same end, Dalla Chiara and di Francia (1993), independent of da Costa, proposed a theory of quasets to make sense of semantic treatment of the language of microphysics (see Dalla Chiara, Guintini, and Krause, 1998).

Following these leads, during the last 20 years or so, came many proposals to spell out an adequate language of microphysics fundamentally different at least in its logical and metaphysical basis from the existing language of labeled Hilbert space – a fragment of functional analysis. A reasonable bulk of these proposals did come from Steven French and Décio Krause.

So the interesting point is that, though the formal structure proposed is solely motivated by the attempt to *avoid identity* in dealing with quantum physical objects, this is (accidentally!) *in apparent metaphysical tuning* with what the Buddhists tried long back, though guided by altogether different motivations to get rid of the eternal existential cycle!

The basic agenda as expressed by Décio Krause recently:

> I am occupied in providing strong arguments favoring metaphysics of non-individual quantum objects. I start taking from granted that the indiscernibility of these entities is necessary to be assumed in quantum physics (as it was in classical physics, as Gibbs' paradox shows, although the situation there is different). Then, I consider the mathematical stuff in which quantum theories are usually written, by realizing – just take any book on QM – that it can be subsumed as being the framework of classical logic and standard mathematics (say, that which can be constructed within a standard set theory like the ZFC system). But the only way to make sense to indiscernible objects within such a framework is by making some mathematical tricks, as QM does when assumes permutation symmetries of some kind (say, symmetric and anti-symmetric state vectors). I try to avoid such a move, by presenting a mathematical theory, namely quasi-set Theory, in which indiscernibility is not made by hand, but can be assumed right at the start. The complications in the logic provide conceptual gain in we being able to express a metaphysics of non-individual entities, which according to me fits better in quantum case.

A brief note by Décio Krause inviting response in *Academia.edu*, September, 2015

So Krause terms the alternative package as metaphysics of nonindividual. This is of course a provocative term from historical point of view. We need not get here into any further details; but what seems to be clear is that the question of failure of individuation and identity can really ensure most fruitful overlap between the foundational issues of physics and philosophy irrespective of Eastern or Western origin. But the immediately next concern is the question that how much of the present physics/quantum mechanics can be meaningfully reproduced on the basis of this alternative metaphysics of nonindividual (quasi-set theory or any nonstandard set theory).

In fact as is well known, mathematical language of classical as well as quantum physics doesn't care (historically) to provide a formal distinction between predicative and existential IS; everything – whatever enters in a Law, IS in some sense of being somewhere (not just in theoretician's mind or dream!). And as an obvious consequence, philosophers of science are well aware that these all blur the traditional epistemic distinction between mathematical and physical objects. In other words, all multiply diverse senses of the undefined term "thing" (*res* in Latin) are vaguely referred to some sense of entity with presumed identity (IS) located/spatially extended somewhere.

But any proposal which amounts to uproot or avoid these ontological presuppositions right from the beginning is supposed to make further sense of an immediate counterfactual question like what should a physical law would look like if *not* committed to the standard sense of IS!! For example, what a Schrodinger equation-like equivalent would look like if based on the metaphysics of nonindividuals from the very beginning ?

Response of Krause himself to this very question:

"A quick answer to your question: yes, I am trying to provide the foundations for a metaphysics of entities devoid of identity conditions, which by historical reasons I call "non-individual". Perhaps it would be better to call them indiscernible entities, but this confuses a little the intended idea.

Schrodinger's equation can be obtained by developing a version of *Fock space* formalism within a theory of *quasi-sets Q*. Once defined the creation and annihilation operators, the Schrodinger's equation can be obtained ....

But of course, someone need to provide the details.

The important thing is that we can avoid identity in dealing with physical objects. *Indiscernibility* is enough."[email response to my query, 5/10/15, for some further hints in this respect, see, Chapter 9, French and Krause, 2006]

In fact, *this* challenge of developing formal systems committed to different variants of metaphysics of nonindividuals constitutes the essence of freedom, though, apparently, from the outset, the question of *picture of Reality* and Nāgārjuna's approach to this seems to be just diagonally opposite to what a physicist would be sympathetic to!

4 Epilogue

Perhaps we may locate this diagonally opposite standpoints as one of the prime reasons why scientific realism, as we know it in the conventional Western sense, had never took prominent shape in India and elsewhere in Asia where Buddhism spread. The existing science as we know it was not possible to be developed here within the *Nāgārjunian spirit of ontological attitude* to *Vikalpa*. Science in its present form rests on presuming *vikalpas* of various orders. But as we have discussed, the metaphysical guidelines of *Mādhyamika* standpoint can surely be exploited to develop alternative formal system for physics to avoid some ontological excess.

However, it must be conceded that the progress in this direction is rather slow. Though some guiding outlines of a quantum theory with nonindividuals as primitive entity had been proposed by Krause et al., an adequate metaphysics of nonindividual is still in its infancy. So a final answer to "how a physics would look like if based on the metaphysical guidelines of *Śūnyavada*" is yet to be worked out in further details.

In fact we are not yet sure whether the conceptual gain would be compatible/ comparable to the cost of admitting greater (nonstandard) logical complexities. But we are surely on the track to answer this question in a formally adequate sense. However to pursue the possibility in further details is something to be developed as a research program, and we leave this topic for further works.

References

1. Post H. Individuality and physics. The Listener. 1963;70:534.
2. Schrodinger E. Science and humanism. Cambridge: Cambridge University Press; 1952.
3. Casellani E, editor. Interpreting bodies, classical and quantum objects in modern physics. Princeton: Princeton University Press; 1998.
4. French S, Krause D. Identity in physics, a historical, philosophical and formal analysis. Oxford: Clarendon Press; 2006.
5. i) Holic F. Neither name, nor number. arxiv:1112.4622v1[quant-ph] 20 Dec 2011. ii) Krause D. The problem of identity and a justification of non-reflexive quantum mechanics, Jan 2011 (available online). iii) Krause D. Entity, but no identity (available online).
6. Wang H. Beyond analytical philosophy. Cambridge, MA: MIT Press; 1985.
7. Kalupahana DJ. Nāgārjuna: the philosophy of the middle way. New York: SUNY; 1986.

Chapter 26
A New Information Theory Explains Śūnya in Samādhi

Alex Hankey

Abstract There is an obvious analogy between the concept of the vacuum state in quantum field theory and that of 'emptiness' or Śūnya. When the mind settles down in meditation to the state of pure consciousness, it has lost all content, all feelings and all emotions, and qualia or qualities also disappear. Some describe the state as fullness, a 'field of all possibilities', while others emphasize its lack of content. As the Bhagavad Gita, II.45, puts it, 'nistraigunyo bhavārjuna', 'be without the three gunas'. Only from a state of purity of the heart and mind can a human fully realize his or her potential and act fully in accordance with all the laws of nature. Our approach shows that such a state is empty of wave functions or quantum fields and that the human mind achieves a state where all such excitations are annihilated.

Keywords Śūnya · Vipassna · Samādhi · Information theory · Quantum field theory · Complexity biology

1 Introduction

Śūnya is a term much favoured by Buddhist philosophers [1] to describe the fundamental state of experience in deep meditation. It was clearly discussed by Nāgārjuna [2] in particular, and the concept has become foundational in Buddhist theories of mind. It distinguishes the Buddhist approach to describing conscious experience from that of the Upaniṣads [3], which comprise the essence of the Vedic philosophical wisdom of ancient India, which most Indologists now regard as having its origins at least 6000 years earlier. It began to flourish as a full-fledged civilization in the cities of Harappa and Mohenjo-daro some 3000 years later around 3500 to 3300 BCE [4], along the banks of the Sapta Sindhu, particularly the Sindhu (Indus) and the Saraswati. The drying up of the Saraswati between 2300 and 1900 BCE [5]

A. Hankey (✉)
SVYASA, Bangalore, Karnataka, India

S. R. Bhatt (ed.), *Quantum Reality and Theory of Śūnya*,
https://doi.org/10.1007/978-981-13-1957-0_26

appears to have been the event that triggered migration to the Gangetic plains. The name 'Meluha' seems to have been associated with that civilization, at least according to the phonetic transcription of trading documents from the Fertile Crescent.

During that vast time period, the events recorded in the Rāmāyaṇa [5] and Mahābhārata [6] may be considered to have taken place. By the time of Kṛṣṇa [7] and the writing of the Bhagavad-Gītā [8], the civilization had developed a highly sophisticated system of philosophy [3] and personal development as recorded in the Purāṇās, the Upaniṣads, and the Ṣat Darśanas [9]. In any Vedic perspective on *Śūnya* [10], it is to these texts that we should turn. They represent the mature wisdom of a civilization already three millennia old, by the time of the Buddha, who some 1500 years after the Saraswati's drying up had caused the eastward migration. The Vedic culture was by then in decay in Bihar and to the east.

The mature Vedic system of practices and wisdom [11] represents the core understanding, which the new approach proposed by Lord Buddha was updating. His was the brilliance and originality that only a fully enlightened, epoch-creating, spiritual master can accomplish. Gautama Siddhartha is a classic example of a soul [12] whose understanding of the human condition in the context of its cosmic potential is supreme and which can communicate the new direction needed for society to adopt in a way that will appeal to all open-minded rulers, intellectuals, and common people alike.

Gautama's was a revival that bypassed the snobbery of the priests and other intellectuals and attempted to finally bring the wisdom and practices developed over the previous millennia to guide all people to spiritual liberation and enlightenment [13], particularly the suffering masses [14], whom the Gurus had largely left to suffer, despite examples and admonitions to the contrary in the Upaniṣads [15].

Having been the heir apparent in a royal house with a level of compassion that cared for all the people, Siddhartha was not concerned with the vanity of learning and exclusive attitudes that characterize the learned but unenlightened who regrettably lack such universal love-based wisdom, even to this day, and whose attitudes so shocked Swami Vivekananda [16] in the late nineteenth century. In order to break with such intellectual snobbery, Lord Buddha emphasized that he was not concerned with philosophical doctrines but with bringing a simple and practical method that could be transmitted to all the people or at least to all those who would listen to him and take his simple but classic message seriously. He taught his Noble Truths [14] to describe the suffering human condition, only too familiar to the common people, and he remedied it with his Eightfold Way [13] implicitly teaching the values of virtuous living based on an inner life enriched by regular spiritual practice.

While Buddha's teachings on meditation have largely become identified with such schools as Zen [17], *Vipassana* [18], and Tibetan systems [19] such as the Six Yogas (i.e. meditative spiritual practices) of Naropa [20], the Chan texts of Chinese Buddhism [21] record many approaches to meditation based on inner development. In one such Chan Sutra, the Buddha is depicted as surrounded by some 26 *arhats* (disciples already at the level of sainthood) and a little boy who asks the key question, 'Respected Sir, please would you tell me how to meditate'? To this question, reminiscent of a similar one posed by a child to Yājñyavalkya in the *Bṛihadāraṇyaka Upaniṣhad* [22], Lord Buddha requests his experienced followers to describe their favourite meditation technique, each speaking in turn. After 25 have spoken comes the turn of the oldest and wisest. This worthy states that he learned his favoured system of meditation when serving the Buddha in a previous age. It consists, he says, in taking an essentially meaningless syllable known of old to have supreme power on the inner level of the meditative mind and letting its repetitions slowly soften until they die away into the mentally vacuous silence and bliss of pure consciousness. In this short description, this greatest of arhats has described a key system adopted by the Vedic civilization of old and used by its highest adherents even to this day. It is the essence of the self-transcending system [23] of Transcendental Meditation [24] brought to both India and the West in the twentieth-century revival of His Holiness Maharishi Mahesh Yogi, by the grace of his master, the Jagad Guru Shankaracharya of Jyotir Math, His Divinity Swami Brahmananda Saraswati Maharaj [25].

Against this background of timeless wisdom and its revival, I intend to here present a new angle on the state of *Śūnya* [26] described by the Buddhist masters [1, 2] and now being linked into concepts of quantum theory and quantum field theory by today's intellectual elite [27]. The present fashion is to link the concept of *Śūnya* to the vacuum of empty space, in its present form of the quantum vacuum state, ultimately derived from Galileo's reversal [28] of Aristotelian doctrine [29]. The problem with this approach, attractive though it seems and beautifully apposite the analogy, is that quantum theory is unable to describe states of mind, because quantum states do not have the double-aspect structure [30] required for a theory of cognitive states, information containing states of self-aware mind.

The approach that I propose to adopt is based on a series of papers [31, 32] that I and my student Rashmi Shetkar [33–36] have published over the last 3 years, based on a theory that I have managed to bring to a state worthy of presentation over the previous decade, since I first presented it in 2006. It is 10 years old and has undergone revisions and updates as a result of discussions with such luminaries as Jonathan Shear, David Chalmers, Brian Josephson, George Sudarshan, and particularly Stuart Hameroff, who stimulated some crucial refinements. Explaining it to students and friends has also helped me clarify both what I think I am attempting to accomplish and the details of how I now approach explaining it. The theory I am proposing is thus a mature and well-considered work that has already been explained in several different contexts, such as its information theoretic aspects, its relationship to Von Neumann's two-part approach to quantum theory [37], light it sheds on Penrose's extensive work [38, 39], and particularly, concerns of Hameroff and his School of Consciousness Studies in Tucson. It presents a new information theory, that of experience information [31], which can describe the information in self-aware cognitive states. After describing that theory in detail, I shall show how it describes a ground state of consciousness [40], one that is mathematically different from, but analogous to, quantum field vacuum states.

2 Experience and Its Foundation

Experience, pointed out David Chalmers [30], presents us with an essentially insoluble problem that he and Jonathan Shear termed the 'hard problem of consciousness'. [41] Self-awareness cannot be described by a system reducible to tiny interacting components, reducible to atoms and molecules, whether they obey the laws of classical physics, or quantum mechanics. Reductionist approaches cannot form the basis of the necessary theory [30], ultimately because we experience ourselves and perceive/conceive of ourselves primarily as wholes. The maturing of the mind in this sense is well-described in the Taittrīya Upaniṣad [42] where the young Bhṛghu, the future Bhṛghu Maharṣi, is sent to meditate for successive periods of time in the forest and asked each time he returns who he is. Initially he identifies with the physical body and then with the physiological sense of energy that controls it and upholds its health, the *Prāṇa*; his next identification is with the mind or at least its aspect that supports thought and emotion, *Manas*; next is the level that controls thought and makes all decisions, the intellect or *Buddhi*.

Going deeper Bhṛghu realizes that his deepest self is an infinite, unbounded field of silence that seems to encompass the universe itself and all possible existences: what we may call the field of *Śūnya*. It is neither a thought nor can it truly be named by a thought, but in the deepest level of meditation, it presents a transcendental level of experience that can subsequently be remembered and discussed. As Bhṛghu proclaimed it 'Aham Viśvam, Bhūvanam Abhyabhavam' – 'I am this All, the essential Being at the basis of all possible existences'.

The Upaniṣads [22] simply designate the experience as 'Tat', an indicator pointing to the abstract level at the foundation of experience, and state that for each of us, 'Aham Tat', 'I am That', adding for the disciple, 'Tat twam asi', 'That thou art' and 'All this is nothing but That'. It is the foundation of all apparently separate individual existences, both subjective and objective, finally concluding, 'That alone is', i.e. only on that basis can all these various entities manifest in the forms they assume and with the qualities, dimensions and properties that each exhibits.

3 Quantum Field Theory

Quantum field theory [43] is remarkable in that it forms the foundation for all known physical theories, a language of description, which seems all-encompassing in its power of accounting for physical appearances and physical laws. The foundation of quantum field theory is the concept of the vacuum state out of which all particles and fields, 'quanta', are manifested or created and to which they will all ultimately return. Electron fields describe electrons; photon fields of light describe electromagnetic phenomena. Proton and neutron fields with their obscure symmetry identified by Princeton's Eugene Wigner [44] describe components of the atomic nucleus. Other quanta such as the graviton and neutrinos, the quark-lepton pair tetrads and others attempt to account for all known phenomena, together with solitons and other exotic animals of material science, including various solutions of different wave equations applying under more or less exotic circumstances. Quantum vacua apply to all such descriptions of nature's display of forms and phenomena in our world of sense perception and its various levels of pleasure and suffering.

So how about the mind? If the mind is not describable by classical physics, if it exists in some describable way rather than as a figment of our own fertile imaginations, then surely it is some kind of quantum phenomenon. What else is there in physical theory that could describe it? Such runs the confused argument that has been propagated by a few confused minds, notably Goswami in his extensive writings on the topic [45]. Quantum field theory *must* be the theory of choice for the mind, but specific forms of it have been proposed as theories of everything [46].

The problem with this is simply stated: it is contradicted by two factors. First quantum field theory is a description of matter, and the mind, if it exists as such, is definitely not a material entity – it is on a *subtle* level, whereas matter is on a gross, physical level. Second, matter retains a definite form by virtue of various degrees of stability. Solids retain form, liquids hold together only fragmenting into drops and droplets under physical duress and even gases have a stable density if not a definite shape. Quantum field theory is a theory of quanta that create stability in the materials to which they apply [32]: the basic constituents of matter, the forces that retain the shape of solids, the forces of cohesion in liquids, and the phonons of sound that retain a uniform stable density in gases. Quantum field theory is a theory of material entities and how they maintain their stable properties that we observe. Similar quanta stabilize atoms and the atomic nucleus. It is probable that the existence of the quanta it supports stabilizes the properties of the vacuum state itself.

Quantum field theory as we know it is not a theory of the mind that we humans experience. The state of *Śūnya* in the mind is a state of awareness free from information content, but it is distinct from states of matter, although paradoxically it is found embodied in physiologies, animal and human and in that sense seems to have some profoundly rich material basis. So what could the basis of the mind be?

4 Instabilities and Their Role in Physics

In order to begin to approach an answer to the question of the basis of mind, we have to come back to some fundamental modern biology. The mind seems to control the body. At any rate that is the way that we use our minds, to send messages to various muscle sets and to execute our desired commands. The mind must therefore stand at the 'locus of control' of the entire consciously controlled part of the physiology [32]. Now this idea instantly distinguishes mind from matter, for modern complexity biology [47] is based on a new universal law of physiology: the locus of control of all physiological systems is a *point of instability*. The physics of mind, if such exists, must be associated with the physics of instability rather than the physics of stability, and this instantly distinguishes the physics that should be used to describe the mind from quantum field theory. For as we have seen, quantum field theory constitutes a universal theory of stability – as we said, of how stable properties of matter are maintained.

A remarkable fact about self-interacting physical systems is that they can become unstable. This possibility was first realized, described and to some extent understood by the great nineteenth-century physicist Van Der Waals [48] when he published his famous equation describing real gases [49], in order to explain the instability observed in carbon dioxide. At a particular value of density, and at a precise temperature (about 33 °C) and a precise pressure (about 30 atmospheres), the density of carbon dioxide becomes unstable, and it begins to fluctuate violently on scales vast compared to its component atoms and molecules, scattering and refracting all light impinging on it, so that it appears opalescent, a phenomenon known as 'critical opalescence'.

What Van Der Waals so brilliantly realized is that under these specific conditions, carbon dioxide has become unstable. It wants to condense from a gas to a liquid, but is not able to do so. Indeed the instability just described occurs at the end point of the boundary line between its liquid form and the corresponding vapour. At temperatures below the critical temperature, two phases exist in the phial containing carbon dioxide at its critical density, one with a low density, the gaseous phase of vapour, and the other the liquid phase of carbon dioxide's liquid form. Above the critical temperature, there is only one phase, but at temperatures close to the critical temperature, the fluid enters a schizophrenic state where it can neither decide to be a gas nor to be a liquid, as it were, and where it fluctuates violently between these two attempted alternatives. Its density has become unstable and fluctuates violently.

One way to explain this is to say that the velocity of sound in the fluid has gone to zero and that the instability has arisen simply because the stabilizing quantum fields of phonons present variously in the stable liquid and the stable gas have vanished [32]. Another way to explain it is to say that the tendency of the CO2 gas molecules to condense into a liquid cannot be prevented by the available entropy in the gas – gaseous states are more disorderly than liquids – and need a corresponding level of entropy to maintain that disorder. Yet another way to explain the phenomenon is to say that the system has reached a state dominated by non-linearities in the physics and that the non-linear processes have destroyed the quantum fields of the stabilizing phonons in the liquid and gas [32]. This novel explanation I shall come back to and explain in more detail later in the article.

One more property of unstable states needs to be added, because it impinges deeply on European science and philosophies' discussion of conscious experience. Motions of component quanta of unstable systems, atoms and molecules in liquids and gases like carbon dioxide become completely unpredictable at instabilities. The celebrated philosopher René Descartes based his opinions of the soul on his vision that the supposed atoms of matter move around deterministically. He proclaimed that the soul is an essentially impotent observer trapped in the flesh [50]. He supposed that on a microscopic level, atoms move in accordance with kinematics principles of motion enunciated by Galileo Galilei (which at a later date Newton [51] expressed as his 'laws of motion'). If all matter is undergoing predictable laws of motion, what room is there for the mind to exert choices? This question has plagued the philosophy of the mind ever since and can only be fully and properly answered by taking recourse to the principles of instability, as I have shown elsewhere in conferences over the last 9 months, particularly the Science of Consciousness Conference in Tucson in April 2016 [52].

The adoption of instability as a condition for the coupling of mind to matter, i.e. to the physiology, is therefore the key to a potentially enormous advance in our understanding of mind and body. But how does instability arise in the physiology? Enter complexity biology.

5 Complexity Biology

Complexity biology with its four component principles (as I like to explain it [32]) constitutes an advance in our understanding of the whole field of biological science as radical as the discovery of the structures of proteins and DNA and the genetic code connecting them in the middle of the twentieth century. Once the genes had been identified as strands of DNA with a sequence of DNA bases to encode the various protein enzymes derived from them, the question naturally arose as to how gene expression was controlled. How is it that particular combinations of genes become expressed in particular cell types? By what means do such combinations get selected so that the resulting cell has its specific properties relevant to the organism as a whole?

The answer to these complicated questions began to be unraveled by a young medical student while studying for his physician's qualification at Stanford University Medical School. Stuart Kauffman [53] began to play with switching mechanisms by which computer models of various loops of genes on the DNA could turn other loops of genes on or switch other loops of genes off. Kauffman rapidly discovered that such switching only seemed to function in a realistic way if the average number of loops influenced by any given loop was precisely two. A number slightly above 2 led to the system manifesting boringly fixed behaviours, while numbers smaller than 2 led to chaotically unstable states continuously blinking on and off.

Kauffman's computer models had hit on a fundamental law of epigenetics – the study of regulation of gene expression: the mutual coordination number of sets of genes being expressed must, on average, be two. The system then functions in a regime called 'the edge of chaos', as Langton later named it. This constitutes the first principle of modern complexity biology: edge of chaos. It also describes an important aspect of the control of all physiological systems including single cells: *physiological systems give variable responses to fixed stimuli.*

This seemingly innocent expression is replete with deep pitfalls and equally deep implications for our understanding of the nature of both physiological systems and biology as a whole. Mechanical systems maintain stable structures, meaning that *mechanisms give fixed responses to fixed stimuli.* The discovery that biological systems do not do so means that physiological systems are *distinct* from mechanical systems. *They do not behave like simple mechanisms.* Here is the rub for Descartes, the slippery banana skin on which his seemingly irrefutable observation falls flat on its face and cannot be recovered.

Because the distributions of physiological responses to sequence of repeated stimuli are fractal distributions of the kind first theorized by the famous mathematician, Benoit Mandelbrot, they are known as fractal distributions and the subject of their retailed study as fractal physiology [54].

Now that we have enunciated the first two of complexity biology's four principles, let us turn to the third, which originally proposed to explain the fractal distributions of response observed to occur in many different systems. In 1987, Per Bak and his colleagues, Tang and Wiesenfeld proposed a principle of 'self-organized criticality' [55]. In essence this stated that physiologically, systems must organize themselves to be at instability, so that responses to stimuli are not fixed and fractal distributions of different response magnitudes result.

Various mechanical models for self-organized criticality were put forward, usually of the nature of a sand pile or an avalanche slope [56]. A particularly appealing one for Western carnivores was that of the egg timer, where a small hourglass shape of tiny coloured sand crystals is allowed to pour through a small funnel into a space below, where a pile of crystals grows up, occasionally slipping in various unpredictable sizes of avalanche (obeying a fractal distribution of relative sizes), as the slope builds up beyond the critical value at which avalanches take place. The critical value of the slope is stabilized, approximately speaking, by being at the 'edge of chaos' beyond which chaotic responses of slips in the pile take place.

Such slopes and responses were well known in the Second World War, when boulder-strewn slopes in Scotland were used to train troops of soldiers to toughen their legs, improve their balance, and hone their fighting skills. The training was very popular with the troops, but being at the edge of chaos, quite large and dangerous slips occasionally occurred.

Personally, I like to refer to the instability at which a physiological system comes to rest as a 'critical instability', in keeping with the terminology introduced to describe the critical point in the gas-liquid system of carbon dioxide studied by Van Der Waals [49]. Critical instabilities are not stable systems and give variable responses to fixed stimuli obeying fractal distributions. Moreover, for a critical point in the physiology to be stably maintained, there must be a dynamic process of stabilization of the values of the parameters where instability occurs. In this way, 'criticality' and 'self-organized criticality' become two separate principles.

In summary, the four principles of complexity biology are edge of chaos, *fractal physiology*, criticality and self-organized criticality [32]. They imply that physiologies behave very differently from mere mechanisms or mechanical systems, and they identify the loci of control of all physiological control systems as at instability. We now apply this to the theory of mind.

6 Criticality-Based Information

As stated, complexity biology comes with the radically new message that the physiology is controlled from critical instabilities. This implies that the mind, being at the locus of control of appropriate parts of physiology, should be located at criticality. If this radically new possibility is true, what kind of mind does this complexity-based theory predict?

A fruitful approach to answering this query turns out to be to consider the kind of information that is supported at instability. Stable systems support digital information [57] at the classical level and quantum information [58] at the quantum level, with various alternative possibilities holding under conditions that yield topological phase transitions, etc. Instability can be regarded as just another condition that radically alters the kind of information supported by a system. So what form does such information take?

Analysis of the structure of information at a critical instability is easily performed by considering an example: fluid flows close to their critical Reynolds number [59], demarcating the transition from smooth laminar flow to chaotic, turbulent flow. The critical point in fluid flow dynamics is another example of edge of chaos, one with the advantage of making a picture of the structure of information at critical instabilities easy to construct.

Below the critical Reynolds number, the flowing fluid can be considered an information field, where the information at each point has a single well-defined value, namely, the velocity vector of the fluid at each point [59]. Above the critical Reynolds number, the information field is chaotic and changing, as vortices form and move through the fluid causing the velocity vectors to change continuously in chaotic ways. At the critical Reynolds number itself, however, vortices want to form, but cannot quite do so. They are present at every point in the fluid in unmanifest potential form, but can never actually manifest and cause disruptive effects on the flow vectors [32].

Instead, their unmanifest presence has a remarkable effect: at each point, the presence of an unmanifest vortex is *to mix vectors that are infinitesimally close to each other into a mixture of flow vectors*. Each simple and unique flow vector below the critical point is thus transformed into an infinite mixture of infinitesimally different flow vectors, all sewn together by an infinitesimal vortex. It is therefore possible to represent the flow vectors as follows [32].

Flow vectors under stable conditions of fluid flow: <--------------, simple vectors.

Flow vectors at instability: <==========**O**, a mixture of vectors, <==========, combined together by an unmanifest vortex loop, **O**.

Here then is a physical and mathematical model for information states at critical instabilities; in this case the flow vectors at critical instabilities of fluid flow, with a structure which we posit can be generalized to other systems at their instabilities, such as the locus of control used by the mind. The next section derives more of its properties from the theory of critical instabilities, showing how to apply it to the mind and thus how to begin to derive the properties of *Śūnya*.

7 Properties of Experience Information

As we shall now show, information of the kind occurring at instabilities derived above seems admirably suited to describe states of the mind when engaged in experience. For this reason, we shall refer to <==========**O** as 'experience information'. The first thing to note is that, unlike any kind of information previously proposed, that at instabilities possesses a 'double-aspect' structure, with one aspect consisting of an *internal feedback loop* with information flowing around it continuously and unchangingly. Such a feedback loop at a macroscopic level would be characterized as having feedback gain, $\mathbf{g} = \mathbf{1}$, a condition identified by Norbert Wiener who was the first to introduce feedback loops into physical theory as *a condition for instability in a physical system.* Wiener's feedback instability demonstrated that instabilities can arise in control theory.

Elsewhere I have shown that this condition constitutes a condition for optimizing sensitivity of regulatory response and therefore for maximizing efficiency in a regulatory control system. From his side, Kauffman and colleagues have shown that criticality is a condition for maximizing information processing and for maximizing information diversity, making it a justifiable and understandable place at which to locate regulatory control.

For all these reasons, it is natural for physiological control systems to centre their function on instabilities, which are feedback instabilities of some aspect of the control system. The consequence is that the information satisfies the requirements proposed by philosopher David Chalmers for any information that can represent information in experience: first, a double-aspect structure, and second, it be non-reductive – not reducible to the properties of component atoms and molecules. Clearly the form <==========**O** has two aspects, the arrow <========== and the loop, **O**. It therefore has the required 'double-aspect' structure. Equally importantly, because the 'infinite mixture of vectors' <========== is at criticality, it is a very highly correlated structure that cannot be reduced, even in principle, to separate components built around some set of separable basis vectors. It is an *irreducible mixture*. The information is therefore non-reductive as Chalmers proposed.

Now to further operative questions: what does the loop represent? And, what kind of information does the information encode? Detailed answers to these questions, with full explanations show that the loop, O, can account for both the sense of 'self' inherent in all experience [41] and for the 'sense of time passing' [60] that, as Immanuel Kant first pointed out in the West [61], seems internal to our sense of 'self'.

In the East, this is implied by the term Asmati, literally 'Amness', that is an inherent aspect of most kinds of Samadhi, while in the West we talk of the sense of 'being' [24, 62], with the emphasis on the –ing or continuity of the sense of continuous existence. The sense of self can more precisely be understood to follow from the presence of the loop [31], because the loop represents a perfectly self-observing system [32], one without information or qualities, and that is precisely the nature of the self understood in Advaita Vedanta with its philosophy of self based on the experience of meditation [63].

A recently published new information theory has the ability to explain the process of meditation and how the attention is naturally drawn to the fundamental state of mind at the 'source of thought' [35]. The new kind of information is a double-aspect information theory of the kind proposed by David Chalmers [30] as necessary for any theory of experience. The new kind of information arises because all loci of control of biological organisms are situated at instabilities [31] and the information by which they operate in cognition and action consists of *the excited states of feedback instabilities* [32]. The structure of the new kind of information is easy to derive from the theory of critical points in fluid flow at the critical Reynolds number making the transition from laminar to turbulent flow [32]. It can then be shown that the information concerning flow is represented by an arrow with an attached loop, <========**O**. The loop represents an internal, self-observing, information loop, while the arrow is a highly correlated mixture of fluid flow vectors [32]. Applying this sophisticated new kind of information to changing states of experience during processes of meditation that culminate in *samadhi*, in which the mind is empty of content, the arrow is reduced to the form, <**O**, where the loop alone remains [36]. This form constitutes the correct physical representation of 'Śūnya' and is analogous to the vacuum state of quantum field theory, though, in this case, it is not the vacuum state of Hilbert space but rather of a Banach space [32], which cannot be expressed as the usual (possibly correlated) combination of weighted sums of quantum basis vectors. We conclude that physical representations of Śūnya require an extension of quantum theory of a kind that the presentation will specify.

8 The Nature of Information in Experience Information

Finally, what kind of information can be encoded in the structure of the arrow, <========, in experience information? One can be certain that it is not digital information, because digital information is reducible to dots and dashes (0s and 1s), while the information contained in the arrow is non-reductive and irreducible. The answer to the problem comes from looking at the structure of the singularity in various kinds of critical points in physical systems [32]. Van Der Waals critical points contain a singularity called a fold catastrophe of the kind proposed by Fields Medalist, René Thom, in his theory of 'structural stability and morphogenesis'. [64] Forms are generated at points where structural stability breaks down, and Thom demonstrated general mathematical properties of the class of mathematical objects where forms can be generated.

A serendipitous point emerges from this. In the West, thoughts have been recognized to contain 'forms' since the time of Plato [65]. Indeed, Immanuel Kant recognized that we perceive objects as wholes and stated that the role of the self is to hold together the parts of an object of perception into a single entity, the 'whole' of the object [61]. Experience information has precisely the structure to incorporate Kant's insight into an embodied reality of perception, in accordance with the thoughts of phenomenology's founders [60, 62]. Not only are Kant's deliberations satisfied so also are those of Husserl in his 'internal sense of time passing' [60] and Heidegger in his *Being and Time* [62].

The state of *Śūnya* is therefore not simply a kind of vacuum state. Rather it is the state of least excitation of a kind of field theory different from familiar quantum fields, a theory of quantum fluctuation fields, with a different kind of potential from a parabola and higher power than two, usually more than four. The state of *Śūnya* is the ground state of a system with a degenerate vacuum state of the kind first studied by K.P. Sinha [66] and ECG Sudarshan and now known to yield unusual phenomena at the boundaries of the domain containing the fields [67].

9 Conclusion

Although it seems attractive and has been suggested by several authors, it is not physically correct to represent the state of Śūnya by an ordinary quantum field theory. The state of Śūnya as proposed by Buddhist philosophers is the state of least excitation of the mind, which must be represented by a system with double-aspect information states supporting experience. Cognitive states of experience, of which the state of *Śūnya* is one, must therefore be represented by the system of excitations of a critical instability with, mathematically speaking, a 'degenerate ground state'. Such systems have completely different properties from quantum fields, in particular they can encode information in the form of 'gestalts' or 'forms'.

References

1. Books LLC. Branches of Buddhism: Hinayana, Pragmatic Buddhism, Schools of Buddhism, Buddhist Modernism, Southern, Eastern and Northern Buddhism. Memphis: General Books LLC; 2010.
2. Potter KH. Nāgārjuna and Mādhyamika Buddhism (from Presuppositions of India's Philosophies). Westport: Greenwood Press; 1976. http://www.advayavada.com/nagarjuna-and-madhyamika-buddhism-potter/.
3. Radhakrishnan S. Principal Upanishads. New Delhi: Harper Collins; 1994.
4. Pruthi RK. Prehistory and Harappan civilization. New Delhi: A.P.H. Publishing; 2004.
5. Danino M. The Lost River: on the Trail of the Saraswati. London: Penguin; 2010.
6. Rajagopalachari C. Ramayana. Mumbai: Bharatiya Vidya Bhavan; 2015.
7. Rajagopalachari C. Mahabharata. Mumbai: Bharatiya Vidya Bhavan; 2010.
8. Subramanian K. Shrimad Bhagavatam. Mumbai: Bharatiya Vidya Bhavan; 2016.

9. Muller M. The six systems of Indian philosophy, Collected works of Max Muller, vol. 16. Charleston: BiblioBazar; 2009.
10. Subhash S. Creation from Śūnya. Anand: Anand Press; 1993.
11. Müller M. Sacred books of the East. Oxford: Oxford UP; 1880.
12. Chogyel T. The life of the Buddha. London: Penguin Books; 2015.
13. Bodhi B. The noble eightfold path: the way to the end suffering. Kandy: Buddhist Publication Soc; 2010.
14. HH The Dalai Lama, (Trans Jinp G.T.) The four noble truths, Thorsons, London, 2013.
15. Radhakrishnan S. Svetasvetara Upanishad. In: Principal Upanishads. New Delhi: Harper Collins; 1994.
16. Chattopadhay R. Swami Vivekananda in India: a corrective life. Delhi: Motilal Banarsidas; 1999.
17. Suzuki DT. Manual of Zen Buddhism. New York: Grove Press; 2007.
18. Hart W. The art of living: Vipassana meditation as taught be S.N. Goenka. Onahlaska: Pariyatti; 2011.
19. Powers J. Introduction to Tibetan Buddhism. Ithaca: Snow Lion; 2007.
20. Six Yogas of Naropaand Niguma. Available at: www.maha-kala.com/archive/naropa/naropa101.htm.
21. Hershock PR. Chan Buddhism. Honolulu: U Hawai'I Press; 2005.
22. Radhakrishnan S. Brihadaranyaka Upanishad. In: Principal Upanishads. New Delhi: Harper Collins; 1994.
23. Travis F, Shear J. Focused attention, open monitoring and automatic self-transcending: categories to organize meditations from Vedic, Buddhist and Chinese Traditions. Consc Cogn. 2010;19(4):1110–8. https://doi.org/10.1016/j.concog.2010.01.007.
24. Yogi MM. Science of being and art of living: transcendental meditation. New York: Plume; 2001.
25. Tiwari R. The whole thing, the real thing: brief biography of Shri Gurudeva. New Delhi: Delhi Photo Company; 1977.
26. Sharma U. Teachings and philosophy of Buddha. New Delhi: Diamond Pocket Books; 2016.
27. See addresses by Chief Guest and Guest of Honour in these Proceedings.
28. Boothroyd J. Galileo Galilei: a life of curiosity. Minneapolis: Lerner Pubs; 2006.
29. Aristotle. (Waterfield R. Trans) Aristotle physics. Oxford: Oxford's World Classics; 1996.
30. Chalmers D. In: Shear J, editor. Explaining consciousness: the hard problem. London: Academic Press; 1997.
31. Hankey A. Complexity biology-based information structures can explain subjectivity, objective reduction of wave packets, and non-computability. Cosmos and History, J Nat Soc Philos. 2014;10(1):237–50.
32. Hankey A. A complexity basis for phenomenology: how information states at criticality offer a new approach to understanding experience of self, being and time. Progr Biophys Mol Biol. 2015;119(3):288–302.
33. Shetkar R, Hankey A. Nagendra H.R. First person accounts of Yoga meditation yield clues to the Nature of Information in Experience. IJOYPPP to be published.
34. Shetkar R, Hankey A, Nagendra HR. Biophysics of meditation. Submitted to the Journal of Consciousness Studies.
35. Shetkar R. Hankey A. Nagendra H.R. Reason for health benefits of deep meditation: self-organized criticality restores regulation to optimal. Eur J Pharm Med Res, 2016,3(5), 435–441.
36. Hankey A, Shetkar R. Self-transcending meditation is good for mental health: why this should be the case. Int Rev Psychiat. 2016;28(3):236–40.
37. Von Neumann J. (Beyer R.T. Trans.) Mathematical foundations of quantum mechanics. Princeton: Princeton U.P.; 1955.
38. Penrose R. The Emperor's New Mind: concerning computers, minds and the laws of physics. Oxford: Oxford University Press; 1999.

39. Penrose R. Shadows of the mind: a search for the missing science of consciousness. Oxford: Oxford University Press; 1994.
40. Domash L. Lectures to Maharishi Internotional University, 1973.
41. Shear J, editor. Explaining consciousness: the hard problem. London: Academic Press; 1997.
42. Radhakrishnan S. Taittariya Upanishad. In: Principal Upanishads. New Delhi: Harper Collins; 1994.
43. Bjorken J. LSC relativistic quantum mechanics. New York: McGraw-Hill Education; 1998.
44. Wigner E. On the consequences of the symmetry of the nuclear hamiltonian on the spectroscopy of nuclei. Phys Rev. 1937;51(2):106.
45. Goswami A, Reed RE, Goswami M. The self-aware universe: how consciousness creates the material world. New York: Penguin-Putnam; 1995.
46. Ellis J. The superstring-theory of everything, or of nothing? Nature. 1986;323:595–8.
47. Waldrop MM. Complexity – the emerging science at the edge of order and chaos. New York: Simon and Schuster; 1993.
48. Kipnis AA, Avelov BE, Rowlinson JS. Van Der Waals and molecular science. Oxford: Clarendon Press; 1996.
49. Johnston DC. Advances in the thermodynamics of the van der Waals fluid. San Rafael: Morgan and Claypool; 2014.
50. Descartes R. Meditations on first philosophy. Blackwell: Start Publishing LLC; 2013.
51. Newton I. The mathematical principles of natural philosophy: principia mathematica philosophiae naturalis. Luxembourg: CreateSpace Publishing; 2016.
52. Hankey A. Free will in a quantum universe. Tucson: Science of Consciousness; 2016.
53. Kauffman S. At home in the universe. The search for the laws self organization and complexity. Oxford: Oxford University Press; 1995.
54. Bassingthwaighte JB, Liebovitch LS, West BJ. Fractal physiology. Oxford: Oxford University Press; 1994.
55. Bak P, Tang C, Wiesenfeld K. Self-organized criticality: an explanotion of the $1/f$ noise. Phys Rev Lett. 1987;59:381.
56. Jensen HJ Self-organized criticality: emergent complex behavior in physical and biological systems. Cambridge Lecture Notes in Physics, 1993.
57. Shannon CE. A mathematical theory of communication. Bell Syst Techn J. 1948;27:379–423 & 623–56.
58. Deutsch D. Quantum theory, the Church-Turing principle and the universal quantum computer. Proc R Soc A. (1985)400 (1818):97–117. Doi: https://doi.org/10.1098/rspa.1985.007.
59. Lesieur M. Turbulence in fluids. 4th ed. Dordrecht: Springer; 2008.
60. Husserl E. The phenomenology of internal time consciousness. Bloomington: Indiana University Press; 1964.
61. Kant I. The critique of pure reason. Pluhar W.S. (trans), Hackett, 1996.
62. Heidegger M. 1927. *Sein und Zeit* (Tübingen): Niemeyer. 19. Aufl. (2006), ISBN-10: 3484701536; ISBN-13: 978–3484701533. *Being and Time.* Trans Macquarrie J. & Robinson E. New York: Harper and Row, 1962. Harper Perennial Modern Thought edition (2008). ISBN-10: 0061575593; ISBN-13**:** 978–0061575594.
63. Adishankaracharya. (Chinmayananda S. trans) Atmabodha. Mumbai: Chinmaya Prakashan; 2009.
64. Thom R. Structural stability and morphogenesis. New York: Addison Wesley; 1989.
65. Plato. (Cornford F.M. Trans) The Republic of Plato. Oxford, 1941. See also, Allen R.E. Trans, Yale U, New Haven Conn, 2006.
66. Sinha KP, Sivaram C, Sudarshan EC. The superfluid vacuum state, time-varying cosmological constant, and nonsingular cosmological models. Found Phys. 1976;6(6):717–26.
67. Hankey A, Rao RN. A new kind of biologically active orientation-sensitive field coupling to complexity based biological regulatory systems? Forsch Komplementmed. 2013;20:314–5. https://doi.org/10.1159/000356797.

MIX
Papier aus verantwortungsvollen Quellen
Paper from responsible sources
FSC® C105338

If you have any concerns about our products,
you can contact us on
ProductSafety@springernature.com

In case Publisher is established outside the EU,
the EU authorized representative is:
Springer Nature Customer Service Center GmbH
Europaplatz 3, 69115 Heidelberg, Germany

Printed by Libri Plureos GmbH
in Hamburg, Germany